T0419275

PROTEIN SCIENCE AND ENGINEERING

PROTEIN STRUCTURE

PROTEIN SCIENCE AND ENGINEERING

Additional books in this series can be found on Nova's website
under the Series tab.

Additional E-books in this series can be found on Nova's website
under the E-books tab.

PROTEIN SCIENCE AND ENGINEERING

PROTEIN STRUCTURE

LAUREN M. HAGGERTY
EDITOR

Nova Science Publishers, Inc.
New York

For permission to use material from this book please contact us:
Telephone 631-231-7269; Fax 631-231-8175
Web Site: http://www.novapublishers.com

Library of Congress Cataloging-in-Publication Data

Protein structure / editor, Lauren M. Haggerty.
p. cm.
Includes index.
ISBN 978-1-61209-656-8 (hardcover)
1. Proteins--Structure. I. Haggerty, Lauren M.
QP551.P69763827 2011
612'.01575--dc22

2011003560

Published by Nova Science Publishers, Inc. † New York

Contents

Preface

To understand the functions of proteins at a molecular level, it is often necessary to determine their three-dimensional structure. A protein may undergo reversible structural changes in performing its biological function. This new book presents current research in the study of protein structure. Topics discussed include enzyme immobilization; structural characteristics of fibrous and globular proteins; mathematical modeling of helical protein structures; three approaches for classifying protein tertiary structures and spectral and fluorescence analysis of protein structure.

Chapter 1 – A large number of diseases are caused by (or at least connected with) aggregation of incorrectly folded (misfolded) proteins. A typical variant in this case is the formation of the so-called amyloid fibrils. Amyloidogenic regions in polypeptide chains are very important because such regions are responsible for amyloid formation and aggregation. Two characteristics (expected probability of formation hydrogen bonds and expected packing density of residues) have been introduced by us to detect amyloidogenic regions in a protein sequence. It has been shown that regions with high expected probability of formation of backbone-backbone hydrogen bonds as well as regions with high expected packing density are mostly responsible for the formation of amyloid fibrils. Moreover, the threshold values obtained from the scales for the prediction of amyloidogenic regions and regions protected from hydrogen/deuterium exchange in protein chain coincide. A comparison of experimentally found amyloidogenic regions with predicted folding nuclei for proteins with experimentally known amyloidogenic regions demonstrates that the average Φ-value is significantly greater in the amyloidogenic regions. Conversely, comparison of experimentally outlined folding nuclei with predicted amyloidogenic regions also demonstrates that the average Φ-value (in this case, the experimental one) is significantly greater in the predicted amyloidogenic regions. This is an indication that amyloidogenic regions are typically incorporated into the native structure early during its formation, i.e. the residues which are crucial for folding are also crucial for misfolding. The modeling of folding of the protein with swapped domains demonstrates that the regions exchanged in the oligomeric form in most cases are also included in the folding nucleus predicted for the monomeric form of the protein. One can hypothesize that if for proteins amyloidogenic regions intersect with the initiation site for folding, then such proteins more probably can misfold in appropriate environment conditions to amyloid fibrils involved *in vivo* in various "amyloid" diseases.

Chapter 2 – Enzyme technology is the application involving modification of enzyme's structure as well as its function. It is currently considered as a useful alternative to conventional process technology in enzyme based industrial as well as analytical fields. Enzyme immobilization is the foremost technique used in enzyme technology, which has gained popularity for the last few decades. It stabilizes structure of the enzymes, thereby allowing their applications under harsh environmental conditions (pH, temperature, organic solvents), and thus enable their uses in non-aqueous enzymology, and in the fabrication of biosensors. Furthermore, it allows repetitive use of enzyme as well as helpful in prevention of product contamination with the enzyme (especially useful in the food and pharmaceutical industries). Recently, the development of techniques for immobilization of multi-enzymes along with cofactor regeneration and retention system are gainfully exploited in developing biochemical processes involving complex chemical conversions. Various methods for enzyme immobilization used in bioreactors and biosensors include adsorption on or covalent attachment to a support, micro-encapsulation, and entrapment within a membrane, film or gel. The ideal immobilization method should employ mild chemical conditions, allow large quantities of enzyme to be immobilized, provide a large surface area for enzyme–substrate contact within a small total volume, minimize barriers to mass transport of substrate and product, and provide a chemically and mechanically robust system. Immobilization helps in the development of continuous processes allowing more economic organization of the operations, automation, decrease of labour and investment to capacity ratio. The present chapter delineates the current status and future potentials of immobilized enzymes in the emerging enzyme based industries with suitable examples.

Chapter 3 – To understand the structure-to-function relationship, life sciences researchers and biologists need to retrieve similar tertiary structures from protein databases and classify them into the same protein fold. The protein structures are stored in the world-wide repository Protein Data Bank (PDB) which is the primary repository for experimentally determined proteins structures. With the technology innovation the number of protein structures increases every day, so, retrieving structurally similar proteins using current structural alignment algorithms may take hours or even days. Therefore, improving the efficiency of protein structure retrieval and classification becomes an important research issue. There are many accurate methods such as SCOP which provide protein structural classification, but they last too long. In this chapter, the authors present three accurate approaches which provide faster classification (min., sec.) of protein tertiary structures. The authors have used two approaches to extract the features of the protein structures. In the authors' voxel based approach, the Spherical Trace Transform is applied to protein tertiary structures in order to produce geometry descriptor. Additionally, some properties of the primary and secondary structure are taken, thus forming better integrated descriptor. In the authors' ray based approach, some modification of the ray descriptor is applied on the interpolated backbone of the protein. Then, the authors use three approaches in order to classify protein molecules. In the authors' first approach, they use the Growing neural Gas (GNG) as a hidden layer and a Radial Basis Function (RBF) as an output layer (which contains separate output node for each class). The authors' second approach uses the well known Support Vector Machine (SVM) method. In the third approach, the elements of the ray descriptor are quantized at 20 levels and then the authors build separate Hidden Markov Model (HMM) for each class. The authors have implemented a system for classifying protein tertiary structures based on these three approaches. The authors' ground truth data contains 6979 protein chains from 150 SCOP domains. 90% of the dataset serves as the training and the other 10% as a test data. The authors examined the classification accuracy of these three approaches according to SCOP hierarchy. The results showed that the ray based descriptor leads to more accurate classifier. Generally, SVM gives the best classification accuracy (98.7%) by using the ray based descriptor. Additionally, the authors compared their HMM approach against the 3D HMM method. The analysis showed that the authors' HMM based approach gives better results than the existing 3D HMM method.

Chapter 4 – Fibrous proteins form coiled coil structure. Long stretches of α-helices form tightly packed arrangements of coiled coil. Helices contain 3.5 residues per turn. Unwinded at the ends, α-helices bind DNA molecules, or dimerizing subunits. In globular proteins, α-helices are shorter. The average number of residues per turn is 3.6. Similarly to coiled coil structure, α-helices include highly packed helix-helix interface region and unwinded at the ends ligand binding region containing heme, metals, and other ligands. Left-handed and right handed, parallel and antiparallel arrangements of α-helices exist. Core regions with similar structure carry common amino acid combinations. Ligand binding regions can also be recognized by a specific amino acid sequence pattern. Besides, α-helices participate in assemblies of protein motifs and assemblies of protein subunits. Distribution of contact areas along helix surface and their link to helix edges can identify the type of motif or type of assembly. Location of the core regions, ligand binding regions, and assembly-related regions on the surface of the α-helix together with their specific combinations are important for the fold of the molecule. In this work, specific interactions of α-helices, their ligands, and oligomerization properties of protein molecules are considered. Specific patterns of amino acid combinations together with their location at the surface of the α-helix are associated with particular fold of the molecule.

Chapter 5 – The c-Myc gene is an important regulator of normal cell growth, differentiation, and programmed cell death, and plays a central role in tumorigenesis. The c-Myc protein contains a potential transactivation domain within its N-terminal region, which contains short acidic, proline-, and glutamine- rich clusters similar to the activation domains of many transcription factors, and is required for the biological activities of c-Myc. However, like the activation domains of most transcription factors, the structural basis for the activity of the c-Myc AD region is not completely understood. In this study, using various structural analytical tools, the authors determined the structural characteristics of the activation domain of c-Myc protein. The authors' analyses show that the c-Myc activation domain possesses all the characteristics of an intrinsically disordered protein, a common phenomenon in many cancer related proteins. In recent years, there have been several indications that relative to prokaryotes, eukaryotic genomes are highly enriched in ID proteins, and it has been predicted that more than two-thirds of proteins associated with cancer-related genes possess long stretches of ID regions. The authors' results suggest that c-myc activation domain is capable of making multiple interactions with its target binding partners, the specificity of which may result in regulation of transcription.

Chapter 6 – The availability of cost-effective DNA sequencing technologies has led to the identification and cataloguing of millions of naturally occurring inherited and somatic human genomic variations. As a result, questions concerning the ultimate phenotypic and functional significance of these variations have been raised. Although some large-scale initiatives have been launched for this purpose, including the Encyclopedia of DNA Elements

(ENCODE) initiative, there is a need for the DNA sequencing and genomics communities to reach out and foster greater collaborative opportunities with the functional genomics community. One area that is ripe for this kind of activity is the functional characterization of coding variations, as structural proteomics and crystallography researchers may both benefit from, and contribute to, an understanding of the molecular and phenotypic influence of such variations. The authors briefly review the motivation for this kind of interaction and draw on publicly available data to showcase its need. The authors also consider how relevant research could be pursued.

Chapter 7 – The authors review recent work on modelling helical parts of protein structure mathematically. The authors use classical calculus of variations and consider mathematical energies dependent on the curvature and torsion of the protein backbone curve. The authors demonstrate classes of mathematical energies for which a given helix, or all helices, are extremisers. A future goal is to find within these energies some which permit less regular protein backbone curves, as also observed in nature.

Chapter 8 – Temperature dependencies of viscosity (η) and conductivity (σ) of blood lipoproteins and A-I apolipoprotein was studied. It has revealed the presence of the anomalous region at temperature 35–38 ± 0.5°C (T_c). Transition heat is very low in human HDL, VLDL and apoA-I, and in LDL it is higher by a factor of 4-5. Some mechanisms of the cortisol interaction with HDL and apoA-I have been studied by infrared (IR) spectroscopy and conductometry. It was shown that the hormone destroy the initial structure of lipoproteins due to formation of new hydrogen bonds and hidrofobic interaction. The hormone has been found to strengthen the tangle → α-helixes and tangle → β-structures transitions and increase the ordering of lipids. Therewith, ΔH_{trans}. rises markedly (13 and more times), and at the same time the anomalous region is shifted by 1-2°C in apoA-I. The anomalous changes of viscosity and conductivity in the physiological temperature range for all lipoprotein fractions and apoA-I seems to be due to the structural phase transition in both proteins and lipids. The structural transition into apoA-I is likely to contain the elements of phase transition of the first and second types. In human and rat VLDL, the smectic → cholesteric phase transition seems to occur. Enthalpy of viscous flow and structural transition in VLDL is higher for rats than for human.

Chapter 9 – This chapter analyzes the advantages of a multiparametric approach to gain exhaustive information related to the structure and folding of globular proteins. The principal bases of the some widespread contemporary physico-chemical methods for studying the structural properties of the globular protein and their changes during the unfolding-refolding reaction are considered. It is discussed what kind of information can be obtained using different physico-chemical methods and what advantages a researcher might have applying several techniques simultaneously.

Chapter 10 – The intrinsic fluorescence of proteins and the fluorescence emitted by fluorescent probes attached to proteins or embedded in lipid layers are extremely sensitive to different environmental parameters (like temperature, viscosity, ionic strength, electric field gradient, pH). The change in the dynamic (spatio-temporal) behaviour of the neighbouring matrix around the fluorophores can affect their fluorescence characteristics. This feature makes the fluorescence techniques to be potentially powerful tools to obtain information about the structural and dynamic parameters of macromolecules and/or supramolecular systems.

In: Protein Structure
Editor: Lauren M. Haggerty, pp. 1-29

ISBN 978-1-61209-656-8

Chapter 1

Misfolded Species Involved Regions Which Are Involved in an Early Folding Nucleus

Oxana V. Galzitskaya*

Institute of Protein Research, Russian Academy of Sciences, Institutskaya str. 4,
Pushchino, Moscow Region, 142290, Russia

Abstract

A large number of diseases are caused by (or at least connected with) aggregation of incorrectly folded (misfolded) proteins. A typical variant in this case is the formation of the so-called amyloid fibrils. Amyloidogenic regions in polypeptide chains are very important because such regions are responsible for amyloid formation and aggregation. Two characteristics (expected probability of formation hydrogen bonds and expected packing density of residues) have been introduced by us to detect amyloidogenic regions in a protein sequence. It has been shown that regions with high expected probability of formation of backbone-backbone hydrogen bonds as well as regions with high expected packing density are mostly responsible for the formation of amyloid fibrils. Moreover, the threshold values obtained from the scales for the prediction of amyloidogenic regions and regions protected from hydrogen/deuterium exchange in protein chain coincide. A comparison of experimentally found amyloidogenic regions with predicted folding nuclei for proteins with experimentally known amyloidogenic regions demonstrates that the average Φ-value is significantly greater in the amyloidogenic regions. Conversely, comparison of experimentally outlined folding nuclei with predicted amyloidogenic regions also demonstrates that the average Φ-value (in this case, the experimental one) is significantly greater in the predicted amyloidogenic regions. This is an indication that amyloidogenic regions are typically incorporated into the native structure early during its formation, i.e. the residues which are crucial for folding are also crucial for misfolding. The modeling of folding of the protein with swapped domains demonstrates that the

* Russia E-mail: ogalzit@vega.protres.ru

regions exchanged in the oligomeric form in most cases are also included in the folding nucleus predicted for the monomeric form of the protein. One can hypothesize that if for proteins amyloidogenic regions intersect with the initiation site for folding, then such proteins more probably can misfold in appropriate environment conditions to amyloid fibrils involved *in vivo* in various "amyloid" diseases.

Introduction

The formation of amyloid fibrils is a case of protein misfolding, in which a protein folds into cross β-structure instead of folding into its native structure. The structures of amyloid fibrils obtained from different proteins are fairly uniform although native structures vary greatly from protein to protein (Rudall, 1952; Nelson et al., 2005). The observed cross β-structure is so similar from protein to protein that even antibodies created to bind amyloid fibers formed by one protein are able to bind to amyloid fibers formed by another protein (Kayed et al., 2003). It is known that amyloidogenic proteins and peptides possess amyloidogenic regions that are critical for fibril formation (Ivanova et al., 2004).

Experimental data demonstrate that amyloid formation depends on the precise amino acid sequence and is a highly specific process. A protein can become non-amyloidogenic after mutations in these regions or after the deletion of an amyloidogenic region, while other mutations in these regions result in increased amyloidogenic propensity. In addition, when synthesized in isolation, these regions (sometimes as small as five residues long (Lopez de la Paz and Serrano, 2004) can sometimes form amyloid fibrils identical to those formed by the whole proteins (Thompson et al., 2000).

It has been shown that only a few residues may be necessary to define amyloidogenic properties (Esteras-Chopo et al., 2005). The sequence STVIIE was identified as the most prone to the formation of amyloid fibrils and was used as a reference to search for amyloidogenic regions in protein chains (Lopez de la Paz and Serrano, 2004). This peptide was used to verify the original hypothesis that amyloidogenic properties are determined by a short region of a protein chain (Esteras-Chopo et al., 2005).

Similar results were reported for mouse β2-microglobulin, which (in contrast to human β2-microglobulin) does not form amyloid structure. Seven amino acid residues of this protein were substituted with their counterparts from human β2-microglobulin (which does form amyloids) in the region that displayed the greatest difference in primary structure between the two proteins (region 83–89). The substitution resulted in the formation of amyloid structure *in vitro* (Ivanova et al., 2004). It should be noted that the peptide involved (NHVTLSQ) formed fibrils, while a peptide that consisted of the same residues joined in an arbitrary order (QVLHTSN) did not (Ivanova et al., 2004). A model of a cross-β spine decorated with the remaining parts of protein molecules was proposed on the basis of the experimental findings.

The formation of amyloid fibrils is a case of protein misfolding, in which a protein folds into cross β-structure instead of folding into its native structure. In addition to proteins that form amyloid fibrils *in vivo* in various "amyloid diseases," there are many other proteins that are not implicated in amyloid diseases but form fibrils *in vitro* (Guijarro et al., 1998; Chiti et al., 1999; Fandrich et al., 2001). There is no sequence homology common to all such proteins or peptides.

Since polypeptide chains can fold into native structures or misfold into amyloid fibrils, there is a competition between the processes of folding and misfolding. Recently it has been shown that transition states of folding and aggregation for human acylphosphatase are structurally unrelated (Chiti et al., 2002), suggesting that the pathways of folding and aggregation for this protein are structurally distinct. For other proteins, partially unfolded species that resemble folding intermediates have been implicated in amyloid formation (Jahn et al., 2006). This suggests that the folding and misfolding free-energy landscapes for these proteins may be more closely related than for other proteins. The identification of an intermediate as the key amyloidogenic precursor of β2-microglobulin fibril elongation under physiological conditions provides direct evidence that the folding and aggregation landscapes for this protein, at least, initially coincide and diverge only at the level of a native-like folding intermediate that contains highly localized structural variations of the native immunoglobulin fold (Jahn et al. 2006). Thus, there is no generally accepted point of view on the coincidence of regions important for amyloid formation and regions important for "normal" folding. Recently, it has been shown that for one half of proteins the correlation between residues important for folding and aggregation intersects, but for the other part such a correlation is absent (Tartaglia and Vendruscolo, 2010). One can hypothesize that if for proteins amyloidogenic regions intersect with the initiation site for folding, then such proteins more probably can misfold in appropriate environment conditions to amyloid fibrils involved *in vivo* in various “amyloid diseases,” for proteins where such intersection is absent. For example, acylphosphatase can form fibrils *in vitro* that are not implicated in amyloid diseases *in vivo*. In the first example (acylphosphatase), such intersection is absent and the protein has not been involved in the disease *in vivo*. But in the second case, such intersection is present and we know that this protein, β2-microglobulin is involved in the disease *in vivo*. In this paper we trace such intersection for different proteins and try to accept or decline this hypothesis.

A crucial event of protein folding is the formation of a folding nucleus, which is a structured part of the protein chain in the transition state. We demonstrate a correlation between locations of residues involved in the folding nuclei and locations of predicted amyloidogenic regions. The average Φ-values are significantly greater inside amyloidogenic regions than outside them. We have found that fibril formation and normal folding involve many of the same key residues, giving an opportunity to outline the folding initiation site in protein chains.

From experimental and theoretical observation we can conclude that there is a pattern in the amino acid sequence, defined by groupings of large and nonpolar side chains (amyloidogenic residues in our case) that lead to restriction of the motion of the polypeptide backbone caused by local hydrophobic cluster formation. It appears that the residues which are crucial for folding are also crucial for misfolding.

During folding, a protein molecule has to overcome a free-energy barrier. The most unstable structure, which corresponds to the top of the barrier (i.e., to the transition state of the folding process), is called the folding nucleus (Fersht, 1995; Fersht, 1997). Since the folding nucleus is unstable, it is not easy to investigate it experimentally. A very laborious experimental method, which is called Φ-analysis (Matouschek et al., 1989), is developed to determine the structure of folding nuclei (then, a related method called Ψ-analysis (Krantz and Sosnick, 2001) is developed). By means of the introduction of point mutations into a protein structure, it is possible to find residues, mutations of which destabilize the transition state as much as for the native state. The Φ-value is calculated for each mutation as follows:

$$\Phi = \Delta_r[F(TS) - F(U)] / \Delta_r[F(N) - F(U)], \quad (1)$$

where $\Delta_r[F(N) - F(U)]$ is the residue r mutation-induced change of the difference in the free energy between the native state N and the unfolded state U, and $\Delta_r[F(TS) - F(U)]$ is the mutation-induced change in the difference between the transition state (TS) and the unfolded state (U) free energies.

Φ-values vary from unity (when the amino acid residue is in the folding nucleus) to zero (when the amino acid residue is not incorporated into the folding nucleus). Φ-values below zero or greater than unity are indicators of the presence of non-native contacts in the transition state; such Φ-values are rare.

Since the experimental method of Φ-analysis is very laborious and there is no sufficient experimental data about folding nuclei in amyloidogenic proteins, it is useful to examine folding nuclei theoretically. At the present time, some methods exist for this purpose (Shakhnovich et al., 1996; Daggett et al., 1996; Onuchic et al., 1996; Galzitskaya and Finkelstein, 1999; Alm and Baker, 1999). Some of them (including our method (Galzitskaya and Finkelstein, 1999; Garbuzynskiy et al., 2004)) are able to predict positions of folding nuclei reasonably well (Galzitskaya and Finkelstein, 1999; Alm and Baker, 1999; Garbuzynskiy et al., 2004; Alm et al., 2002; Galzitskaya et al., 2005; Garbuzynskiy et al., 2005).

The goal of this work consists in comparison of those amino acid residues which are crucial for folding and misfolding processes of the same proteins. The understanding of peculiarities of the folding process of amyloidogenic proteins can probably shed light on why these proteins are amyloidogenic. As the experimental data on both folding nuclei and amyloidogenic regions in the same proteins are scarce, we compared experimentally found amyloidogenic regions with predicted folding nuclei for proteins with experimentally known amyloidogenic regions, and experimentally outlined folding nuclei with predicted amyloidogenic regions for proteins with experimentally known folding nuclei positions. We demonstrate that: most of the experimentally determined amyloidogenic fragments intersect with the regions incorporated into the predicted folding nuclei of amyloidogenic proteins, and also most of the predicted amyloidogenic regions intersect with experimentally outlined folding nuclei. In average, Φ-values for amino acid residues in amyloidogenic regions are significantly greater than Φ-values for amino acid residues in non-amyloidogenic regions. Thus, amyloidogenic regions are important not only for the process of amyloid formation but also for the process of "normal" folding. It appears that amyloidogenic regions readily form either the folding nucleus (during "normal" folding) or amyloid fibrils (during misfolding), thus playing a crucial role in the competition between folding and misfolding. This can probably also indicate that this part of the folding nucleus can be involved in amyloid formation. Moreover, the modeling of folding of the protein with swapped domains demonstrates that the regions exchanged in the oligomeric form in most cases are also included in the folding nucleus predicted for the monomeric form of the protein.

One can hypothesize that if for proteins amyloidogenic regions intersect with the initiation site for folding, then such proteins more probably can misfold in appropriate environment conditions to amyloid fibrils involved *in vivo* in various "amyloid" diseases.

Results and Discussion

Intersection of Experimentally Determined Amyloidogenic Regions with the Predicted Folding Nuclei

To investigate folding/unfolding behavior of amyloidogenic proteins, we have constructed a database of globular proteins with experimentally revealed amyloidogenic regions. From literature data, we selected those globular proteins in which the position of amyloidogenic regions is known from experimental data. The database now includes seven proteins (Table 1). The size of the proteins varies from 98 (acylphosphatase) to 253 (human prion protein) amino acid residues. Although each of these proteins can form amyloid fibrils (*in vivo* or *in vitro*), the globular states of these proteins are rather different. Though some amyloidogenic proteins in their native globular states are β-structured (for example, β2-microglobulin), the other proteins are α-helical (for example, myoglobin) or possess large unstructured regions (for example, human prion protein). The number and size of the experimentally revealed amyloidogenic regions also vary (see Table 1). Altogether, the seven amyloidogenic proteins contain 12 amyloidogenic regions which cover almost a quarter of the length of the proteins (24% of amino acid residues).

Using a theoretical method, we searched for the position of the folding nuclei in the amyloidogenic proteins during their "normal" folding process from unfolded state into their folded native structure. Using the method which (as we have previously shown (Galzitskaya and Finkelstein, 1999; Garbuzynskiy et al., 2004; Galzitskaya et al., 2005; Garbuzynskiy et al., 2005)) is able to predict the position of folding nuclei reasonably well (correlation coefficient between predicted and experimentally known Φ-values reaches 65% if the native 3D structure which is used in the modeling is of high quality, see Garbuzynskiy et al., 2004), we modeled the behavior of each protein molecule at the point of thermodynamic equilibrium (i.e. under the conditions when unfolded and native states are equally stable; under these conditions, all the other states are unstable). The behavior of the protein molecule is modeled as a simplified sequential reversible unfolding of its known native 3D structure (see Materials and Methods). The obtained network of folding/unfolding pathways is further analyzed by the dynamic programming method (Finkelstein and Roytberg, 1993) to find transition states and (correspondingly) folding nuclei. The involvement of a residue in the folding nucleus (or rather, into the ensemble of the folding nuclei) is reflected in its Φ-value, which is unity when this residue has all of its native contacts in the transition state and zero when this residue has no contacts in the transition state. The analysis of the obtained predictions have shown that 43% of amino acid residues of the proteins have Φ-values larger than 0.5 (more than 50% of native contacts are formed) and can be attributed to folding nucleus. It should be mentioned that such a large size of the folding nucleus is assumed to be typical for the mid-transition point (our modeling was done just in these conditions) while far from the equilibrium (that is, in pure water etc.) the folding nucleus should be typically smaller (Finkelstein and

Table 1. Proteins with experimentally determined amyloidogenic regions

Name of the protein	PDB entry	Number of amino acid residues		Experimentally investigated amyloidogenic regions	*In vivo* and *in vitro* or *in vitro* only
		in the protein	in the used 3D structure (region[a])		
Acylphosphatase	1aps[b]	98	98 (1–98)	16–31 (Chiti et al., 2002); 87–98 (Chiti et al., 2002)	*in vitro*
β2-microglobulin	1im9	99	99 (1–99)	20–41 (Kozhukh et al., 2002); 59–71 (Jones et al., 2003); 83–89 (Ivanova et al., 2004)	*in vivo* and *in vitro*
Gelsolin	1kcq	104	104 (158–261)	52–62 (Maury and Nurmiaho-Lassila, 1992)	*in vitro*
Transthyretin	1bm7	127	114 (10–123)	10–19 (Chamberlain et al., 2000); 105–115 (Jaroniec et al., 2002)	*in vivo* and *in vitro*
Lysozyme	1931	130	129 (1–129)	49–64 (Krebs et al., 2000)	*in vivo* and *in vitro*
Myoglobin	1wla	153	153 (1–153)	7–18 (Picotti et al., 2007); 101–118 (Fandrich et al., 2003)	*in vitro*
Human prion	1qm0	253	143 (125–228)	169–213 (Lu et al., 2007)	*in vivo* and *in vitro*

[a]The numeration (in brackets) corresponds to that in the used PDB entry.

[b]Experimentally, amyloidogenic regions were revealed for human acylphosphatase. 3D structure of this protein is unknown. However, there is a 3D structure for horse acylphosphatase sequence of which is highly similar to that of the human protein (sequence identity 95%).

Badretdinov, 1997); this assumption is confirmed experimentally (when several sets of Φ-values corresponding to different conditions are published: for instance, in Ternström et al., 1999).

We tested whether the theoretically found folding nuclei intersect with experimentally found amyloidogenic regions. It appears that 8 of 12 amyloidogenic regions are situated in folding nuclei where Φ-values are large. For several proteins, the regions with the largest Φ-values coincide with the amyloidogenic regions (see Figure 2 below).

For amino acid residues in amyloidogenic regions, the average Φ-value is 0.58±0.02 while amino acid residues in non-amyloidogenic regions have the average Φ-value that is significantly smaller (0.43±0.01) (here and below, the shown error is the error of the average which is calculated as $\sigma/\sqrt{n}$ where σ is the standard deviation of the distribution, and n is the number of points). Thus, in amyloidogenic regions, an average amino acid residue has more than 50% of its contacts formed in the transition state (i.e., in the folding nucleus by our definition). The p-value obtained with Student's t-test (that is, the probability that the observed difference is accidental) is $2*10^{-11}$ that confirms that the difference between the average Φ-values of amino acid residues in amyloidogenic and in non-amyloidogenic regions is significant.

The distributions of Φ-values are shown in Figure 1. It is clear that the distributions of Φ-values in amyloidogenic and in non-amyloidogenic regions are different. It should be noted that a folding nucleus usually contains more amino acid residues (on average, 43% of amino acid residues) than are present in amyloidogenic regions (on average, 24% of amino acid residues), so the folding nucleus is typically larger.

Thus, we demonstrate that theoretically found folding nuclei intersect with experimentally found amyloidogenic regions. The average Φ-values are significantly greater inside amyloidogenic regions than outside them. It means that amyloidogenic regions tend to be present in the folding nucleus of a protein when the protein folds into its native structure.

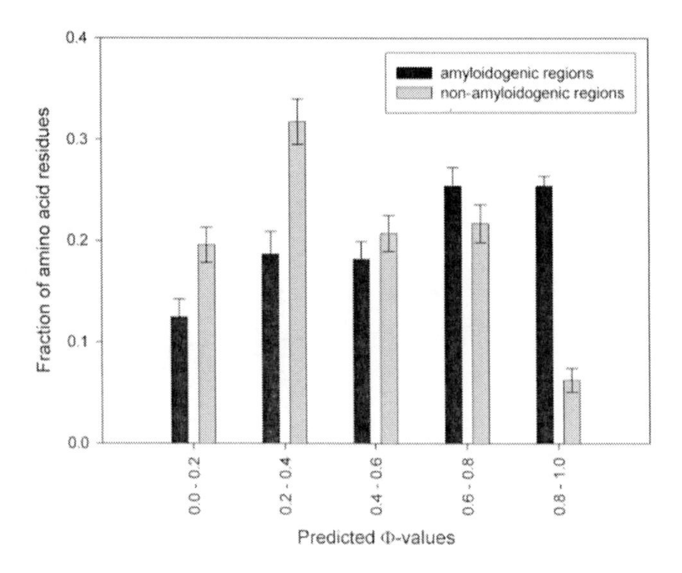

Figure 1. Distribution of predicted Φ-values in experimentally shown amyloidogenic (black bars) and non-amyloidogenic (gray bars) regions of seven investigated proteins.

A Description of Globular Proteins with Experimentally Determined Amyloidogenic Regions

The profiles of Φ-values and the positions of amyloidogenic regions for individual proteins are shown in Figure 2. Totally, amyloidogenic regions correspond to regions with large predicted Φ-values for five of the seven proteins and the difference in the average Φ-values is statistically reliable for each of these five proteins (see the average Φ-values inside amyloidogenic fragments and outside of them for each protein in its profile – as well as probability that this difference in the averages could be obtained by chance).

As it has been mentioned in the introduction the transition states of folding and aggregation for human acylphosphatase are structurally unrelated (Chiti et al., 2002), suggesting that the pathways of folding and aggregation for this protein are structurally distinct. In our case we correctly predict both amyloidogenic regions (19–25 and 91–98). Moreover, the regions with high predicted Φ-values correspond to these regions. The average Φ-value inside amyloidogenic regions is higher (0.59) than in the non-amyloidogenic regions (0.38).

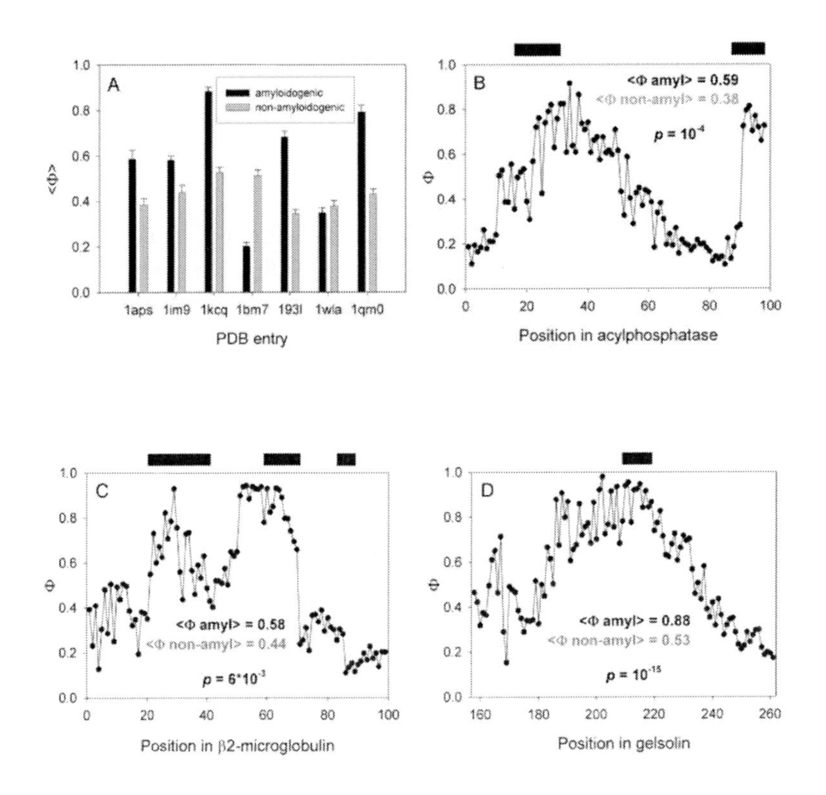

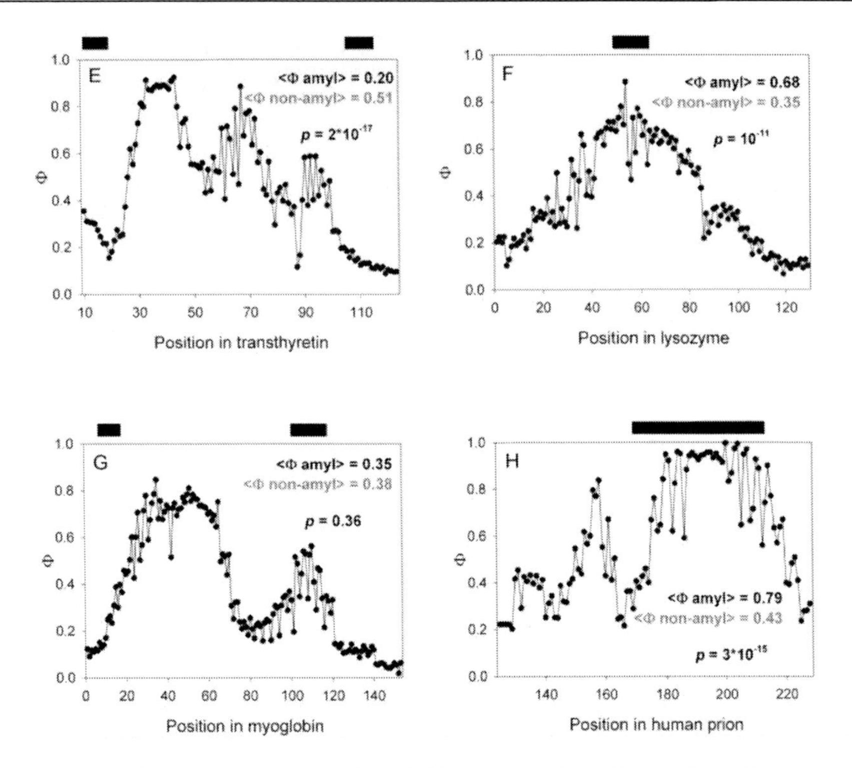

Figure 2. Φ-value profiles for the investigated amyloidogenic proteins. The positions of experimentally found amyloidogenic regions are shown as rectangles above each plot. For each protein, the average Φ-value over amyloidogenic and over non-amyloidogenic region(s) as well as probability to obtain the difference between these averages by chance (evaluated by two-tailed *t*-test) is given.

It has been found experimentally that the following sequences play a dominant role in the amyloidogenesis of β2-microglobulin: residues 20–41 (Kozhukh et al., 2002), residues 59–71 (Jones et al., 2003), and residues 83–89 (Ivanova et al., 2004). All predicted regions are consistent with the experimental data except for fragment 83–89. It should be noted that both large fragments have high Φ-values inside amyloidogenic regions. The average Φ-value inside amyloidogenic regions is higher (0.58) than in the non-amyloidogenic regions (0.44).

For gelsolin one region (52-62) has been found experimentally is responsible for amyloid formation (Maury and Nurmiaho-Lassila, 1992). The average Φ-value inside amyloidogenic regions is higher (0.88) than in the non-amyloidogenic regions (0.53).

The most amyloidogenic peptide fragments from transthyretin (TTR) have been demonstrated in two regions: residues 10–19, which encompass the A strand of the inner β sheet structure that readily forms amyloid fibrils when dissolved in water at low pH (Chamberlain et al., 2000, MacPhee and Dobson, 2000); and residues 105–115, which adopt an extended β strand conformation that is similar to that found in the native protein (Jaroniec et al., 2002). We predicted correctly these important regions (11–17 and 105–113) and one additional region with strong expected packing density (28-34). Namely, the last region corresponds the residues with high predicted Φ-values. The average predicted Φ-value inside amyloidogenic regions is lower (0.20) than in the non-amyloidogenic regions (0.51).

The experimentally found amyloidogenic fragment of lysozyme (one amyloidogenic fragment for lysozyme (residues 26–123 (Frare et al., 2006) or, by another data, residues 49–64 (Krebs et al., 2000)), which has been specifically implicated in amyloidogenic conversion, is a part of the β domain in the native structure of the protein. Our predictions for lysozyme are consistent with experimental results; however, three additional fragments (25–33, 76–82, and 107–114) are also predicted. These regions cover the above mentioned fragments 26-123. The average Φ-value inside amyloidogenic regions is higher (0.68) than in the non-amyloidogenic regions (0.35).

Two amyloidogenic fragments for myoglobin (residues 7–18 according to Picotti et al., 2007, and 101–118 according to Fandrich et al., 2003) have been found experimentally. For apomyoglobin, the primary folding initiation site has been identified at residues 103–115, which encompass the major part of the G helix (Matheson et al., 1978). Moreover, it has been demonstrated that interactions among the A, G, and H helices of apomyoglobin might be important for folding initiation (Gerritsen et al., 1985). The results of our search for residues involved in the folding initiation sites coincide with these sites: 8–16, 27–33, 67–74, and 100–116. The results are consistent with the experimental results that implicated these regions in equilibrium intermediates of apomyoglobin and the earliest kinetic steps in the folding pathway (Nishimura et al., 2005). It should be noted that this has been shown experimentally by using the fragments of the whole protein, but yet we do not know what regions are responsible for amyloid formation for the whole protein. The average predicted Φ-value inside amyloidogenic regions is lower (0.35) than in the non-amyloidogenic regions (0.38).

In the partially unstructured human prion protein, we predict region detected by experiment (169-213 (Lu et al., 2007)). The first α-helix near the N-terminus showed lower protection to solvent exchange compared to the other helices (Liu et al., 1999). The N-terminus of the molecule appears to be predisposed to adopt multiple conformations that may feature in the transition from PrP^{C} to the toxic PrP^{Sc} form (Liu et al., 1999). From the Figure 2 H one can see that the C-terminus of the molecule has higher Φ-values than the N-terminus. Moreover, the average Φ-value inside amyloidogenic regions is higher (0.79) than in the non-amyloidogenic one (0.43). It has been shown that this fast-folding protein has a transition state that is not compact (*m* value analysis gives a β_t value of only 0.3) but contains a developing nucleus between helices 2 and 3 (Hart et al., 2009). This is in agreement with our prediction (see Figure 2 H). There are two residues with large Φ -values (V180 and M206), both of which are squarely on the interface between helix 2 and helix 3. These residues form the majority of the contact area in the central region of this interaction. The fact that a mutation in this nucleus had a negligible effect on stability but still led to formation of aberrant conformations during folding implies an easily perturbed folding mechanism (Hart et al., 2009). It is notable that in inherited forms of human prion disease, where point mutations produce a lethal dominant condition, 20 of the 33 amino acid replacements occur in the helix-2/3 sequence (Hart et al., 2009).

Thus, we have demonstrated that amyloidogenic regions are often predicted to be part of the folding nuclei in amyloidogenic proteins. Therefore, we can hypothesize that amyloidogenic regions often play a crucial role not only in amyloid fibril formation but also in the process of "normal" folding of amyloidogenic proteins into their native structure, since amyloidogenic regions compose part of the folding nucleus in these proteins.

It should be mentioned here that all proteins considering in this chapter except for one (acylphosphortase) are involved in disease. Practically for all these proteins we observe intersection between regions important for folding and misfolding. This confirm our hypothesis that if regions which are important for folding intersect with the regions which are responsible for amyloid formation, then such proteins more probably can misfold in appropriate environment conditions to amyloid fibrils involved *in vivo* in various "amyloid" diseases.

Intersection of Predicted Amyloidogenic Regions and Protected From Hydrogen/Deuterium Exchange with Experimentally Outlined Folding Nuclei

Further, conversely, we have compared predicted amyloidogenic regions with experimentally found folding nuclei for those 20 proteins where folding nuclei have already been outlined experimentally. Prediction of amyloidogenic regions was made by previously described (Galzitskaya et al., 2006a,b,c) method (see also Materials and Methods) which predicts amyloidogenic regions using only amino acid sequence. For each amino acid residue, the method predicts the number of expected contacts and those regions within which all residues have large number of expected contacts are predicted as amyloidogenic regions. As it was demonstrated previously (Galzitskaya et al., 2006c), this method is able to predict amyloidogenic regions.

The comparison of the degree of the involvement into the folding nucleus (reflected in experimental Φ-values) of residues in the predicted amyloidogenic and non-amyloidogenic regions have demonstrated again that there is a reliable difference. In the predicted amyloidogenic regions, the average over Φ-values is 0.41 ± 0.02 while in the predicted non-amyloidogenic regions, the average over Φ-values is 0.33 ± 0.01. Student's *t*-test gives probability 4×10^{-3}; thus, the above difference is statistically reliable. The distribution of experimentally measured Φ-values is shown in Figure 3. It is obvious that the distributions in amyloidogenic and in non-amyloidogenic fragments are different. The fraction of residues from the amyloidogenic regions with large Φ-values is greater compared to the non-amyloidogenic regions. This result is valid for both cases (predicted Φ-values *vs.* experimental amyloidogenic regions and experimental Φ-values *vs.* predicted amyloidogenic regions, see Figure 1 and Figure 3). In Table 2, the predicted amyloidogenic regions for each of 20 investigated proteins are shown as well as those experimental Φ-values which are above 0.50. One can see that only for four of the 20 proteins we do not observe an intersection between folding nuclei and amyloidogenic regions. It should be underlined that, while the fraction of all known experimental Φ-values in the predicted amyloidogenic regions is 25%, the fraction of large Φ-values within amyloidogenic fragments is 33% (those Φ-values are shown in bold in Table 2). The obtained fraction (33%) remains the same if to consider as the large experimental Φ-values those ones which are above 0.45 or above 0.40. Since this difference is not occasional, the method of prediction of amyloidogenic regions gives us opportunity to predict the location of nucleation centers from protein sequence.

In addition one can suggest that regions protected from the hydrogen/deuterium exchange will be responsible for aggregation. It has been demonstrated earlier that the scale of probability of formation of hydrogen bonds (donors scale) can be used for both the prediction as protection from hydrogen/deuterium exchange and for the prediction of amyloidogenic/aggregation regions (Dovidchenko et al., 2009; Garbuzynskiy et al., 2010). For these 20 proteins the regions protected from the hydrogen/deuterium exchange have been predicted by the program elaborated by us (see Table 2). One can see that for 17 proteins, we obtained intersection with predicted amyloidogenic regions. Moreover, additional predicted regions by the second scale (probability formation of hydrogen bonds) also intersect with residues with high phi-values (such residues are highlighted by bold and italic in Table 2, the last column).

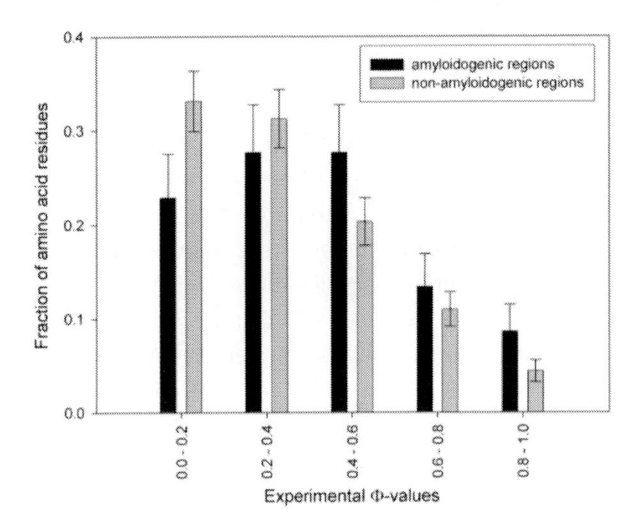

Figure 3. Distribution of experimental Φ-values in predicted amyloidogenic (black bars) and non-amyloidogenic (gray bars) regions of 20 investigated proteins.

It should be noted here that thresholds obtained from the optimization of two methods for prediction of protected regions from hydrogen/deuterium exchange and amyloidogenic regions are the same (0.69). Although, the testing has been done on different databases (in the first case the database includes 14 proteins with known experimental data about protection of amino acid residues from hydrogen/deuterium exchange, the second database includes 144 amyloid forming peptides and 263 non-amyloid forming peptides) but using the same scale of probability of formation of hydrogen bonds.

It is interesting to analyze the profiles of real contacts calculated from the 3D structure and expected contacts calculated from the amino acid sequence. Figure 4 demonstrates such a comparison for all 20 proteins considered in this paragraph. Also we added the profile of probability of hydrogen bond formation which can tell us about such properties as protection from the hydrogen/deuterium exchange and amyloidogenic regions.

Thus, both comparisons (predicted folding nuclei *vs.* experimentally known amyloidogenic fragments and experimentally known folding nuclei *vs.* predicted amyloidogenic fragments and regions protected from the hydrogen/deuterium exchange) indicate that the location of folding nuclei intersects with the location of amyloidogenic regions. In other words, nucleation centers for folding and for misfolding often intersect.

For 20 proteins considered we have no information about involvement of these proteins in some disease. But we may hypothesize that if for proteins amyloidogenic regions intersect with the initiation site for folding, then such proteins more probably can misfold in appropriate environment conditions to amyloid fibrils involved *in vivo* in various “amyloid diseases,” for proteins where such intersection is absent. Because globular proteins are highly amyloidogenic, their primary defense against aggregation is fast and effective folding.

Table 2. Experimentally found Φ–values, predicted regions which are protected from hydrogen/deuterium exchange, and predicted amyloidogenic regions

	Protein or domain name	Size Number of Φ-values	Regions protected from hydrogen/deuterium exchange	Predicted amyloidogenic regions	Residues with experimental Φ-values larger than 0.50 (Φ-value)[a]
1	IgG binding domain of streptococcal protein G, 1pgb	56 (20)	**3 —— 8**	1–7[b]	32(0.55); 46(0.96); 47(0.67); 49(0.84)
2	B-domain of staphylococcal protein A, 1bdc	60 (15)	**13 —— 21**	14–21	**14(0.65)**; **17(1.0)**; **18(1.0)**; 32(0.57)
3	SH3 domain of α-spectrin, 1shg	62 (13)	**7 —— 14**	8–14; 31–35; 40–44[b]	47(0.63); 52(0.58); 53(0.61); 55(0.53)
4	SH3 domain of src tyrosine kinase transforming protein, 1srm	64 (26)	1 —— 5	9–16; 41–47	30(0.62); 32(0.55); 35(0.77); **44(0.54); 47(0.95)**; 48(0.72); 50(0.86); 53(0.68); 55(0.56); 56(0.71)
5	DNA-binding protein Sso7d, 1bf4	64 (15)	**20 —— 27** 54 —— 59	22–26; 30–34[b]	**25(0.65);** 45(0.59); ***57(0.54); 59(0.60)***
6	Chymotrypsin inhibitor (CI2), 2ci2	64 (34)	18 —— 22 **44 — 52**	29–35; 39–52	18(0.70); **49(0.53)**

Table 2. Continued

	Protein or domain name	Size Number of Φ-values	Regions protected from hydrogen/deuterium exchange	Predicted amyloidogenic regions	Residues with experimental Φ-values larger than 0.50 (Φ-value)[a]
7	IgG binding domain of streptococcal protein L, 2ptl	64 (38)	**8——13** 58——62	7–13	4(0.51); **7(0.62); 8(0.53)**; 14(0.85); 21(0.75)
8	Colicin E9 immunity protein, 1imp	86 (19)	**15——21** **34——38**	15–22; 34–40	13(0.98); **15(0.57); 16(0.52)**
9	Colicin E7 immunity protein, 1cei	87 (19)	15——20 **35——40**	34–44	3(0.52); 7(0.56); ***16(0.72); 18(0.52)***; 27(0.56); **37(0.51); 44(1.0)**; 68(0.86); 77(0.86); 78(0.78)
10	Ig repeat 27 of titin, 1tiu	89 (22)	10——15 **56——66** 83——88	56–64	23(0.82); 49(0.66); **56(0.52); 58(0.79)**; **60(0.67)**; 71(0.63);-75(0.63)
11	Barstar, 1btb	89 (22)	71——75 84——89	35–56	5(0.63); 9(0.72); 18(0.69); 25(0.68); **36(0.70)**; **37(0.59); 50(0.77)**; 68(0.52); ***72(0.81);*** 77(0.90); 79(0.63); ***85(0.51)***
12	3rd Fn3 repeat of tenascin (short form), 1ten	90 (25)	**18——22** 70——74	19–26	36(0.53); 48(0.67); ***70(0.54)***

	Protein or domain name	Size Number of Φ-values	Regions protected from hydrogen/deuterium exchange	Predicted amyloidogenic regions	Residues with experimental Φ-values larger than 0.50 (Φ-value)[a]
13	10th FN3 module of fibronectin, 1ttf	94 (16)	**28——35** **70——75**	28–35; 69–75	8(0.55); **34(0.61)**; 36(0.52); **72(0.51); 74(0.65)**; 92(0.79)
14	Spliceosomal protein U1A, 1urn	96 (10)	**30——34** **40——44** **54——59** 93——97	9–15; 30–45; 53–59	**44(0.60)**
15	Ribosomal protein S6, 1ris	97 (10)	25——29 **44——48** **59——65** 77——81 86——92	1–9; 45–51; 59–65	**6(0.52)**
16	Human muscle acylphosphatase, 1aps[c]	98 (17)	**18——23** 33——38 85——89	19–25; 91–98	11(0.93); 47(0.54); 54(0.98); **94(0.76)**
17	FK-506 binding protein (FKBP12), 1fkb	107 (22)	**102—106**	96–107	23(0.52); 27(0.85); 63(0.51); **101(0.57)**
18	Barnase, 1rnb	110 (19)	1-5 31-35	87–98; 103–109	13(0.60); 15(0.80); 57(1.0); **87(0.80)**; **90(0.9)**; **91(1.0); 96(0.40)**

Table 2. Continued

	Protein or domain name	Size Number of Φ-values	Regions protected from hydrogen/deuterium exchange	Predicted amyloidogenic regions	Residues with experimental Φ-values larger than 0.50 (Φ-value)[a]
19	Villin 14T domain 1, 2vil	126 (18)	**45——49** **61——65** 89——94 112—117	18–24; 42–49; 58–66	**23(0.65)**; 28(0.58); **44(0.85)**; **46(0.69); 48(0.62);** 73(0.69); 77(0.52); 78(0.56); 81(0.75); 84(0.68)
20	CheY, 3chy	128 (9)	5——9 **17——26** 66——71 93——98	16–22; 51–57; 80–86	35(0.75); 37(0.60); 41(0.68)

[a]If several Φ-values are known for a single residue (several different mutations were made), the Φ-values were averaged.

[b]For these proteins, amyloidogenic regions are revealed only with window size of five residues. The size of the sliding window for the rest proteins was seven residues. In the case of prediction of regions protected from hydrogen/deuterium exchange the size of window was five residues.

[c]Experimentally, amyloidogenic regions were revealed for human acylphosphatase. 3D structure of this protein is unknown. However, there is a 3D structure for horse acylphosphatase sequence of which is highly similar to that of the human protein (sequence identity 95%).

Table 3. Prediction of amyloidogenic regions for proteins with swapped regions

	SCOP class, name, and number of residues	PDB file with swapped structure	Predicted amyloidogenic regions	Location of swap region	PDB file with non-swapped structure
1	a, PECTIN METHYLESTERASE INHIBITOR, 153	1x8z	20-27 70-74 140-148	1-31	1x91
2	a, CALBINDIN D9K, 76	1ht9	30-35 68-72	45-76	1ksm
3	b, T-CELL SURFACE ANTIGEN CD2, 94	1a6p	16-20 72-26 89-94	1-41	1hng
4	b, NERVE GROWTH FACTOR RECEPTOR TRKA, 109	1wwa	14-22, 32-36 45-51, 62-67 93-98	1-19	1www
5	b, CYANOVIRIN-N, 101	1l5b	53-57	1-51	2ezm (CASP3)
6	c, SPORULATION RESPONSE REGULATOR SPO0A, 130	1dz3	1-7, 51-59 65-69, 80-85 101-106 114-120	108-130	1qmp
7	d, BARNASE, 110	1yvs	1-6, 12-16 33-37, 92-98	1-37	1a2p
8	d, IMMUNOGLO-BULIN G BINDING PROTEIN G, 56	1q10	1-7 31-35	38-56	1pgb
9	d, CYSTATIN A, 98	1n9j	52-57 64-69 80-84	1-48	1nb5
10	d, PROTEGRIN-3 PRECURSOR, 101	1lxe	3-12 24-30	1-38	1n5h
11	d, PROTEIN-TYROSINE PHOSPHATASE YOPH, 136	1k46	40-44 64-68 115-119	1-29	1m0v
12	d, RIBONUCLEASE A, 124	1f0v	105-109 116-120	114-124	1kf3
13	d, RIBONUCLEASE A, 124	1a2w	105-109 116-120	1-20	1kf3
14	d, MAJOR PRION PROTEIN (human), fragment 119-226, 108	1i4m	18-23, 29-34 56-60, 62-66 93-99	75-108	1qm2
15	d, SH2 DOMAIN, 114	1fyr	12-16 59-63 70-77 98-102	76-114	1bmb
16	d, CATABOLITE REPRESSION HPR, 87	1mol	20-24 48-53	1-12	1k1c

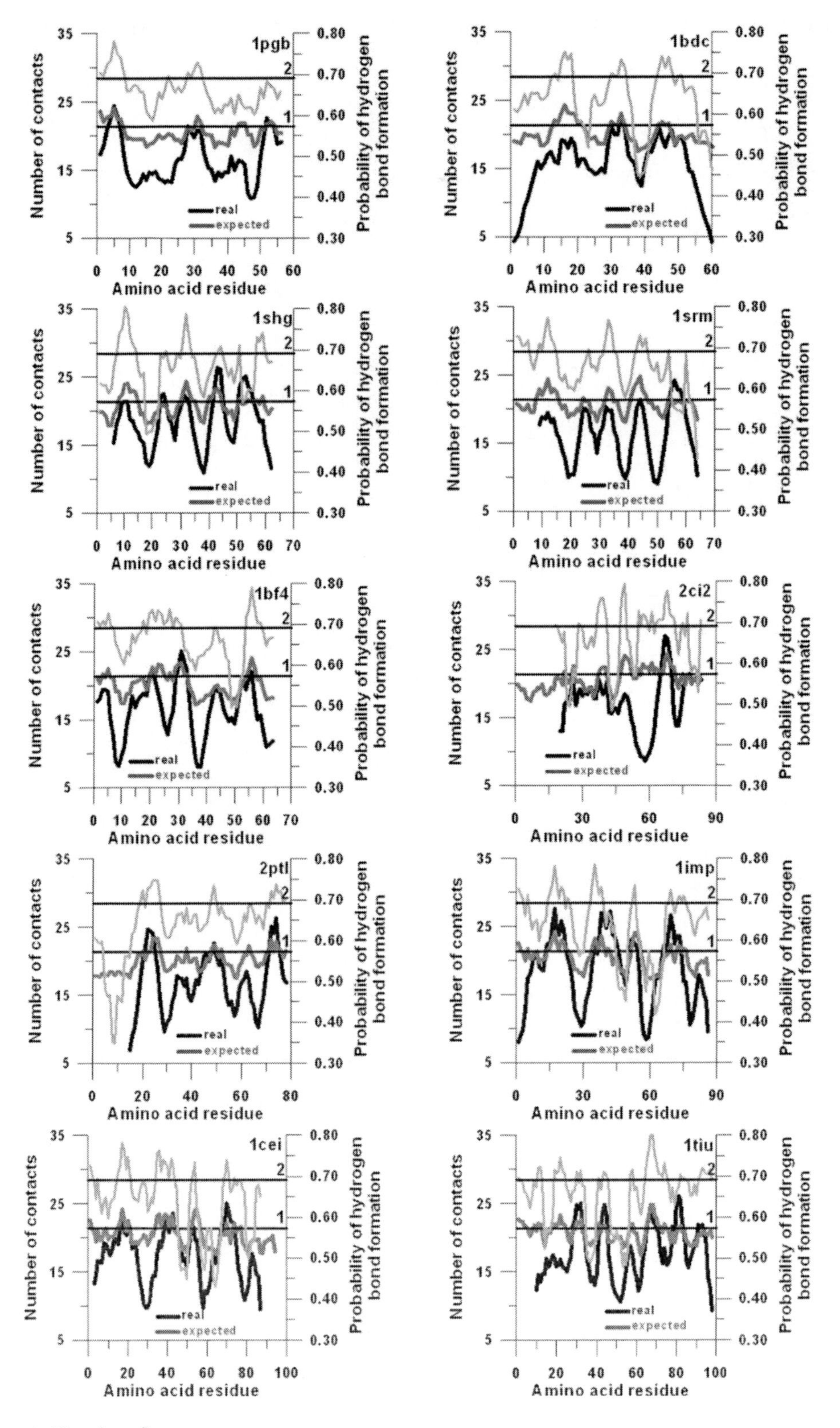

Figure 4. (Continued).

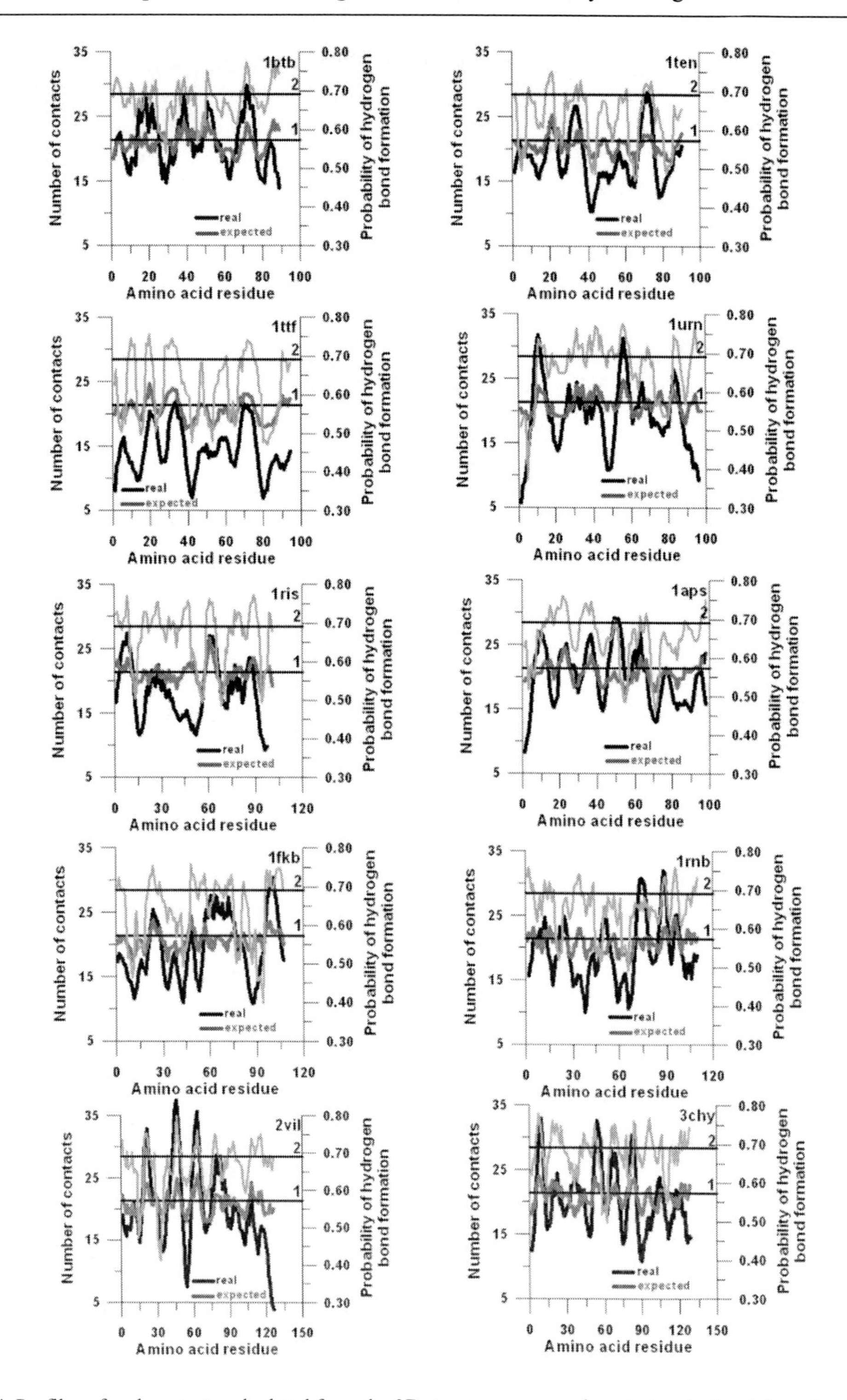

Figure 4. Profiles of real contacts calculated from the 3D structure, expected contacts calculated from the amino acid sequence, and probability of hydrogen bond formation for all 20 proteins considered in this paragraph. Horizontal line 1 corresponds the threshold for prediction of amyloidogenic regions using the contact scale. Horizontal line 2 corresponds the threshold for prediction of amyloidogenic regions using the scale of probability of hydrogen bond formation.

Modeling of Folding of the Proteins with Swapped Domains

Eisenberg and his colleagues (Bennett et al., 1994, 1995) have proposed a mechanism for protein oligomerization, 3D domain swapping. In a domain swapping oligomer, one segment of a monomeric protein is replaced by the same segment from another chain. Some cases of amyloid formation can be explained through a domain swapping mechanism where the swapped segment is either a beta-hairpin or other conformation. It has been shown how the swapping of beta-hairpin form twisted beta-sheet conformation, explaining the remarkable high stability of the protofibril *in vitro* (Sinha et al., 2001).

One can hypothesize that regions which are responsible for aggregation and protein folding will often appear as swapping regions in the proteins with swapped domains. For this purpose we collected 16 pairs for which we know 3D structures of swapped domains in the oligomeric form and 3D structures of the monomeric form. For these proteins, we calculated the amyloidogenic/aggregation regions using our program (FoldAmyloid) and for monomeric form we calculated Φ-values for each residue. From Table 3 one can see that for 10 out of 16 proteins, swapping regions intersect with the predicted amyloidogenic regions (such pdb files are highlighted by bold, third column). Moreover, for 10 out of 16 proteins, high Φ-values belong to residues from the swapping regions (such pdb files are highlighted by bold, the last column). By modeling the folding of proteins with swapped domains we again observe that regions, which are responsible for swapping, are also responsible for folding and misfolding. It should be noted that in all cases the swapping region is at the termini of the protein chain.

Naturally occurring variants of cystatin C, gelsolin, immunoglobulin, transthyretin, lysozyme, prion, and α_1-antitrypsin are known. For some of these proteins only the proteins with single mutations polymerize and the wild-type does not, as in the cases of lysozyme and gelsoline. In α_1-antitrypsin both the wild-type and the mutant form fibrils. Such a swapping mechanism is attractive as it involves a universal mechanism in proteins, critical for their function. For cystatin C (1a67), one of the regions identified as amyloidogenic (58AGVNYF63/59GVNYFL64) is located at the long β-sheet formed upon domain-swapping, as shown by the crystal structure of its dimmer (Pastor et al., 2008). The putative role of this region in the oligomerization of cystatin C has been discussed (Janowski et al., 2001). There is more than one potential candidate motif which may undergo swapping to yield amyloids for transhtyretin (1bmz) and serum amyloid P component (1sac), and for immunoglobulin (1bmz) (Sinha et al., 2001).

Summarizing all cases, one can hypothesize that if for proteins amyloidogenic regions intersect with the initiation site for folding, then such proteins more probably can misfold in appropriate environment conditions to amyloid fibrils involved *in vivo* in various "amyloid" diseases.

Materials and Methods

Creation of the Database of Amyloidogenic Proteins

We collected globular proteins for which amyloidogenic regions have been experimentally determined and the three-dimensional structure of the native state is resolved and located in the Protein Data Bank (PDB) (Berman et al., 2000). If 3D structure of a protein is resolved more than once, we chose the X-ray structure with the best resolution. A list of the selected proteins along with their PDB entries and experimentally determined amyloidogenic regions are presented in Table 1.

The Database of Proteins with Experimentally Outlined Folding Nucleus

Into this database (Garbuzynskiy et al., 2004; Garbuzynskiy et al., 2005), we collected the proteins where folding nuclei are experimentally outlined (that is, where Φ-values which reflect the extent of involvement of a residue into the folding nucleus are known). We excluded those Φ-values which were below zero or greater than unity (which have no interpretation in terms of native-like contacts) as well as those Φ-values which were measured based on non-single point mutations or mutations which were not changes from a larger residue to a smaller one (Fersht, 1999). List of the experimentally found Φ-values as well as the corresponding mutations can be found at http://phys.protres.ru/resources/phi_values.htm. Experimentally found Φ-values (> 0.5) and predicted amyloidogenic regions can be found at http://phys.protres.ru/resources/phi values amyloid.html/.

Prediction of Amyloidogenic Regions in Proteins

For prediction of amyloidogenic regions, we used the previously described (Galzitskaya et al., 2006a,b,c) method of prediction of regions with large number of contacts per residue in protein sequences. For each amino acid residue in the protein sequence, an expected number of contacts per residue is attributed according to the type of the residue (which is the average number of contacts at a distance below 8 Å for the given type of residue in 3D structures of proteins (Galzitskaya et al., 2006a,b)). Then, the values are averaged with a sliding window of seven (or five) residues (Galzitskaya et al., 2006c). In the obtained profile, those regions inside which all residues have values above a threshold (which is 21.4 expected contacts per residue) are predicted as amyloidogenic if the size of such a region is not smaller than the sliding window (Galzitskaya et al. 2006a,b,c). Thus, the predicted amyloidogenic regions are regions which have a large number of expected contacts per residue.

We tried to predict amyloidogenic regions not only by our method but by other methods which (as their authors state) can predict amyloidogenic regions. But at first, we decided to test the quality of the predictions of these methods (as well as our method) on those seven proteins for which there are experimentally known amyloidogenic fragments. We used TANGO (dis.embl.de) and ZYGGREGATOR (http://www-vendruscolo.ch.cam.ac.uk/zyggregator.php). The predictions which we have done using all three methods are shown in the Table 4.

One can see that TANGO does not find the most of experimentally revealed amyloidogenic fragments (only two of 12 experimental amyloidogenic regions are correctly predicted, and one (for myoglobin) is in the wrong place). Moreover, the predictions of TANGO (made by this server) do not reproduce the results reported in the article describing this method (Fernandez-Escamilla et al., 2004). ZYGGREGATOR works better for this set of proteins (seven of 12 amyloidogenic regions are correctly predicted, that is, intersect with real amyloidogenic fragments) but ZYGGREGATOR predictions are still worse compared to our method (we predict correctly 10 of 12 experimentally revealed amyloidogenic fragments).

Since both methods do not give good predictions of amyloidogenic regions for proteins for which the quality of predictions can be evaluated, we could not add predictions done by these methods to our manuscript.

Theoretical Search for Folding Nuclei

We model the process of reversible unfolding of the protein "normal" (native) globular structure into the unfolded state at the point of thermodynamic equilibrium (where the free energies of the folded and the unfolded states are equal). Under these conditions, only the completely folded and completely unfolded states are observed while all other states (misfolded, partially folded etc.) are destabilized. The pathways of folding and unfolding coincide and all dead-ends are destabilized, so protein chain behavior can be approximated as a reversible folding/unfolding process.

We start from a known tertiary structure of the investigated protein in the native state and stepwise unfold its amino acid residues. A link of five residues is unfolded at a single step (this size of link was shown to be optimal for predictions of folding nuclei in globular proteins) (Garbuzynskiy et al., 2004). Finally, we obtain the completely unfolded state. The unfolded residues lose all contacts that existed in the tertiary structure but gain the entropy of the coil (except for the entropy which is expended on fixation of unfolded loops that protrude from the folded part of the structure). Thus, any amino acid residue in our model can exist in one of two states: folded, where it has all of its native contacts with the other folded residues, and unfolded, where it has no contacts but has more entropy.

Considering all the ways of unfolding the protein structure, we obtain a network of protein folding/unfolding pathways. The free energy of any state is calculated as follows:

$$F(I) = \varepsilon \times n_I - T[\eta_I \times \sigma + \sum_{loops \in I} S_{loop}], \quad (2)$$

where n_I is the number of native atom-atom contacts in the native-like part of I (n_I does not include contacts of neighbor residues, which also exist in the coil; in this work, we consider two atoms to form a contact if their centers are separated by less than 6Å, which is the value of this parameter that is optimal for folding nuclei predictions (Garbuzynskiy et al., 2004)); ε is the energy of one contact (all contacts are assumed to be equal in energy); η_I is the number of residues in the unfolded part of I; T is the temperature; σ is the entropy difference between the coil and the native state of a residue (we take $\sigma = 2.3R$ (Privalov, 1979), where R is the gas constant). The sum Σ is taken over all closed unfolded loops that protrude from the native-like part of I. For a loop between fixed residues k and l, S_{loop} is calculated as follows (Finkelstein and Badretdinov, 1997):

$$S_{loop} = -{}^{5}/_{2}R \ln|k-l| - {}^{3}/_{2}R\,(r^{2}_{kl} - a^{2}) / (2Aa|k-l|), \quad (3)$$

Table 4. Comparison of predictions of amyloidogenic regions by different methods

Name of the protein	PDB entry	Experimentally investigated amyloidogenic regions	Regions predicted		
			by our method	by Tango	by Zyggregator
Acylphosphatase	1aps	16-31 (Chiti et al., 2002); 87-98 (Chiti et al., 2002)	19 - 25; 91 - 98	-	19 - 20; 36 - 37
β2-microglobulin	1im9	20-41 (Kozhukh et al., 2002); 59-71 (Jones et al., 2003); 83-89 (Ivanova et al., 2004)	24 - 30; 60 - 70	61 - 69	24 - 26; 63 - 66
Gelsolin	1kcq	52-62 (Maury and Nurmiaho-Lassila, 1992)	1 - 7	-	30; 32 -33; 40 - 41
Transthyretin	1bm7	10-19 (Chamberlain et al., 2000); 105-115 (Jaroniec et al., 2002)	10 - 17; 28 - 34; 105 - 113	107 - 111	28; 93 - 94; 117 - 118
Lysozyme	1931	49-64 (Krebs et al., 2000)	25 - 31; 54 - 65; 106 - 113; 121 - 129	-	25 - 26; 28 - 29; 31; 55 - 56
Myoglobin	1wla	7-18 (Picotti et al., 2007); 101-118 (Fandrich et al., 2003)	8 - 14; 27 - 33; 100 - 116; 134 - 140	66 - 72	10 - 12; 14; 67 - 70; 104; 109
Human prion	1qm0	169-213 (Lu et al., 2007)	136 - 142; 159 - 166; 176 - 183	-	126 - 128; 177 - 179; 190; 212 - 213; 215

where r_{kl} is the distance between the C_α atoms of residues k and l, a = 3.8Å is the distance between the neighbor C_α atoms in the chain, and A is the persistence length for a polypeptide (according to Flory (Flory, 1969), we take A = 20Å). The term $-{}^5/_2R \ln|k - l|$ is the most significant term in this equation; the coefficient $-{}^5/_2$ (rather than Flory's value $-{}^3/_2$) follows from the condition that a loop cannot penetrate inside the globule (Finkelstein and Badretdinov, 1997).

Calculation of Φ-Values

Since in our model all native contacts have equal energy and non-native contacts are absent, the Φ-value is a fraction of native contacts formed in the transition state compared to the native state (Garbuzynskiy et al., 2004):

$$\Phi = \langle \Delta_r(n_I) \rangle_{I\#} / \Delta_r(n_0), \quad (4)$$

where $\Delta_r(n_0)$ is number of contacts deleted by the mutation in the given amino acid residue r in the native state of the protein while $\langle \Delta_r(n_I) \rangle_{I\#}$ is the number of contacts deleted by the mutation averaged over all transition states found with our method. We consider all mutations on Gly; if a residue in the wild-type protein is already Gly, we take its probability to be structured in the transition state ensemble (Garbuzynskiy et al., 2005).

The statistical reliability of differences between the average Φ-values (namely, the probability that the difference was obtained by chance) was evaluated using two-tailed t-test with unequal dispersions.

Creation of a Database pf Swapped Proteins

First, all entries where a word "swap" was met were selected from the PDB database. Multi-domain proteins and proteins with split domains were excluded. Large proteins (larger than 200 amino acid residues) were also excluded. There were several cases when a PDB entry contained both swapped and non-swapped chains; we strongly preferred such cases. 16 such pairs have been found (see Table 3).

Predicting Protection of Amino Acid Residues From Hydrogen-Deuterium Exchange Using Amino Acid Sequence Only

For the prediction of the extent of protection of amino acid residues from hydrogen-deuterium exchange by the protein amino acid sequence, we collected statistics of intraprotein contacts and of hydrogen bonds in the known spatial structures of proteins (Dovidchenko et al., 2009). For this purpose, we composed a database of spatial structures (based on the SCOP database). The database included proteins or protein domains with the amino acid sequence identity below 25% and belonging to the most widespread four structural classes (all-α proteins, all-β proteins, α/β proteins, and α+β proteins). The obtained database includes 3769 proteins and protein domains. For each of the 20 types of amino acid residues we calculated the average (for a given type) number of intraprotein contacts as well as the probability of intraprotein hydrogen bond formation by the amide group of the main chain of amino acid residues of the given type. Upon prediction of number of contacts and probability of hydrogen bond formation by protein amino acid sequence, these average values were attributed to each residue of the given type. The average number of intraprotein contacts was minimal for glycine (17.11 contacts) and maximal for tryptophane (28.48 contacts). Large hydrophobic amino acid residues had (on average) more intraprotein contacts compared to small and hydrophilic residues. The probability of hydrogen bond formation by the NH-group of a residue varied from 0 (in the case of proline in which NH-group is absent) to 0.84 (for leucine). It should be noted that hydrophilic amino acid residues had smaller probabilities of hydrogen bond formation inside the main chain of the protein compared to hydrophobic residues which probably reflects a larger involvement of hydrophobic residues in regular secondary structure elements.

During prediction of protection of amino acid residues from hydrogen-deuterium exchange, residues which had the number of predicted contacts above a cutoff value were predicted as protected, and residues which had a smaller number of predicted contacts were predicted as non-protected. Similarly, residues which had the probability of hydrogen bond formation above a cutoff value were predicted as protected, and residues which had smaller probability of hydrogen bond formation were predicted as non-protected. The results of the prediction of protection of a residue from hydrogen-deuterium exchange were the best if the cutoff value was equal to 22 intraprotein contacts per residue. Consequently, residues of tryptophane, phenylalanine, tyrosine, isoleucine, leucine, methionine, valine, and cysteine were predicted as protected residues while the other types of residues were predicted as non-protected ones in the optimal case. Thus, mainly large hydrophobic residues are predicted as protected from hydrogen-deuterium exchange. A similar situation is observed in the case of predicting protection of amino acid residues by the predicted probability of hydrogen bond formation. Residues of leucine, isoleucine, valine, methionine, cysteine, phenylalanine, tyrosine, alanine, and tryptophane are predicted as protected ones, while the other types of residues are predicted as non-protected. The best quality of predictions is in the case when the cutoff value is 0.69 (amino acid residues which have probability of hydrogen bond formation greater than 0.69 are predicted as protected from hydrogen-deuterium exchange, and the other residues are predicted as non-protected ones). It is interesting that the set of amino acid residues predicted as protected ones is practically the same in both variants of prediction. The only exception is alanine which is predicted as non-protected in predictions by contacts and as protected in predictions by hydrogen bonds.

Acknowledgments

The author is grateful to S. Garbuzinsky and A.V. Glyakina for assistance in preparation of this paper.

Funding

This work was supported by the programs "Molecular and Cellular Biology" (01200959110) and "Fundamental Sciences to Medicine", by the Russian Foundation for Basic Research (11-04-00763), by the "Russian Science Support Foundation", and by a grant from the Federal Agency for Science and Innovation (#02.740.11.0295).

References

Alm, E. and Baker, D. 1999. Prediction of protein-folding mechanisms from free-energy landscapes derived from native structures. *Proc. Natl. Acad. Sci. USA* 96: 11305–11310.

Alm, E., Morozov, A.V., Kortemme, T., and Baker, D. 2002. Simple physical models connect theory and experiment in protein folding kinetics. *J. Mol. Biol.* 322: 463–476.

Bennett, M.J., Choe, S., Eisenberg, D. 1994. Domain swapping: entangling alliances between proteins. *Proc Natl Acad Sci U S A* 91(8): 3127-3131.

Bennett, M.J., Schlunegger, M.P., Eisenberg, D. 1995. 3D domain swapping: a mechanism for oligomer assembly. *Protein Sci.* 4(12): 2455-2468.

Berman, H.M., Westbrook, J., Feng, Z., Gilliland, G., Bhat, T.N., Weissig, H., Shindyalov, I.N., and Bourne, P.E. 2000. The Protein Data Bank. *Nucl. Acids Res.* 28: 235–242.

Chamberlain, A.K., MacPhee, C.E., Zurdo, J., Morozova-Roche, L.A., Hill, H.A., Dobson C.M., and Davis J.J. 2000. Ultrastructural organization of amyloid fibrils by atomic force microscopy. *Biophys. J.* 79: 3282–3293.

Chiti, F., Taddei, N., Baroni, F., Capanni, C., Stefani, M., Ramponi, G., and Dobson, C.M. 2002. Kinetic partitioning of protein folding and aggregation. *Nat. Struct. Biol.* 9: 137–143.

Chiti, F., Webster, P., Taddei, N., Clark, A., Stefani, M., Ramponi, G., and Dobson, C.M. 1999. Designing conditions for *in vitro* formation of amyloid protofilaments and fibrils. *Proc. Natl. Acad. Sci. USA* 96: 3590–3594.

Daggett, V., Li, A., Itzhaki, L.S., Otzen, D.E., and Fersht, A.R. 1996. Structure of the transition state for folding of a protein derived from experiment and simulation. *J. Mol. Biol.* 257: 430–440.

Dovidchenko, N.V., Lobanov, M.Yu., Garbuzynskiy, S.O. and Galzitskaya, O.V. 2009. Prediction of amino acid residues protected from hydrogen-deuterium exchange in a protein chain. *Biochemistry* (Moscow), 74(8), 1091-1102.

Esteras-Chopo, A., Serrano, L., and Lopez de la Paz, M. 2005. The amyloid stretch hypothesis: Recruiting proteins toward the dark side. *Proc. Natl. Acad. Sci. USA* 102: 16672–16677.

Fandrich, M., Fletcher, M.A., and Dobson C.M. 2001. Amyloid fibrils from muscle myoglobin. *Nature* 410: 165–166.

Fandrich, M., Forge, V., Buder, K., Kittler, M., Dobson, C.M., and Diekmann, S. 2003. Myoglobin forms amyloid fibrils by association of unfolded polypeptide segments. *Proc. Natl. Acad. Sci. USA* 100: 15463–15468.

Fernandez-Escamilla, A.M., Rousseau, F., Schymkowitz, J., Serrano, L. 2004. Prediction of sequence-dependent and mutational effects on the aggregation of peptides and proteins. *Nature Biotechnology*, 22:1302-1306.

Fersht, A.R. 1995. Characterizing transition states in protein folding: an essential step in the puzzle. *Curr. Opin. Struct. Biol.* 5: 79–84.

Fersht, A.R. 1997. Nucleation mechanisms in protein folding. *Curr. Opin. Struct. Biol.* 7: 3–9.

Fersht, A.R. 1999. *Structure and mechanism in protein science: a guide to enzyme catalysis and protein folding.* W.H. Freeman and Company, New York.

Finkelstein, A.V. and Badretdinov, A.Ya. 1997. Rate of protein folding near the point of thermodynamic equilibrium between the coil and the most stable chain fold. *Fold. Des.* 2: 115–121.

Finkelstein, A.V. and Roytberg, M.A. 1993. Computation of biopolymers: a general approach to different problems. *Biosystems*, 30: 1–19.

Flory, P.J. 1969. *Statistical Mechanics of Chain Molecules*, Interscience, New York.

Frare, E., Mossuto, M.F., Polverino de Laureto, P., Dumoulin, M., Dobson, C.M., Fontana, A. 2006. Identification of the core structure of lysozyme amyloid fibrils by proteolysis. *J. Mol. Biol.*, 361: 551–561.

Galzitskaya, O.V. and Finkelstein, A.V. 1999. A theoretical search for folding/unfolding nuclei in three-dimensional protein structures. *Proc. Natl. Acad. Sci. USA* 96: 11299–11304.

Galzitskaya, O.V., Garbuzynskiy, S.O., and Finkelstein, A.V. 2005. Theoretical study of protein folding: outlining folding nuclei and estimation of protein folding rates. *Journal of Physics: Condensed Matter* 17: S1539–S1551.

Galzitskaya, O.V., Garbuzynskiy, S.O., and Lobanov, M.Yu. 2006a. A search for amyloidogenic regions in protein chains. *Mol. Biol. (Moscow)* 40: 821–828.

Galzitskaya, O.V., Garbuzynskiy, S.O., and Lobanov, M.Yu. 2006b. Is it possible to predict amyloidogenic regions from sequence alone? *Journal of Bioinformatics and Computational Biology* 4: 373–388.

Galzitskaya, O.V., Garbuzynskiy, S.O., and Lobanov, M.Yu. 2006c. Prediction of amyloidogenic and disordered regions in protein chains. *PLoS Computational Biology* 2: e177.

Garbuzynskiy, S.O., Finkelstein, A.V., and Galzitskaya, O.V. 2004. Outlining folding nuclei in globular proteins. *J. Mol. Biol.* 336: 509–525.

Garbuzynskiy, S.O., Finkelstein, A.V., and Galzitskaya, O.V. 2005. On the prediction of folding nuclei in globular proteins. *Mol. Biol. (Moscow)* 39: 906–914.

Garbuzynskiy, S.O., Lobanov, M. Yu. and Galzitskaya, O. V. 2010. FoldAmyloid: a method of prediction of amyloidogenic regions from protein sequence. *Bioinformatics* 26(3): 326-332.

Gerritsen, M., Chou, K.C., Nemethy, G., Scheraga, H.A. 1985. Energetics of multihelix interactions in protein folding: Application to myoglobin. *Biopolymers*, 24(7):1271–1291. Erratum 24:2177, 2005.

Guijarro, J.I., Sunde, M., Jones, J.A., Campbell, I.D., and Dobson, C.M. 1998. Amyloid fibril formation by an SH3 domain. *Proc. Natl. Acad. Sci. USA* 95: 4224–4228.

Hart, T., Hosszu, L.L., Trevitt, C.R., Jackson, G.S., Waltho, J.P., Collinge, J., Clarke, A.R. 2009. Folding kinetics of the human prion protein probed by temperature jump. *Proc Natl Acad Sci U S A.* 106: 5651-5656.

Ivanova, M.I., Sawaya, M.R., Gingery, M., Attinger, A., and Eisenberg, D. 2004. An amyloid-forming segment of beta2-microglobulin suggests a molecular model for the fibril. *Proc. Natl. Acad. Sci. USA* 101: 10584–10589.

Jahn, T.R., Parker, M.J., Homans, S.W., and Radford, S.E. 2006. Amyloid formation under physiological conditions proceeds via a native-like folding intermediate. *Nat. Struct. Mol. Biol.* 13: 195–201.

Janowski, R., Kozak, M., Jankowska, E., Grzonka, Z., Grubb, A., Abrahamson, M. & Jaskolski, M. 2001. Human cystatin C, an amyloidogenic protein, dimerizes through three-dimensional domain swapping. *Nature Struct. Biol.* 8: 316–320.

Jaroniec, C.P., MacPhee, C.E., Astrof, N.S., Dobson, C.M., and Griffin, R.G. 2002. Molecular conformation of a peptide fragment of transthyretin in an amyloid fibril. *Proc. Natl. Acad. Sci. USA* 99: 16748–16753.

Jones, S., Manning, J., Kad, N.M., and Radford, S.E. 2003. Amyloid-forming peptides from beta2-microglobulin—Insights into the mechanism of fibril formation *in vitro*. *J. Mol. Biol.* 325: 249–257.

Kayed, R., Head, E., Thompson, J.L., McIntire, T.M., Milton, S.C., Cotman, C.W., and Glabe, C.G. 2003. Common structure of soluble amyloid oligomers implies common mechanism of pathogenesis. *Science* 300: 486–489.

Kozhukh, G.V., Hagihara, Y., Kawakami, T., Hasegawa, K., Naiki, H., and Goto, Y. 2002. Investigation of a peptide responsible for amyloid fibril formation of beta2-microglobulin by *Achromobacter* protease I. *J. Biol. Chem.* 277: 1310–1315.

Krantz B.A. and Sosnick T.R. 2001. Engineered metal binding sites map the heterogeneous folding landscape of a coiled coil. *Nature Struct. Biol.* 8: 1042–1047.

Krebs, M.R., Wilkins, D.K., Chung, E.W., Pitkeathly, M.C., Chamberlain, A.K., Zurdo, J., Robinson, C.V., and Dobson, C.M. 2000. Formation and seeding of amyloid fibrils from wild-type hen lysozyme and a peptide fragment from the beta-domain. *J. Mol. Biol.* 300: 541–549.

Liu, H., Farr-Jones, S., Ulyanov, N.B., Llinas, M., Marqusee, S., Groth, D., Cohen, F.E., Prusiner, S.B. and James, T.L. 1999. Solution structure of Syrian hamster prion protein rPrP(90-231). *Biochemistry* 38, 5362-5377.

Lopez de la Paz, M. and Serrano, L. 2004. Sequence determinants of amyloid fibril formation. *Proc. Natl. Acad. Sci. USA* 101: 87–92.

Lu, X., Wintrode, P.L, and Surewicz, W.K. 2007. Beta-sheet core of human prion protein amyloid fibrils as determined by hydrogen/deuterium exchange. *Proc. Natl. Acad. Sci. USA* 104: 1510–1515.

MacPhee, C.E., Dobson. C.M. 2000. Chemical dissection and reassembly of amyloid fibrils formed by a peptide fragment of transthyretin. *J. Mol. Biol.* 297: 1203–1215.

Matheson, R.R. Jr, Scheraga, H.A. 1978. A method for predicting nucleation sites for protein folding based on hydrophobic contacts. *Macromolecules*, 11:819–829.

Matouschek, J.T., Kellis, J., Serrano, L., and Fersht, A.R. 1989. Mapping the transition state and pathway of protein folding by protein engineering. *Nature* 340: 122–126.

Maury, C.P. and Nurmiaho-Lassila, E.L. 1992. Creation of amyloid fibrils from mutant Asn187 gelsolin peptides. *Biochem. Biophys. Res. Commun.* 183: 227–231.

Nelson, R., Sawaya, M.R., Balbirnie, M., Madsen, A.O., Riekel, C., Grothe, R., and Eisenberg, D. 2005. Structure of the cross-beta spine of amyloid-like fibrils. *Nature* 435: 747–749.

Nishimura, C., Lietzow, M.A., Dyson, H.J., Wright, P.E. 2005. Sequence determinants of a protein folding pathway. *J. Mol. Biol.* 351(2): 383–392.

Onuchic, J.N., Socci, N.D., Luthey-Schulten, Z., and Wolynes, P.G. 1996. Protein folding funnels: the nature of the transition state ensemble. *Fold. Des.* 1: 441–450.

Pastor, M.T., Kümmerer, N., Schubert, V., Esteras-Chopo, A., Dotti, C.G., López de la Paz, M., Serrano L. 2008. Amyloid toxicity is independent of polypeptide sequence, length and chirality. *J. Mol. Biol.* 375: 695–707.

Picotti, P., de Franceschi, G., Frare, E., Spolaore, B., Zambonin, M., Chiti, F., de Laureto, P.P., and Fontana, A. 2007. Amyloid fibril formation and disaggregation of fragment 1-29 of apomyoglobin: insights into the effect of pH on protein fibrillogenesis. *J. Mol. Biol.* 367: 1237–1245.

Privalov, P.L. 1979. Stability of proteins: small globular proteins. *Adv. Protein Chem.* 33: 167–241.

Rudall, K.M. 1952. The proteins of the mammalian epidermis. *Adv. Protein Chem.* 7: 253–290.

Shakhnovich, E., Abkevich, V., and Ptitsyn, O. 1996. Conserved residues and the mechanism of protein folding. *Nature* 379: 96–98.

Sinha, N., Tsai, C.J., Nussinov, R. 2001. A proposed structural model for amyloid fibril elongation: domain swapping forms an interdigitating beta-structure polymer. *Protein Eng.* 14: 93-103.

Tartaglia, G.G. and Vendruscolo, M. 2010. Proteome-level interplay between folding and aggregation propensities of proteins. *J. Mol. Biol.* 402: 919-928.

Ternström, T., Mayor, U., Akke, M., and Oliveberg, M. 1999. From snapshot to movie: phi analysis of protein folding transition states taken one step further. *Proc. Natl. Acad. Sci. USA* 96: 14854–14859.

Thompson, A., White, A.R., McLean, C., Masters, C.L., Cappai, R., and Barrow, C.J. 2000. Amyloidogenicity and neurotoxicity of peptides corresponding to the helical regions of PrP(C). *J. Neurosci. Res.* 62: 293–301.

In: Protein Structure
Editor: Lauren M. Haggerty, pp. 31-50
ISBN 978-1-61209-656-8

Chapter 2

Enzyme Immobilization: A Breakthrough in Enzyme Technology and Boon to Enzyme Based Industries

Alka Dwevedi and Arvind M. Kayastha [*]
School of Biotechnology, Faculty of Science, Banaras Hindu University, Varanasi 221005, India

Abstract

Enzyme technology is the application involving modification of enzyme's structure as well as its function. It is currently considered as a useful alternative to conventional process technology in enzyme based industrial as well as analytical fields. Enzyme immobilization is the foremost technique used in enzyme technology, which has gained popularity for the last few decades. It stabilizes structure of the enzymes, thereby allowing their applications under harsh environmental conditions (pH, temperature, organic solvents), and thus enable their uses in non-aqueous enzymology, and in the fabrication of biosensors. Furthermore, it allows repetitive use of enzyme as well as helpful in prevention of product contamination with the enzyme (especially useful in the food and pharmaceutical industries). Recently, the development of techniques for immobilization of multi-enzymes along with cofactor regeneration and retention system are gainfully exploited in developing biochemical processes involving complex chemical conversions. Various methods for enzyme immobilization used in bioreactors and biosensors include adsorption on or covalent attachment to a support, micro-encapsulation, and entrapment within a membrane, film or gel. The ideal immobilization method should employ mild chemical conditions, allow large quantities of enzyme to be immobilized, provide a large surface area for enzyme–substrate contact within a small total volume, minimize barriers to mass transport of substrate and product, and provide a chemically and mechanically robust system. Immobilization helps in the development of continuous processes allowing more economic organization of the operations,

[*] Corresponding Author: Tel: +91-542-2368331 Fax: +91-542-2368693 E-mail: kayasthabhu@gmail.com

automation, decrease of labour and investment to capacity ratio. The present chapter delineates the current status and future potentials of immobilized enzymes in the emerging enzyme based industries with suitable examples.

Introduction

Enzymes are the versatile class of bio-catalysts which enhance rate of a reaction without being substantially consumed during the reaction. They are precise and display high regio-, chemo- and enantio-selectivity while operating under mild conditions. There are many enzymes known which are commercially available. Their prices can vary significantly, depending on the degree of difficulty in isolating them or if they are readily available from recombinant sources. Enzyme technology is a branch of biotechnology involving development of bulk and high added-value products utilizing enzymes as biocatalysts in a sustainable manner. Enzymes are being implemented in production of various products for human needs (e.g. food, animal feed, pharmaceuticals, fine and bulk chemicals, fibers, hygiene and environmental technology), as well as in a wide range of analytical purposes, especially in diagnostics.

Genetic engineering has been one of the tools for enzyme technology but could not be as successful as anticipated. It suffers from various drawbacks, most importantly; post-translational modifications, problem of inclusion bodies, costly, tedious, time consuming and requisite for expertise. Immobilization has been the foremost enzyme technology being used these days due to its simplicity, decrease labour and cost-efficacy. It leads to physical confinement or localization of enzymes in a certain defined region of space with retention of their catalytic activities, less sensitivity towards their environment with insistent usability [Pierre, 2004; Sheldon, 2007]. Enzyme immobilization has been first adopted at the "enzyme engineering conference" held at Henniker, New Hampshire, USA in 1971. The first industrial application of immobilized enzyme was reported in 1967 in Japan; development of columns containing immobilized aminoacylase from *Aspergillus oryzae* for the synthesis of racemic DL-amino acid from the corresponding optically active enantiomers. Around 1970s, in England, immobilized penicillin acylase was used to prepare 6-amino penicillanic acid from penicillin G or V. In the same year, immobilized glucose isomerase was used for converting glucose into fructose in USA. These successful industrial applications prompted extensive research in enzyme technology leading to a steady increase in the number of industrial process based on sophisticated, immobilized-enzyme reactors [Hartmier, 1983].

Immobilization allows practical utilization of enzymes, particularly in industries while operating under mild conditions. The parameters which determine the success or failure of immobilization depends on the methodology used. It helps in the development of continuous processes allowing more economic organization of the operations and automation [Mateo *et al.*, 2007]. It allows availability of the product in greater purity required in food processing and pharmaceutical industry since contamination could cause serious toxicological, sensory, or immunological problems. It leads to greater control over enzymatic reaction as well as high volumetric productivity with lower residence time. This has great significance in the food industry, especially in the treatment of perishable commodities as well as in other applications involving labile substrates, intermediates or products [Persson *et al.*, 2002]. Immobilization procedures that would assure the desired properties to a chosen bio-system for a chosen application are established through experimental optimization. The retention of enzyme activity should be determined for the efficacy of the immobilization procedure. For the enzyme system to be characterized during immobilization, following properties are relevant: specific activity, the Michaelis-Menten constant K_m, the optimum pH and temperature, and the activation energy. The major emphasis should be the stability of enzyme including thermal and storage stabilities as well as reusability in various practical applications [Krajewska, 2009]. Here we present recent developments for immobilized enzymes used in variety of fields. A brief overview of the evolution of the use of immobilized enzymes with increasing efficacy, have been discussed along with the possible developments.

Enzyme Immobilization, Proficient Tool of Enzyme Technology

Enzymes catalyze the most complex chemical processes under the mildest experimental and environmental conditions. They are excellent industrial catalyst for the sustainable development, *viz.* green chemistry, production of clean fuels from renewable sources, improvement of the utilization of natural resources, food chemistry, very precise biosensors etc. Enzymes have few limitations to be used for industrial purpose: unstable; soluble; undergo inhibitions especially by metal ions; poorly selective on non-natural substrates, etc. Therefore, most enzymes have to be greatly amended before industrial implementation. Enzymes may be immobilized by using physical and chemical methods leading to its reusability for very long times. The utilization of necessary immobilization techniques improves various physico-chemical properties of enzymes and the possibilities of using them as industrial catalysts [Hanefeld *et al.*, 2008].

- Simply fixing enzymes onto porous supports prevents them from aggregation, proteolysis, interaction with hydrophobic interfaces, etc.
- The multi-point covalent attachment of the enzyme (e.g., through its region containing the highest density in Lys residues) on highly activated supports (e.g., with epoxy or glyoxyl groups) promotes an intense rigidification of its three dimensional structure.
- Multi-subunit immobilization of multimeric enzymes promotes stabilization under conditions where dissociation of subunits is the main mechanism of inactivation.
- Selectivity of enzymes can be greatly improved by promoting multipoint covalent immobilization on different enzyme regions (e. g., in the proximity of the active center). This is achieved by introducing Cys residues in different regions of the enzyme surface using disulfide-epoxy supports.
- In many instances, enzymes may be immobilized, stabilized and purified in just one step using tailor made supports (e.g., epoxy matrices activated with metal ions).
- Enzymes can be purified, immobilized, and hyper-activated and stabilized *via* interfacial activation on hydrophobic supports, *viz.* lipases recognizing their natural substrates like, hydrophobic oil interfaces.

- Enzymes can be reversibly immobilized on ionized polymeric layers attached to the support surfaces. Ionic exchange of enzymes on these supports is very intense leading to its stability and the supports can be re-used after enzyme inactivation.

Types of Enzyme Immobilization

A. Physical

During physical methods of immobilization there is no covalent bond formation, rather dependent on physical forces (like, Electrostatic, Protein-Protein etc). In principle, completely reversible i.e., original enzyme can be regenerated. Details of various physical methods are given below.

1) Entrapment within Cross-Linked Polymers

This technique is also called lattice entrapment, inclusion, occlusion etc. Method involves the formation of a highly cross-linked network of a polymer in the presence of an enzyme. Enzyme molecules are physically entrapped within the polymer lattice and cannot permeate out of gel matrix. Appropriately sized substrate and product molecules can transfer across to insure a continuous transformation. Thus, this method can be applied to selective enzyme systems. Usually with this method no change in the intrinsic properties of the enzyme are anticipated. Method has several advantages: simplicity, different physical shapes, no chemical modification and small amounts of enzymes required. However, this method suffers from some disadvantages e.g., leakage or leaching of small sized substrate/products, operationally good immobilization requires delicate balance of experimental factors (e.g., good mechanical properties *versus* activity). Important examples include urease immobilization onto polyacrylamide and alginate [Das *et al.,* 1998; Nakarani and Kayastha, 2007].

2) Adsorption

The method consists of addition of enzyme to a surface-active adsorbent, and removal of any non-adsorbed enzyme by washing. The absorbents usually require special pre-treatment in order to insure good adsorption. The adsorbents may be organic or inorganic in nature. It is dependent on experimental variables like, pH, temperature, nature of the solvent, ionic strength, concentration of enzyme and adsorbent. In principle, it should be a completely reversible process; a change in pH, ionic strength should lead to desorption. In practice, however, this does not happen. Irreversible binding is not a problem, if the immobilized enzyme is still active and is to be used in a continuous process. Commonly employed adsorbents include alumina, anion and cation-exchange resins, celluloses, collagen etc. Method has several advantages: simplicity, large choice of carriers, mild process, simultaneous purification and immobilization (e.g. Asparginase on CM-cellulose). Method suffers from some disadvantages e.g. attaining optimum conditions is often a matter of trial and error; If strong binding does not occur, desorption will occur and this can be a nuisance, especially when operating at high substrate concentrations [Hage *et al.*, 1986; Marquez etal., 2008; Alloue *et al.*, 2008].

3) Microencapsulation

Microenapsulated enzymes are formed by enclosing enzyme within spherical semi permeable polymer membranes, having diameter of 1-100 μm. It can be done using permanent or non-permanent membranes. Permanent membranes are made of chemicals like cellulose nitrate, polystyrene, while non-permanent membranes are made of liquid surfactant. Main advantage of this method is extremely large surface area for contact of substrate and enzyme within a relatively small volume [Das *et al.,* 1997]. Some disadvantages include that the size of substrate/product should be sufficiently small and high concentrations of enzyme are required.

B. Chemical

Chemical methods of immobilization involve formation of at least one covalent bond (or partially covalent). In reality, more than one covalent bond is involved. Such a method is usually irreversible. There are basically two types of chemical methods which are as follows:

1) Cross-Linking

Method involves formation of covalent bonds between enzyme molecules by means of bi- or multi-functional reagents, leading to formation of multiple covalent bonds. Though many multifunctional reagents are involved only glutaraldehyde has found extensive use [Das and Kayastha, 1998; Kayastha and Srivastava, 2001; Tripathi *et al*., 2007; Kumar *et al*., 2009; Kumari and Kayastha, 2011]. This is probably because it reacts with proteins readily under mild conditions. Figure 1 shows attachment of α-amylase using glutaraldehyde onto Amberlite MB-150 beads leading to its colour change on immobilization. Multifunctional reagents can be used not only to link enzyme molecules to polymer but to link enzyme molecules to each other. Major disadvantages of this method: difficulty in controlling the reaction, the need for large quantities of enzyme, much of which is lost by involvement of the active site in bond formation, gelatinous nature of the final product.

2) Covalent Attachment

Covalent attachment takes place *via* non-essential amino acid (other than active site groups) to a water-insoluble, functionalized supports (carrier, matrix, polymer), which are either organic or inorganic polymers. Bond formation is also termed as fixation, linkage, binding, bonding, coupling, grafting etc. Supports are either *synthetic*: acrylamide-based, methacrylic acid-based, styrene based etc; or *natural*: glass, Sephadex, Agarose, Sepharose etc. Majority of support requires activation (as they do not possess reactive groups but have only hydroxyl, amide, amino, carboxyl groups) before they can be used for immobilization. Some of the reactions used are diazotization, schiff base (imine bond) formation etc. [Kayastha *et al.,* 2003; Reddy and Kayastha, 2006].

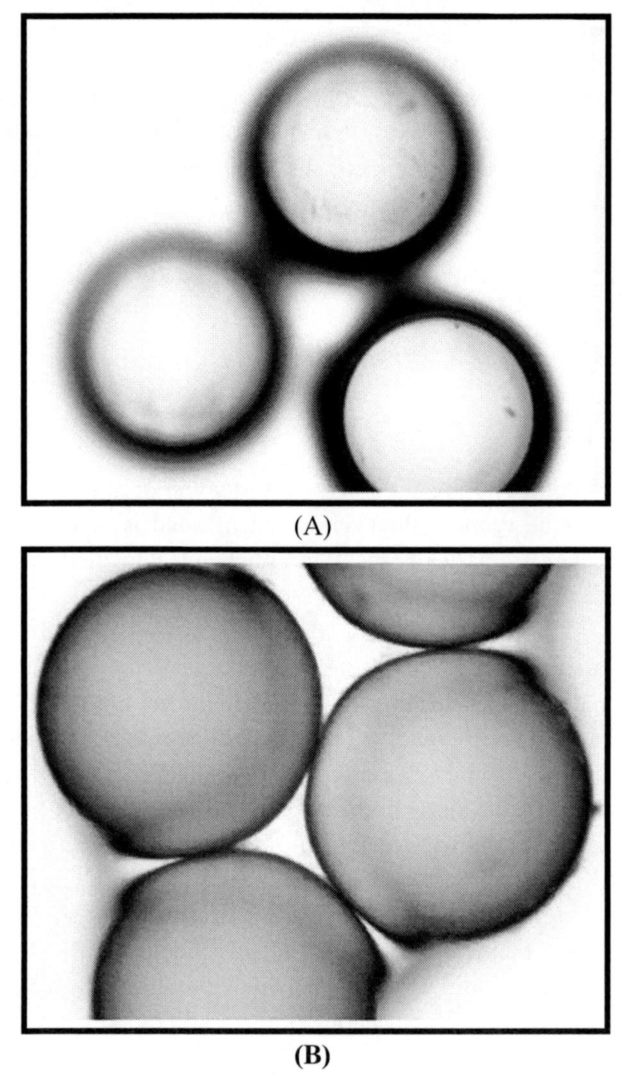

Picture has been adapted from Tripathi *et al.*, 2007.

Figure 1. Microscopic images of Amberlite MB 150 without enzyme (A) α-amylase coupled to glutaraldehyde activated Amberlite MB 150 (B) using Nikon light microscope.

Effect of Enzyme Immobilization on its Kinetic Properties

Enzyme immobilization leads to change in its microenvironment that may be drastically different from that existing in free solution. The new microenvironment may be a result of the physical and chemical nature of the support matrix alone, or it may result from interactions of the matrix with substrates or products involved in the enzymatic reaction. Following are the various parameters which are affected by enzyme immobilization due to change in its microenvironment:

Optimum pH

Optimum pH of the enzyme is determined by the pK_a of amino acids present at its active site. There is change in optimum pH either the enzyme is present on the surface of the matrix or entrapped inside it. It is either broadened or shifts to acidic or basic side with respect to soluble enzyme depending on the type of immobilization chosen, charge and surface area of the matrix [Tripathi *et al.*, 2007; Kumar *et al.*, 2009; Dwevedi and Kayastha, 2009a; Dwevedi and Kayastha., 2009b]. In case of charged matrices, change in optimum pH profile is defined by following equation:

$$\Delta p\mathrm{H} = p\mathrm{H}_i - p\mathrm{H}_o = 0.43\frac{\varepsilon\varphi}{kT}$$

where, $p\mathrm{H}_i$ represents pH in vicinity of the immobilized enzyme and $p\mathrm{H}_o$ represents pH of the bulk solution, ε represents the positive charge on the proton and φ represents the average electrostatic potential, k is Boltzmann's constant and T is the temperature. When the support is cationic (positively charged), the protons in the vicinity of the support are repelled. Hence, the local (i.e., in the vicinity of the enzyme) pH is higher that that of the bulk. Thus, there would be shift in optimum pH towards acidic side while vice versa in case of anionic support.

An immobilized enzyme preparation having high loading (that is a large quantity of enzyme activity per unit of polymer especially when the support has very high surface area) leads to substrate diffusion limitation. Figure 2 shows scanning electron micrograph of immobilized β-galactosidase from *Pisum sativum* (*Ps*BGAL) onto Amberlite MB-150 beads having diameter of 5 μm leading to broadening of optimum pH when lactose was the substrate [Dwevedi and Kayastha, 2009b]. As the enzyme has high intrinsic specific activity (occurs during immobilization), the substrate concentration gradient through the particle becomes steep and consequently the substrate may not penetrate to the centre of immobilized enzyme particle (enzyme present on the surface or inside the matrix). With the constraint of pH (below or above optimum pH), the substrate concentration gradient becomes less steep. Thus, allow the substrate to penetrate further into the immobilized enzyme particle having high intrinsic specific activity, due to increased enzyme concentration during immobilization [*Zulu* effect, Trevan, 1980; Portaccio *et al.*, 2003]. Two factors therefore work antagonistically on the reaction rate, the change in pH reducing the rate while the rise in effective enzyme concentration tends to increase the rate, thereby moderating the effect of the pH change. Therefore, due to high local enzyme concentration and substrate limitation leads to broadening of optimum pH.

Optimum Temperature

Immobilization of enzyme to the matrix led to a barrier towards free movement of enzyme. Therefore, gain in kinetic energy by the immobilized enzyme due to an increase in temperature is limited to much extent leading to an increase in optimum temperature. Displacement of optimum temperature for immobilized enzymes has been observed in many cases, but the extent of displacement would differ with type of matrix and interactions between the enzyme and the matrix. Further broadening of optimum temperature has been observed in various cases due to high intrinsic specific activity (Zulu effect) as observed in optimum pH [Reddy *et al.*, 2004; Tripathi *et al.*, 2007; Kumar *et al.*, 2009; Dwevedi and Kayastha, 2009a; Dwevedi and Kayastha., 2009b].

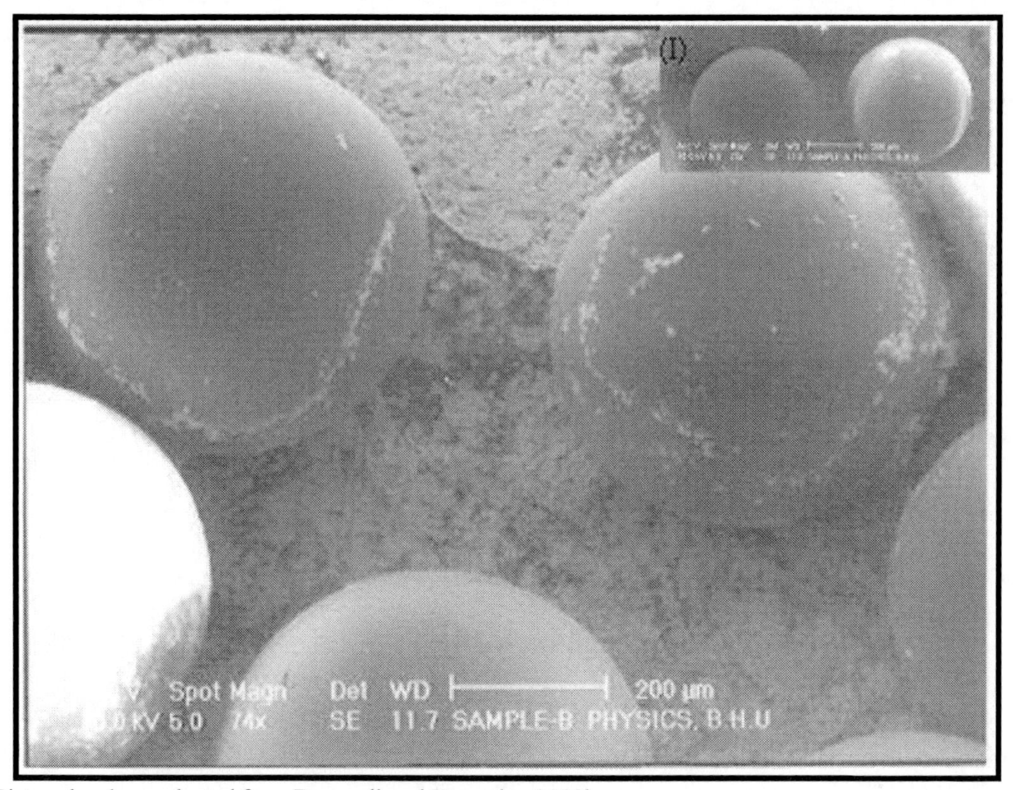

Picture has been adapted from Dwevedi and Kayastha, 2009b.

Figure 2. Scanning electron micrograph (SEM) of *Ps*BGAL (*β*-galactosidase from *Pisum sativum*) immobilized on Amberlite MB-150 beads and its control (inset) with scale bar of 200 μm.

Kinetic Parameters

They are dependent on following factors [Trevan, 1980]:

- Enzyme must be attached to the matrix in correct conformation (most appropriately *via* residues away from active site).
- Diffusion barrier (*internal* and *external*): Internal diffusion barrier is present in case of immobilized enzyme system where enzyme is present inside the matrix. Here, substrate has to diffuse inside the matrix to get hydrolyzed to its products. On the other hand, external diffusion barrier (when enzyme is present on the surface of the matrix) is a result of the thin, unstirred layer of solvent that surrounds the polymer particle, called the 'Nernst layer'. Solutes diffuse in this layer by a combination of

passive molecular diffusion and convection. Higher the concentration of solute greater would be the external diffusion barrier. The thickness of this layer is affected (within limits) by the speed at which the solvent around the immobilized enzyme particle stirred. Increasing the stirring rate will reduce this external diffusion layer as represented by following equation;

$$\alpha = \frac{V_{\max}}{K_m}$$

where α is the diffusion constant

- Partitioning effects arise due to polyionic matrices leading to interaction between matrix and the ionic solute. If partition coefficient $p = S_i/S_o$, where S_i is the concentration of ions around enzyme and S_o is the concentration of ions in bulk phase. Thus, Michaelis-Menten equation becomes:

$$v = \frac{V_{\max} \times S_o \times p}{K_m + S_o \times p}$$

and thus,

$$K_{m(app.)} = \frac{K_m}{p}$$

Commercial Implications

1) Enzymatic Synthesis of Aspartame

Chemically aspartame is L-aspartyl-L-phenylalanine methyl ester dipeptide. It is a common artificial sweeter, which is 150-200 times sweeter compared to sugar on g/g basis. In industry, it is synthesized chemically. Recently, an enzymatic method has been developed and is competing with the chemical method. Immobilized thermolysin was used as a biocatalyst in one of the processes in which *N*-(benzloxycarbonyl)-L-aspartate and L-phenylalanine methyl ester were coupled enzymically in ethylacetate and the product treated with acid to produce aspartame. The new sweetener serves mainly as a sucrose substitute in soft drinks such as diet Coke and diet Pepsi and many sugar free candies and chewing gums [United States Patent No. 6617127].

2) Enzymic Production of L-Aspartic acid and L-Malic Acid

Immobilized aspartase from *E. coli* is used to catalyze the interaction of fumaric acid with ammonia to yield L-aspartic acid, while immobilized fumarase from *Brevibacterium flavum* is used to transform fumaric acid to malic acid. Many tons of L-aspartic or malic acid is obtained each month from 1000 L columns with the immobilized cells containing aspartase. L-malic acid is widely used as a food additive in fruit and vegetable juices, carbonated soft drinks, jams and sweets. L-aspartic acid is in demand for the production of aspartame [United States Patent No. 6238895].

3) Production of 6-Amino Penicillanic Acid by Immobilized Penicillin Amidase

The penicillin compounds, such as ampicillin and amoxicillin, are prepared by acylation of 6-amino penicillanic acid (6-APA) derived from penicillin V or G. The high stability of immobilized enzyme is illustrated by the fact that 300 g of dry polymer and 20 g of protein can be used nearly 1000 times in batch reactions, in each of which 1600-1800 kg of penicillin G is split to 6-APA in a quantitative yield. It is estimated that more than 100 tons of immobilized penicillin amidase is now used globally per year [United States Patent No. 4113566].

4) Stereochemical Resolution of Racemic Amino Acids by Immobilized Aminoacylase:

Optically active L-amino acids can be obtained from their corresponding, chemically synthesized DL-forms by the use of an enzymic procedure. Immobilized aminoacylase bound to DEAE-Sephadex has been frequently used [Katchalski-Katzir, 1993].

5) Immobilized Glucose Isomerase in the Production of High Fructose Corn Syrup

Glucose isomerase catalyses the transformation of glucose to fructose; fructose is sweeter than glucose at an equilibrium state where approximately equimolar amounts of glucose and fructose are present. Production of high fructose corn syrup from starch requires three enzyme-mediated stages:

- Liquifaction of the starch by α-amylase
- Saccharification by glucoamylase
- Isomerization of the glucose formed by glucose isomerase

The product which is 55% enriched fructose syrup is obtained at a price 20% lower than that of sucrose, based on sweetening power. The popular new sweetener serves mainly as a sucrose substitute in soft drinks such as Coke and Pepsi [United States Patent Nos. 3694314 and 3817832].

6) Enzymic Synthesis of Acrylamide

Rhodococcus rhodochrous J1, an organism rich in nitrile hydratase (NHase), could transform acrylonitrile into acrylamide (Scheme I), with 100% substrate conversion and a yield of 656 g of acrylamide per litre of rea ction mixture. In the industrial process developed, resting cells of *R. rhodochrous* J1 are entrapped in a polyacrylamide gel, enabling continuous reuse [Kobayashi *et al.*, 1992].

Scheme I: $CH = CH - C \equiv N + H_2O \xrightarrow{\text{NHase}} CH_2 = CH - CONH_2$

7) Immobilized Lactase in the Hydrolysis of Lactose in Milk

The enzyme lactase (β-galactosidase) splits lactose into its monosaccharide components, glucose and galactose [Biswas *et al.,* 2003]. People who suffer from lactase deficiency become ill when they drink milk, which contains 4.5% lactose. The problem can be circumvented by enzymatic hydrolysis of the lactose prior to milk consumption. Immobilized β-galactosidase reactors can be used for continuous hydrolysis of lactose in dairy products [Dwevedi and Kayastha, 2009a; Dwevedi and Kayastha 2009b]. Lactose tends to crystallize at low temperatures. Its crystallization in ice creams can be prevented by prior treatment of the milk with immobilized β-galactosidase.

Biosensors

A biosensor is an analytical device for the detection of an analyte that combines a biological component with a physico-chemical detector component. It has three distinct types of components:

- Biological component (e.g., enzymes, antibodies etc.).
- Detector element (works in a physicochemical way; optical, piezoelectric, electrochemical, etc.) that transforms the signal resulting from the interaction of the biological component into another signal that can be more easily measured and quantified.
- Signal processors that are responsible for the display of the results in a user-friendly way. This sometimes accounts for the most expensive part of the sensor device.

The application of biosensors as analytical tools is a growing research topic in areas such as environmental surveillance, batch food analysis and clinical monitoring, and is beginning to impact on quality-of-life issues. The choice of biosensor design for a particular application is governed by diverse factors: the chemical nature of the analytical medium (e.g., hydrophobic *versus* hydrophilic); the sample size (e.g., intracellular and extracellular monitoring *versus* batch analysis); the time resolution and recording duration required; and the concentration of the target analyte relative to the corresponding interference compounds for the chosen technique (electrochemical, optical, gravimetric, tonometric, thermal, magnetoelastic, etc.). For *in-vivo* monitoring, *viz.* in the brain during behavior, implantable biosensors showing good biocompatibility, sensitivity, selectivity and stability in the strongly lipophilic environment are needed, and amperometric enzyme-based devices incorporating a permselective polymer have been applied successfully in many neurochemical studies. Following are important parameters to be taken into consideration in the development of biosensors:

- Enzyme kinetic parameters
- Permeability and permselectivity parameters
- Permeability characteristics of basic designs
- Enzyme concentrations

Enzyme immobilization plays a crucial role in the synthesis of biosensors based on enzymatic reaction. The attachment of enzymes to the surface of the sensor (be it metal, polymer or glass) requires activation of the surface. This can be done by polylysine, aminosilane, epoxysilane or nitrocellulose in case of silicon chips/silica glass. The bound enzymes may be fixed by layer by layer or three dimensional lattices can be used to chemically or physically entrap them. In case of physical entrapment, most commonly hydrogel *viz.* glassy silica generated by polymerization of silicate monomers (added as tetra alkyl orthosilicates) or acrylate hydrogel which polymerize upon radicle initiation are used for entrapping enzymes along with other stabilizing polymers, such as PEG [Gupta and Chaudhury, 2007; Liao *et al.*, 2008]. Enzymes may be covalently attached through functional groups such as amino or carboxylic groups. For instance, urease can be immobilized on a co-polymer film of poly (*N*-3-aminopropyl pyrrole) and polypyrrole, to fabricate a urea biosensor [Rajesh *et al.*, 2005], cholesterol oxidase can be bound on polyaniline-carbon nanotube composite film for cholesterol biosensor [Dhand *et al.*, 2008] and glucose oxidase (GOD) biosensor can be fabricated by incorporating GOD into graphite paste modified with tetracyanoquinodimethane [Pandey *et al.*, 1992]. One biosensor can be converted into other by sequential enzymatic reactions, for example glucose biosensor to lactose biosensor [Dwevedi *et al.*, 2009]. A novel glucose biosensor has been developed [Reddy *et al.*, 2005] which has integrated potentiometric as well as amperometric detection of enzyme catalyzed reaction. Figure 3, show the diagrammatic representation of glucose biosensor made. The construction involved immobilizing of glucose oxidase onto the platinized outer tip surface using the precipitation of electrodeposition paint with direct entrapment of the enzyme in the polymer film. Products of enzyme substrate reaction are targeted in a dual-detection mode on one hand with the covered platinum layer at the tip region as amperometric detector and on the other hand with a proton-selective liquid membrane based potentiometric sensor inside the open pipet tip.

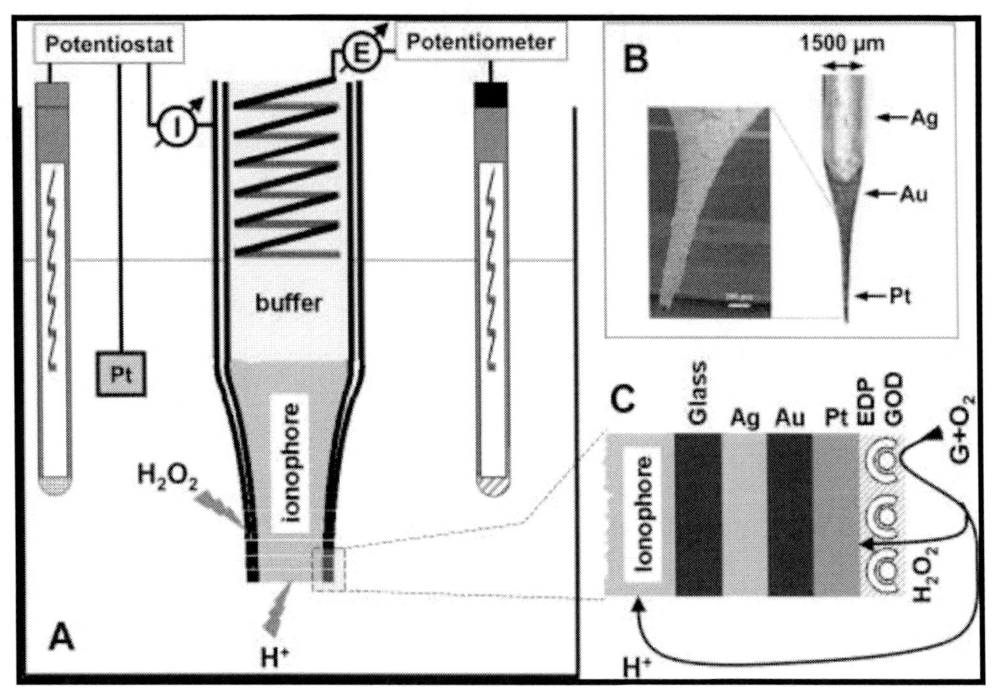

Picture has been adapted from Reddy *et al.*, 2005.

Figure 3. Bifunctional needle-type glucose micro-biosensor with an amperometric and potentiometric detection of enzyme-catalyzed glucose oxidation available at the tip of a pulled micropipette (A). Scanning electron micrograph of a metallized pipet tip used as platform for enzyme immobilization and amperometric detection of H_2O_2 (B). Schematic of the cross section through the tip of a complete glucose microbiosensor showing the multilayer approach for establishing the dual-detection mode (C).

Recently, gold nanoparticles have been given prominent status towards development of biosensors due to their very high surface area [Mao *et al.*, 2009; Ryu *et al.*, 2010]. It is known that metal bears shine due to reflection of light by electron clouds present around nuclear core. Further, when metal absorbs resonant wavelength it causes the electron cloud to vibrate, dissipating the energy. This process usually occurs at the surface of a material and is therefore called surface plasmon resonance (oscillations of the electron cloud present at the surface). It can be explained as there are certain wavelengths for metal where photons are not reflected, but instead are absorbed and converted into surface plasmon resonance (electron cloud vibrations). For example, in case of gold metal, these wavelengths occur in the infrared portion of the spectrum (wavelength > 800 nm). In case of gold nanoparticles (having very high surface area), they experience surface plasmon resonance in the visible portion of the spectrum. Therefore, a certain portion of visible wavelengths will be absorbed, while another portion will be reflected. The portion reflected will lend the material a certain color. Small nanoparticles absorb light in the blue-green portion of the spectrum (~ 400-500 nm) while red light (~700 nm) is reflected, yielding a deep red color. As particle size increases, the wavelength of surface plasmon resonance related absorption shifts to longer wavelengths. This means that red light is now adsorbed, and bluer light is reflected, yielding particles with a pale blue or purple color. As particle size continues to increase toward the bulk limit, surface plasmon resonance wavelengths move into the infra-red portion of the spectrum and most visible wavelengths are reflected. This gives the nanoparticles clear or translucent color.

These properties have been used to create biosensors. Individual small gold nanoparticles appear red; however, when particles aggregate together the plasmon resonances can combine. The particle would appear as one large particle rather than two separate ones. Plasmon resonance associated-absorption wavelengths would shift from blue to red, and reflected light would shift from red to blue. Therefore, particle color would change from red to blue on aggregation. For example, in case of DNA biosensor for diagnosis of pathogens based on colour change brought by nanoparticles on aggregation [Parab *et al.*, 2010]. The aggregation of gold nanoparticles has also been employed in the synthesis of pregnancy kits [Rojanathanes *et al.*, 2008]. Nanoparticle is also an important matrix for enzyme immobilization, helpful for enzyme based biosensors. High surface area leads to tremendous increase in intrinsic specific activity contributed by *Zulu* effect. This results in increased catalysis as well sensitivity of immobilized enzyme [Liu *et al.*, 2005; Zhu *et al.*, 2007; Pandey *et al.*, 2007; Dwevedi *et al.*, 2009].

Test strips are one of the simplest forms of biosensors. They are single-use strips of cellulose coated with the appropriate enzyme and suitable reagents. They are dipped into the sample solution and the presence of the substrate or analyte in the sample is detected by a change in the colour of the strips, which can be compared with a coloured chart to estimate the approximate amount of analyte, unlike true biosensors, which are very sensitive and give a true quantitative value [Reddy *et al.,* 2004]. A wide variety of test strips based on enzymes are commercially available. Efforts are in the direction of immobilizing more than one enzyme (co-immobilization) for assay of more than one analyte in the biological fluids e.g. glucose and cholesterol can be analyzed together in case of diabetics who may also be suffering from hypertension and heart diseases.

Immobilization by Cross-Linking of Enzymes among Themselves without Any Support

Cross linked enzyme aggregates (CLEAs) involves covalent binding between enzymes molecules both intramolecularly as well as intermolecularly without any requirement of support. Aggregates can be formed with respect to all the five classes of enzymes (oxidoreductase, transferases, isomerases, lyases, and hydrolases). These insoluble aggregates have higher thermal stability as compared to the corresponding free-soluble enzymes. These amorphous preparations can be molded into various shapes and forms depending upon individual applications (e.g., use in fluidized bed reactor, enzyme electrode design, and separation of enzyme inhibitors). CLEAs are prepared by aggregating enzymes using precipitants like acetone, ammonium sulfate, ethanol or 1, 2-dimethoxyethane. Aggregation is followed by cross-linking using glutaraldehyde and addition of additives. In case of lipase from *Burkholderia cepacia*, enzyme is cross-linked with bovine serum albumin as protecting reagent in the presence of dextrin [Sheldon, 2007].

The synthesis of CLEAs from penicillin G amidase (an industrially important enzyme) is used in the synthesis of semi-synthetic penicillin and cephalosporin antibiotics. The free enzyme has limited thermal stability and a low tolerance to organic solvents, which makes it an ideal candidate for stabilization as a CLEA. Indeed, penicillin G amidase CLEAs, prepared by precipitation using ammonium sulfate proved to be effective catalysts for the synthesis of ampicillin [Bruggink *et al.*, 2003; Mateo *et al.*, 2006]. Figure 4 shows transmission electron micrograph of CLEAs prepared using trypsin from porcine pancreas prepared using compressed CO_2 at 4.00 MPa followed by cross-linking using 5% glutaraldehdye [Chen *et al.*, 2006]. Remarkably, the productivity of the CLEA was much higher than that of the free enzyme from which it was made. The method of preparation of CLEAs is exquisitely simple, amenable to rapid optimization, low costs therefore finds short time to reach to the market. It is applicable to a wide variety of enzymes, including crude preparations, affording stable, recyclable catalysts with high retention of activity and tolerance to organic solvents. The technique is also applicable to the preparation of combi-CLEAs containing two or more enzymes, which can be advantageously used in catalytic cascade processes. CLEAs have also been used in micro-channel reactors (device in which chemical reactions take place in a confinement with typical lateral dimensions below 1 mm). Concisely, CLEAs have wide applications in industrial bio-transformations as well as other areas requiring immobilized enzymes [Honda *et al.*, 2006].

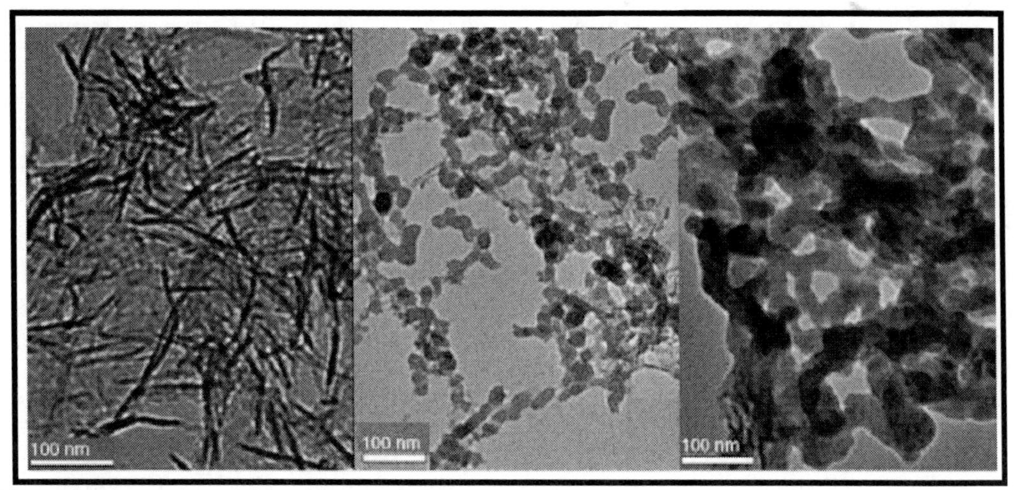

Picture has been adapted from Chen *et al.*, 2006.

Figure 4. Transmission electron micrograph of cross-linked enzyme aggregates (CLEAs) of trypsin using CO_2-expanded reverse micelle solutions. (a) Reverse micelle solution: 20 mmoL L^{-1}, trypsin: 0.5 mg mL^{-1}; (b) Reverse micelle solution: 40 mmoL L^{-1}, trypsin: 0.5 mg mL^{-1}; (c) Reverse micelle solution: 40 mmoL^{-1}, trypsin: 1.0 mg mL^{-1}.

Structure-Based Development of Immobilization

Recent research has demonstrated the combined use of experimental and computational methods can provide rational guidelines for the selection of optimal polymeric supports as well as for the choice of immobilization technique. For example, the surface polarity of lipases (from *Candida antarctica*, *Thermomyces lanuginosus*), penicillin G acylase were analyzed with the GRID computational method, to establish the hydrophobic and hydrophilic areas on the enzyme surface [Goodford, 1985]. A map of regions suitable for the establishment of interactions with different supports was created. This is of particular importance for enzyme orientation upon binding to the support as different areas will interact differently with the support. Most lipases face their hydrophobic interface to adopt the active conformation when attached to the hydrophobic support. In case of hydrophilic carrier, lipases require the active site of enzyme to face the reaction medium to assume an inactive conformation. It was found that lipases have many lysine residues near to its active site. Therefore, immobilization *via* covalent attachment would lead to enzyme deactivation. Here, genetic engineering can be utilized to remove all lysine residues close to the active site. For example, in case of lipase from *Thermomyces lanuginosus*, only one lysine residue opposite to the active site can be retained. This is then utilized for the selective covalent immobilization on matrices like Eupergit®, Accurel etc. Thus all the active sites of the enzyme are oriented toward the reaction mixture and higher activity of the enzyme would be observed [Basso *et al.*, 2007]. Ideally, no lysine residue located in proximity of the opening of the active site should be involved in the covalent binding otherwise lead to enzyme inactivation. In such cases alternative options such as adsorption or entrapment should be pursued [Hanefeld *et al.*, 2009].

Glycosylation greatly changes the polarity of the enzyme and the sugar moieties can also be used for covalent attachment [Dwevedi *et al.*, 2009]. These days approach is moving towards using enzymes which are glycoproteinaceous in nature. It can be achieved by using eukaryotic enzyme sources or glycosylating the prokaryotic enzymes artificially. *N*-glycosylation is more helpful than *O*-glycosylation during enzyme immobilization constituting an additional hydrophilic region causing considerable variation of the surface properties of the enzyme. Introduction of the sugar moiety in penicillin G acylase can be exploited to achieve a more active and stable enzyme using suitable carrier. By choosing a polymer like Sepabeads® having aldehyde functionalities (obtained *via* pre-activation of amino groups with glutaraldehyde) and a hydrophilic nature, two targets could be achieved [Hanefeld *et al.*, 2009]:

- Favourable orientation of the glycosylated penicillin G acylase due to hydrophilic interactions between the mannoses of glycan moiety and the polymer.
- Stabilization of the enzyme due to the formation of cyclic acetals between the mannose units of glycan moiety and the functionalized polymer.

Summary and Conclusion

Immobilization of enzymes has been largely a trial and error approach and progress is being made towards targeted immobilization of enzymes. Recent advances in the design of materials with specified pore sizes and surface functionality has enabled more precise control of the immobilization process with retention of catalytic activity and stability. Simulation of the surface characteristics of the target enzyme can be used to aid in the design of appropriate support materials still this approach has not found much awareness. As the structure and mechanism of more enzymes become available, more controlled immobilization methods will be generated. It should be noted that, even in those cases where a three dimensional structure of the enzyme is unavailable, structural models can be built up by using homology modelling methods. In particular, successful industrial applications of biocatalysts require systems that are not only stable and active, but are low in cost and can undergo repeated re-use. Enzyme immobilization is slowly turning into a well-understood science part of it still remains an art.

Future Perspectives

Past experience clearly suggests us that we must learn to work with impure enzyme preparations, like the whole cells or the cell homogenates [D'Souza *et al.*, 1999]. A significant future is expected for development of processes using multienzyme systems, specially those involving cofactor regeneration for the production of high value compounds. The initial industrial success in this direction is encouraging and may hold a vast potential for the future. The immobilized enzyme technology may also help in the future for integrating bioprocess with downstream processing with an effort to increase the productivity while minimizing product recovery cost. Immobilized enzyme technology may also be useful in non-aqueous enzymology, not only in terms of stabilization of the biocatalysts but also in the development of continuous bioreactors. One of the rapidly emerging areas in the field of sensors is the development of biosensors for industrial process control wherein immobilized enzyme technology plays a pivotal role. Thus, there are interesting possibilities within the field of immobilized enzymes and it is imminent that in the future many applications will be replaced by immobilized systems and many more new systems will become technically as well as commercially feasible.

References

Alloue, W. A.; Destain, J.; El Medjoub, T.; Ghalfi, H.; Kabran, P. and Thonart, P. (2008) Comparison of *Yarrowia lipolytica* lipase immobilization yield of entrapment, adsorption, and covalent bond techniques. *Appl. Biochem. Biotechnol.* 150, 51-63.

Basso, A.; Braiuca, P.; Cantone, S.; Ebert, C.; Linda, P.; Spizzo, P.; Caimi, P.; Hanefeld, U.; Degrassi, G. and Gardossi, L. (2007). *In Silico* analysis of enzyme surface and glycosylation effect as a tool for efficient covalent immobilisation of CalB and PGA on Sepabeads®. *Adv. Synth. Catal.* 349, 877-886.

Basu, I.; Subramanian R.V.; Mathew, A.; Kayastha, A. M.; Chadha, A. and Bhattacharya, E. (2005). Solid state potentiometric sensor for the estimation of tributyrin and urea. *Sens. Actuat. B. Chem.* 107, 418-423.

Biswas, S.; Kayastha, A. M. and Seckler, R. (2003). Purification and characterization of a thermostable *β*-galactosidase from kidney beans (*Phaseolus vulgaris* L.) cv. PDR 14. *J. Plant Physiol.* 160, 327-337

Bruggink, R.; Schoevaart, T. and Kieboom, A. P. G. (2003). Concepts of nature in organic synthesis: Cascade catalysis and multi-step conversions in concert. *Org. Proc. Res. Dev.* 7, 622-640.

Chen, J.; Zhang, J.; Han, B.; Li, Z.; Li, J. and Feng, X. (2006). Synthesis of cross-linked enzyme aggregates (CLEAs) in CO_2-expanded micellar solutions. *Colloids Surf. B: Biointerf.* 48, 72-76.

D'Souza, S. F. (1999). Immobilized enzymes in bioprocess. *Curr. Sci.* 77, 69-86.

Das, N. and Kayastha, A. M. (1998). Immobilization of urease from pigeonpea (*Cajanus cajan* L.) on flannel cloth using polyethylenimine. *World J. Microbiol. Biotechnol.* 14, 927-929.

Das, N.; Kayastha, A. M. and Malhotra, O. P. (1998). Immobilization of urease from pigeonpea (*Cajanus cajan* L.) on polyacrylamide gels and calcium alginate beads. *Biotechnol. Appl. Biochem.* 27, 25-29.

Das, N.; Prabhakar, P.; Kayastha, A. M. and Srivastava, R. C. (1997). Enzyme entrapped inside the reverse micelle in the fabrication of a new urea sensor. *Biotechnol. Bioeng.* 54, 329-332.

Dhand, C.; Arya, S. K.; Datta, M. and Malhotra, B. D. (2008). Polyaniline–carbon nanotube composite film for cholesterol biosensor. *Anal. Biochem.* 383, 194-199.

Dwevedi, A. and Kayastha, A. M. (2009a). Optimal immobilization of β-galactosidase from Pea (*Ps*BGAL) onto Sephadex and chitosan beads using response surface methodology and its applications. *Bioresour. Technol.* 100, 2667-2675.

Dwevedi, A. and Kayastha, A. M. (2009b). Stabilization of β-galactosidase (from peas) by immobilization onto Amberlite MB-150 beads and its application in lactose hydrolysis. *J. Agric. Food Chem.* 57, 682-688.

Dwevedi, A.; Singh, A. K.; Singh, D. P.; Srivastava, O. N. and Kayastha, A. M. (2009). Lactose nano-probe optimized using response surface methodology. *Biosens. Bioelectron.* 25, 784-790

Goodford, P. J. (1985). A computational procedure for determining energetically favourable binding sites on biologically important macromolecules. *J. Med. Chem.* 28, 849-857.

Gupta, R. and Chaudhury, N. K. (2007). Entrapment of biomolecules in sol-gel matrix for applications in biosensors: problems and future prospects. *Biosens. Bioelectron.* 22, 2387-2399.

Hage, D. S.; Walters, R. R. and Hethcote, H. W. (1986). Split-peak affinity chromatographic studies of the immobilization-dependent adsorption kinetics of protein A. *Anal. Chem.* 58, 274–279.

Hanefeld, U.; Gardossi, L. and Magner, E. (2009). Understanding enzyme immobilization. *Chem. Soc. Rev.* 38, 453-468.

Hartmier, W. (1983). *Immobilized biocatalysts: An introduction*, Springer Verlag, Berlin.

Honda, T.; Miyazaki, M.; Nakamura, H. and Maeda, H. (2006). Facile preparation of an enzyme-immobilized microreactor using a cross-linking enzyme membrane on a microchannel surface. *Adv. Synth. Catal.* 348, 2163-2171.

Katchalski-Katzir, E. (1993). Immobilized enzymes: learning from past successes and failures. *Trends Biotechnol.* 11, 471-478.

Kayastha, A. M. and Srivastava, P. K. (2001). Pigeonpea (*Cajanus cajan* L.) urease immobilized on glutaraldehyde activated chitosan beads and its analytical applications. *Appl. Biochem. Biotechnol.* 96, 41-53.

Kayastha, A. M.; Srivastava, P. K.; Miksa, B. and Slomkowski, S. (2003). Unique activity of ureases immobilized on poly (styrene-co-acrolein) microspheres. *J. Bioact. Compat. Polym.* 18, 113-124.

Kobayashi, M.; Nagasawa, T. and Yamada, H. (1992). Enzymatic synthesis of acrylamide: a success story not yet over. *Trends Biotechnol.* 10, 402-408.

Krajewska, B. (2009). Ureases II. Properties and their customizing by enzyme immobilizations: A review. *J. Mol. Catal. B: Enzym.* 59, 22-40.

Kumar, S.; Dwevedi, A. and Kayastha, A. M. (2009). Immobilization of soybean (*Glycine max*) urease on alginate and chitosan beads showing improved stability: Analytical applications. *J. Mol. Catal. B: Enzym.* 58, 138-145.

Kumari, A. and Kayastha, A. M. (2011). Immobilization of soybean (*Glycine max*) α-amylase onto Chitosan and Amberlite MB-150 beads: Optimization and characterization. *J. Mol. Catal. B: Enzym.* 69, 8-14.

Liao, K. C.; Hogen-Esch, T.; Richmond, F. J.; Marcu, L.; Clifton, W. and Loeb, G. E. (2008). Percutaneous fiber-optic sensor for chronic glucose monitoring *in vivo*. *Biosens. Bioelectron.* 23, 1458-1465.

Liu, G.; Lin, Y.; Ostatná, V. and Wang, J. (2005). Enzyme nanoparticles-based electronic biosensor. *Chem. Commun.* 27, 3481-3483.

Mao, X.; Ma, Y.; Zhang, A.; Zhang, L.; Zeng, L. and Guodong Liu, G. (2009). Disposable nucleic acid biosensors based on gold nanoparticle probes and lateral flow Strip. *Anal. Chem.* 81, 1660-1668.

Marquez, L. D. S.; Cabral, B. V.; Freitas, F. F.; Cardoso, V. L. and Ribeiro, E. J. (2008). Optimization of invertase immobilization by adsorption in ionic exchange resin for sucrose hydrolysis. *J. Mol. Catal. B: Enzym.* 51, 86-92.

Mateo, B.; Chmura, A.; Rustler, S.; van Rantwijk, F.; Stolz, A. and Sheldon, R. A. (2006) Synthesis of enantiomerically pure (*S*)-mandelic acid using an oxynitrilase–nitrilase bienzymatic cascade: a nitrilase surprisingly shows nitrile hydratase activity. *Tetrahed. Symm.* 17, 320-323.

Mateo, C.; Palomo, J. M.; Fernandez-Lorente, G.; Guisan, J. M. and Fernandez-Lafuente, R. (2007). Improvement of enzyme activity, stability and selectivity via immobilization techniques. *Enzyme Microb. Technol.* 40, 1451-1463.

Nakarani, M. and Kayastha, A.M. (2007). Kinetics and diffusion studies in urease-alginate biocatalyst beads. *Orient. Pharm. Exp. Med.* 7, 79-84.

Pandey, P.; Singh, S. P.; Arya, S. K.; Gupta, V.; Datta, M.; Singh, S. and Malhotra, B. D. (2007). Application of thiolated gold nanoparticles for the enhancement of glucose oxidase activity. *Langmuir* 23, 3333-3337.

Pandey, P. C.; Kayastha, A. M. and Pandey, V. (1992). Amperometric enzyme sensor for glucose based on graphite paste-modified electrodes. *Appl. Biochem. Biotechnol.* 33, 139-144.

Parab, H. J.; Jung, C.; Lee, J.-H. and Park, H. G. (2010). A gold nanorod-based optical DNA biosensor for the diagnosis of pathogens. *Biosens. Bioelectron.* 26, 667-673.

Persson, M.; Mladenoska, I.; Wehtje, E. and Adlercreutz, P. (2002). Preparation of lipases for use in organic solvents. *Enzyme Microb. Technol.* 31, 833-841.

Pierre, A. C. (2004). The sol-gel encapsulation of enzymes. *Biocatal. Biotransform.* 22, 145-170.

Portaccio, M.; De Luca, P.; Durante, D.; Rossi, S.; Bencivenga, U.; Canciglia, P.; Lepore, M.; Mattei, A.; De Malo, A. and Mita, D. G. (2003). *In vitro* studies of the influence of ELF electromagnetic fields on the activity of soluble and insoluble peroxidase. *Bioelectromagnetics* 24, 449-456.

Rajesh; Bisht, V.; Takashima, W. and Kaneto K. (2005). An amperometric urea biosensor based on covalent immobilization of urease onto an electrochemically prepared copolymer poly (*N*-3-aminopropyl pyrrole-co-pyrrole) film. *Biomaterials* 26, 3683-3690.

Reddy, K. R. C. and Kayastha, A. M. (2006). Improved stability of urease upon coupling to alkylamine and arylamine glass and its analytical use. *J. Mol. Catal. B: Enzym.* 38, 104-112.

Reddy, K. R. C.; Srivastava, P. K.; Dey, P. M. and Kayastha, A. M. (2004). Immobilization of pigeonpea (*Cajanus cajan*) urease on DEAE-cellulose paper strips for urea estimation. *Biotechnol. Appl. Biochem.* 39, 323-327.

Reddy. K.R.C.; Turcu, F.; Schulte, A.; Kayastha, A. M. and Schuhmann, W. (2005). Fabrication of a potentiometric/amperometric bifunctional enzyme microbiosensor. *Anal. Chem.* 77, 5063-5067.

Rojanathanes, R.; Sereemaspun, A.; Pimpha, N., Buasorn, V.; Ekawong, P.; Wiwanitkit, V. (2008). Gold nanoparticle as an alternative tool for a urine pregnancy test. *Taiwan J. Obstet. Gynecol.* 47, 296-299.

Ryu, S.-W.; Kim, C.-H.; Han, J.-W.; Kim, C.-J.; Jung, C.; Hyun Gyu Park, H. G. and Choi, Y-.K (2010). Gold nanoparticle embedded silicon nanowire biosensor for applications of label-free DNA detection. *Biosens. Bioelectron.* 25, 2182-2185.

Sheldon, R. A. (2007). Cross-linked enzyme aggregates (CLEA®S): stable and recyclable biocatalysts. *Biochem. Soc. Trans.* 35, 1583-1587.

Sheldon, R. A. (2007). Enzyme immobilization: the quest for optimum performance. *Adv. Synth. Catal.* 349, 1289-1307.

Srivastava, P. K.; Kayastha, A. M. and Srinivasan (2001). Characterization of gelatin immobilized pigeonpea urease and preparation of a new urea biosensor for urea assay. *Biotechnol. Appl. Biochem.* 34, 55-62.

Swati, M.; Kumar, S.; Reddy, K.R.C. and Kayastha, A. M. (2007) Immobilization of urease from pigeonpea (*Cajanus cajan*) on agar tablets and its application in urea assay. *Appl. Biochem. Biotechnol.* 142, 291-297.

Trevan, M. D. (1980). Effect of immobilization on enzymatic activity. In Immobilized Enzymes, An Introduction and Applications in Biotechnology; Trevan, M. D., Ed.; Wiley & Sons: New York, pp. 11-55.

Tripathi, P.; Kumari, A.; Rath, P. and Kayastha, A. M. (2007). Immobilization of α-amylase from mung beans (*Vigna radiata*) on Amberlite MB 150 and chitosan beads: A comparative study. *J. Mol. Catal. B: Enzym.* 49, 69-74.

Wiseman, A. (1985) *Handbook of Enzyme Technology*, Ellishorwood, Chichester, England.

Zhu, Y.; Zhu, H.; Yang, X.; Xu, L. and Li, C. (2007). Sensitive biosensors based on (dendrimer encapsulated Pt nanoparticles)/enzyme multilayers. *Electroanalysis* 19, 698-703.

Reviewed by Professor P.M. Dey, School of Biological Sciences, Division of Biochemistry, Royal Holloway, University of London, Egham, Surrey TW20 0EX, U.K.

In: Protein Structure
Editor: Lauren M. Haggerty, pp. 51-69
ISBN 978-1-61209-656-8

Chapter 3

Three Approaches for Classifying Protein Tertiary Structures

Georgina Mirceva**[*] **and Danco Davcev
Ss. Cyril and Methodius University in Skopje, Faculty of Electrical Engineering and Information Technologies, Skopje, Macedonia

Abstract

To understand the structure-to-function relationship, life sciences researchers and biologists need to retrieve similar tertiary structures from protein databases and classify them into the same protein fold. The protein structures are stored in the world-wide repository Protein Data Bank (PDB) which is the primary repository for experimentally determined proteins structures. With the technology innovation the number of protein structures increases every day, so, retrieving structurally similar proteins using current structural alignment algorithms may take hours or even days. Therefore, improving the efficiency of protein structure retrieval and classification becomes an important research issue. There are many accurate methods such as SCOP which provide protein structural classification, but they last too long. In this chapter, we present three accurate approaches which provide faster classification (min., sec.) of protein tertiary structures. We have used two approaches to extract the features of the protein structures. In our voxel based approach, the Spherical Trace Transform is applied to protein tertiary structures in order to produce geometry descriptor. Additionally, some properties of the primary and secondary structure are taken, thus forming better integrated descriptor. In our ray based approach, some modification of the ray descriptor is applied on the interpolated backbone of the protein. Then, we use three approaches in order to classify protein molecules. In our first approach, we use the Growing neural Gas (GNG) as a hidden layer and a Radial Basis Function (RBF) as an output layer (which contains separate output node for each class). Our second approach uses the well known Support Vector Machine (SVM) method. In the third approach, the elements of the ray descriptor are quantized at 20 levels and then we build separate Hidden Markov Model (HMM) for each class. We have implemented a system for classifying protein tertiary structures based on these three

[*] georgina@feit.ukim.edu.mk

approaches. Our ground truth data contains 6979 protein chains from 150 SCOP domains. 90% of the dataset serves as the training and the other 10% as a test data. We examined the classification accuracy of these three approaches according to SCOP hierarchy. The results showed that the ray based descriptor leads to more accurate classifier. Generally, SVM gives the best classification accuracy (98.7%) by using the ray based descriptor. Additionally, we compared our HMM approach against the 3D HMM method. The analysis showed that our HMM based approach gives better results than the existing 3D HMM method.

Keywords: PDB, SCOP, protein classification, SGNG, SVM, HMM.

1. Introduction

Proteins are one of the most important molecules in the living organisms, since they play a vital structural and functional role in their cells. They are constructed of several polypeptide chains of amino acids, which fold into complex tertiary structures. The way these chains fold in the 3D space are very important in order to understand the function of the protein molecule. To understand the structure-to-function relationship, life sciences researchers and biologists need to retrieve similar structures from protein databases and classify them into the same protein fold. Therefore, the 3D representation of a residue sequence and the way this sequence folds in the 3D space are very important. The protein tertiary structures are stored in the world-wide repository Protein Data Bank (PDB) [1], [2] that is the primary repository for experimentally determined proteins structures. It was created in 1971 at the Brookhaven National Laboratories in the USA.

With the technology innovation the number of protein structures increases rapidly; therefore the necessity of a fast and completely automated protein classification is obvious. This classification must be biased towards functional and evolutionary properties of the proteins. At present, the Structural Classification of Protein (SCOP) database [3], which is manually constructed by human experts, is believed to maintain the most accurate structural classification. The SCOP database describes the structural and evolutionary relationships between proteins. Since the existing tools for comparing secondary structure elements cannot guarantee 100 percent success in the identification of protein structures, SCOP uses experts' experience to carry out this task. The Evolutionary relationship of proteins is presented as they are classified in hierarchical manner. The main levels of the hierarchy are "Family" (based on the proteins' evolutionary relationships); "Superfamily" (based on some common structural characteristics); and "Fold" (based on secondary structure elements). The deepest level at the hierarchy is the domain, so if we can predict the domain, we know all upper levels. Under the hierarchical configuration of the SCOP database, proteins with globally or locally similar structures are usually grouped into the same SCOP fold. The manual classification provides reliable results, but it is labour intensive. Therefore, the gap between protein holdings of PDB and SCOP databases continues to grow [4], see Fig. 1. Hence, developing an efficient and accurate classifier of protein structures will have a vital impact on effectively classifying the newly-discovered structures.

The CATH (Class, Architecture, Topology, and Homologous superfamily) database [5], which is held at the UCL University of London, contains hierarchically classified structural elements (domains) of the proteins stored in the PDB database. The CATH system uses automatic methods for the classification of domains, as well as experts' contribution, where automatic methods fail to give reliable results. For the classification of structural elements, five main hierarchical levels are used: "Class" (is determined by the percentage of secondary structure elements and their packing); "Architecture" (describes the organization of the secondary structure elements); "Topology" (provides a complete description of the whole schema and the way the secondary structure elements are connected); "Homologous Superfamily" (structural elements that have at least 35% amino-acid sequence identity belong to the same Homologous Superfamily); and "Sequence" (at this last level of hierarchy, the structures of the same Homologous Superfamily are further classified according to the similarity of their amino acid sequences). The CATH database is constructed by applying the Secondary Structure Alignment Program (SSAP) [6]. SSAP (Secondary Structure Alignment Program) utilizes a two-layer dynamic programming technique to align two proteins. In this way the optimal structural alignment of two proteins is determined.

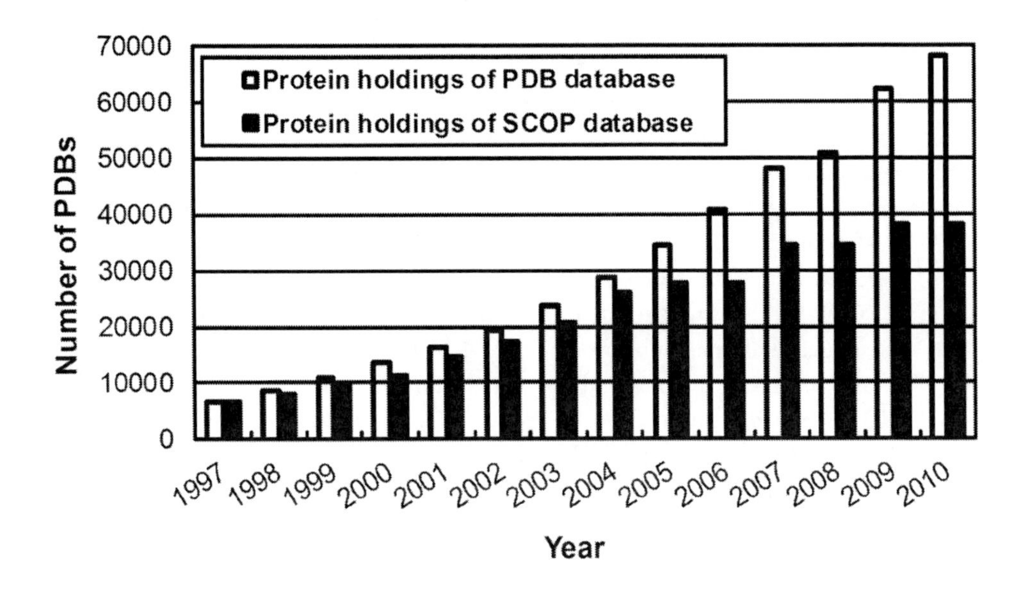

Figure 1. Number of protein holdings in the PDB and SCOP databases.

The FSSP (Families of Structurally Similar Proteins) database [7] was created according to the DALI classification method [7], [8] and is held at the European Bioinformatics Institute (EBI). It provides a fully automated classification of protein structures. The similarity between two proteins is based on the best alignment of their secondary structures. The DALI algorithm first calculates the distance matrix of the protein. The distance matrix is a symmetrical matrix whose elements are the 3D Euclidean distances between each pair of C_α atoms. In order to simplify the alignment in later stages, the distance matrix is decomposed into hexapeptide-hexapeptide fragments, named contact patterns. Then, a Monte Carlo algorithm is applied in order to find similar contact patterns and assembles pairs of contact patterns. Since Monte Carlo is a randomized algorithm, it cannot guarantee convergence with the globally optimal solution. The evaluation of a pair of proteins is a highly time consuming task, so the comparison between a macromolecule and all the macromolecules of the database requires days. Therefore, one representative protein for each class is defined that way that no two representatives have more than 25% amino-acid sequence identity. The unclassified protein is compared only to the representative protein of each class that requires an entire day.

All these sophisticated protein classification methods, are very time consuming, especially SCOP, where most of the phases are made manually. Therefore the number of released proteins in PDB which are not classified by SCOP increases every day, as shown in Fig. 1. Thus the necessity of fast, accurate and completely automated protein classification methods is obvious.

One way to determine the protein similarity is to use sequence alignment algorithms like Needleman–Wunch [9], BLAST [10], PSI-BLAST [11] etc. These algorithms offer a fast and efficient recognition of the overlapping subsequences in two protein structures which leads to detection of closely related protein structures. However, these methods cannot recognize proteins with remote homology.

Instead of sequence alignment methods, structure alignment methods like CE [12], MAMMOTH [13] and DALI [14] can be used to detect and highlight distant homology relations between protein structures. In general, these methods are very precise and efficient and they have high degree of successful mapping of the existing structures in new proteins. Structure alignment methods perform comparison of the query protein against all proteins in order to find the most similar existing protein to the novel protein structure. Having in mind that structure alignment methods are quite cost expensive, the speed of the classification with these methods is always questioned. For example, CE takes 209 days [12] to classify 11.000 novel protein structures.

Also, there are numerous methods which combine sequence and structure alignment of the proteins. SCOPmap [15] uses a pipelined architecture for the classification. SCOPmap is based on four sequence alignment methods: BLAST [10], PSI-BLAST [11], RPS-BLAST and COMPASS [16] and two structure alignment methods: VAST [17] and DaliLite [18]. FastSCOP [19] is another, more efficient system than SCOPmap which is based on 3D-BLAST [20] and MAMMOTH [13]. 3DBLAST is a structure alignment method that is used as a pre-processing filter to produce the top 10 scores. Afterwards, the top 10 results are used by MAMMOTH in order to find the most similar protein structures to the query structure.

In contrary of classification of protein molecules by applying alignment techniques, the classification based on mapping of the protein structure in the high-dimensional feature space is found as very promising. After proper mapping of the protein structures in the feature space, some classification method can be used. There are many algorithms which can be used for this purpose, like the Naive Bayesian classifier, the nearest neighbour classifier, decision trees, neural networks, support vector machines, Hidden Markov Model and so on. In [21], the distance matrix of the protein molecule is calculated, and then the local and global features are extracted from the distance matrix histograms. The descriptor consists of 24 local and 9 global features. The classification is based on the E-predict algorithm [21]. In [22], a protein descriptor is extracted from the features of the protein sequence in order to avoid complex structure comparison. The protein descriptor contains information for the number of different amino acids, the hydrophobicity, the polarity, the Van der Waals volume, and the secondary structure. By using this protein descriptor, proteins are classified by using a Naive Bayes classifier and boosted C4.5 decision trees [22]. ProCC [23] first decomposes the protein structures into multiple SSE triplets. The algorithm then extracts 10 features from a SSE triplet based on the spatial relationships of SSEs such as distances and angles. R*-Tree is utilized to index the feature vectors of SSE triplets. Similarly, a query protein is decomposed into multiple SSE triplets, which are searched against the R*-Tree. For each database protein, a weighted bipartite graph is generated based on the matched SSE triplets of the retrieval results. A maximum weighted bipartite graph matching algorithm is used for computing the overall similarity score between the query protein and the database protein. Once the algorithm finds the top *k* similar database proteins, K-NN [24] and SVM [25] techniques are adapted to classify the query protein into a known fold. When the classifier cannot assign a class label to the query protein with enough confidence, the algorithm employs a clustering technique to detect new protein folds. The ProCC takes about 9 minutes to compare a query structure with 2733 database proteins.

The Naive Bayes classifier is not appropriate to be used in this research, since it assumes that the features are independent that is not truth in our case. The nearest neighbour classifier used in [26] makes comparison of the unclassified protein with all proteins in the database, which can last too long. This can be avoided by applying some algorithm that uses only several database proteins in the comparison, or by building a profile for each class.

In this chapter we use three approaches for protein classification. Our first approach is based on the Growing Neural Gas (GNG) neural network [27]. GNG is a representative of the Self-organizing Maps (SOMs) [28]. First, we would like to explain some common properties shared by the neural networks of this type. The network structure is a graph consisting of a set of nodes and a set of edges connecting the nodes. Each node has an associated position (reference feature vector) in the input space. Adaptation, during training, is done by moving the position of the nearest (winning) node and its topological neighbours in the graph toward the input signal. There are two kinds of SOMS: Static SOMs and Growing SOMs. The Static SOMs have a pre-defined structure which is chosen a priori and does not change during the parameter adaptation. On the other hand, growing SOMs have no pre-defined structure, which is modified by successive additions (and possibly deletions) of nodes and edges, therefore it can better adapt to the distribution of the training data. In [29], some neural networks of this type are presented. Some of them use Competitive Hebbian Learning (CHL) [30] to build up a topology during the self-organization. The principle of CHL is to create an edge between the winning and the second winning node at each adaptation step.

The Growing Neural Gas (GNG) model [27] does not impose any explicit constraints on the graph topology. The graph is generated and updated by CHL. The topology of the GNG network reflects the topology of the input data distribution.

However, the Self-organizing Maps provide unsupervised learning, thus some modification should be made in order to use them for protein classification – supervised learning. There are several approaches for supervised learning based on GNG. The Supervised Growing Neural Gas (SGNG) [31] described by Bernd Fritzke combines the Growing Neural Gas (GNG) [27] as a hidden layer, and RBF [32] as an output layer. The RBF layer contains separate output node for each class, and each GNG node is connected with an edge with each of the output nodes. In [33], we introduced the SGNG method for protein classification.

Instead of using many nodes in GNG which are close in the feature space and correspond to belonging of the protein into the same class, a Support Vector Machines (SVM) method [25] can be used. Our second approach is based on SVM. In [34], nine different protein classification methods are used, including support vector machines (SVMs) with four different kernel functions, SVM - pairwise and SVM -Fisher. With a SVM classifier, only the border representatives are considered, thus leading to simpler model and a faster classifier. However, with SGNG each GNG node influences the output of each RBF node according to the corresponding weight. Therefore with SGNG the feature space is not "simply" divided that way that each point in the feature space corresponds only to one class, as in the case of using SVM.

Our third approach is based on Hidden Markov Model (HMM), which in contrary to the previous approaches builds a separate model (profile) for each class. There are several methods [35], [36] for protein classification which use a Hidden Markov Model (HMM) for alignment of protein sequences (protein primary structures). Alexandrov and Gerstein [37] introduced the HMM for classifying protein tertiary structures. In [38], the three-dimensional structure is converted into one-dimensional representation, and then a HMM is applied in order to classify protein tertiary structures. For a given state, the average Root-Mean-Square deviation (RMSd) from the C_α atoms of the fragments to their centroid is used to measure the structural dispersion of each state. To reconstruct the protein 3D structures from their description as a series of states, and to keep some comparison possible, the building procedure employed by Kolodny et al. [39] is used. In [40], it is shown that a HMM method based on the protein tertiary structure is more accurate (10% higher precision) than a similar method based on the protein sequence (primary structure). This is due to the fact that the tertiary structure cares much more information than the primary structure.

In this chapter we present three approaches for classification of protein structures. First, we extract some features of the protein structures, by applying our protein voxel [41], [42] and ray based descriptors [42]. After proper positioning of the protein structures, the Spherical Trace Transform is applied to them to produce descriptor vectors, which are rotation invariant. Additionally, some features of the primary and secondary structure of the protein are taken, thus forming better integrated descriptor. Also, a modified ray based descriptor is applied on the interpolated backbone of the protein molecule. The extracted feature vectors are used for proper mapping of the protein structures in the feature space. Then, classification methods are used. In this chapter we use the Supervised Growing Neural Gas (SGNG) [31], Support Vector Machines (SVM) [25] and Hidden Markov model (HMM) [43], [44] for protein classification.

The rest of the chapter is organized as follows, in section 2 our protein classification approaches are presented, some experimental results and discussions are given in section 3, while section 4 concludes the chapter a gives directions for future work.

2. Our Protein Classification Approaches

In this chapter, we present three approaches for classification of protein structures. The information about protein structure is stored in PDB files which contain information about the primary, secondary and tertiary structures of the proteins. First, we extract some features of the protein structures, thus forming the protein voxel and ray based descriptors. These feature vectors are used for mapping the protein structures in the feature space.

Then, classification methods are used in order to classify each newly protein in the corresponding protein domain in the SCOP hierarchy. We use the Supervised Growing Neural Gas (SGNG) [31], Support Vector Machines (SVM) [25] and Hidden Markov model (HMM) [43], [44] for protein classification.

2.1. Protein Voxel Based Descriptor

In [41], [42] our protein voxel based descriptor is presented. First, we extract the features of the tertiary structure of the protein by using the voxel based algorithm for 3D object retrieval [45]. Since the exact position of each atom and its radius are known, it may be represented by a sphere. The surface of each sphere is triangulated thus forming the 3D-mesh model of the protein structure. After triangulation, we perform voxelization. Voxelization transforms the continuous 3D-space into 3D discrete voxel space. First, we divide the continuous 3D-space into voxels. Depending on the positions of the polygons of the 3D-mesh model, to each voxel a value is attributed equal to the fraction of the total surface area of the mesh which is inside that voxel. The information contained in the voxel grid is processed further to obtain both correlated information and more compact representation of the voxel attributes as a feature. We apply the 3D Discrete Fourier Transform (3D-DFT) to obtain a spectral domain feature vector which provides rotation invariance of the descriptors. Since the 3D-DFT input is a real-valued array, the symmetry is present among the obtained coefficients, so the feature vector is formed from all non-symmetrical coefficients. This vector presents the geometrical features of the protein molecule. Additionally, some features of the primary and secondary structure of the protein molecules are extracted [46], thus forming the attribute-based descriptor vector. In this way we form better integrated protein descriptor.

2.2. Protein Ray Based Descriptor

The C_α atoms form the backbone of the protein molecule. There are some residues that hang up on the C_α atoms, which are not important in the classification. Our previous analyses [42] showed that by taking into account only the C_α atoms of the protein and extracting some suitable descriptor, we can get better results. Since proteins have distinct number of C_α atoms, we have to find a unique way to represent all proteins with descriptors with same length. We approximate the backbone of the protein with fixed number of points, which are equidistant along the backbone. Finally, we use modification of the ray descriptor [45]. In the ray descriptor, the surface of the 3D object is triangulated thus forming the mesh model, and then the elements of the descriptor are calculated as the Euclidean distances between the points of the mesh model to the centre of the mass. In this approach, we approximate the backbone by interpolating to fixed number of C_α atoms, and then the elements of the descriptor are calculated as Euclidean distances from the approximated points to the centre of mass.

2.3. Supervised Growing Neural Gas (SGNG)

In [33], we introduced the Supervised Growing Neural Gas (SGNG) for protein classification. SGNG [31] is described by Bernd Fritzke as an algorithm which combines the Growing Neural Gas (GNG) [27] as a hidden layer, and RBF [32] as an output layer. The RBF layer contains separate output node for each class. The SGNG network topology is shown in Fig. 2.

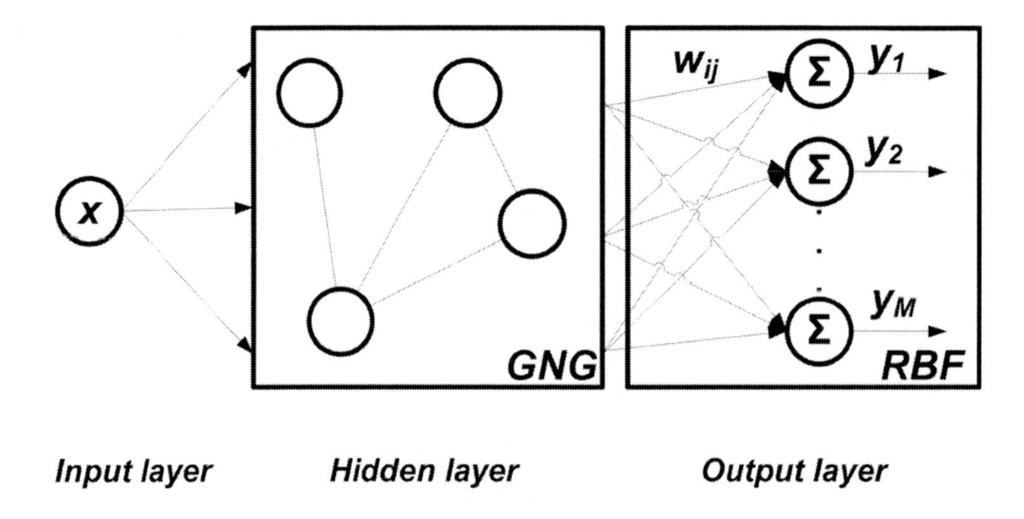

Figure 2. The topology of the Supervised Growing Neural Gas (SGNG).

Let assume that we want to classify n-dimensional feature vectors into M classes. The output layer has one output node for each class. In testing, the classification decision is made according to the output node with the greatest output value. The training phase is done by presenting pairs of input and expected output vectors. As in GNG, we start with two randomly selected nodes connected by an edge. The edge has no weight, since it is not part of the actual RBF network, but represents the fact that two nodes are neighbours. The neighbour information is maintained by application of the Competitive Hebbian learning (CHL) [30]. The adaptation of the hidden nodes is performed as in GNG, the winner node s is moved some fraction of the distance to the input, while the neighbours of s are moved some smaller fraction of their distance to the input.

Let the input vector be denoted as x, the desired output vector denoted as d, and the output vector of the RBF layer is denoted as y. The standard deviation of the j-th RBF node is denoted as σ_j, the output from the hidden node i is denoted as z_i, w_{ij} is the weight from the hidden node i to the output node j and let η be the learning rate.

First, we create two randomly positioned GNG nodes, connected by an edge with *age*=0 and set their errors to 0. Present an input vector x, with expected output vector d. Find the two nodes s and t nearest to x, with reference vectors w_s and w_t. The L_2 norm is used as a distance measure. Evaluate the net using the input vector x. Adjust the weights by applying the delta rule (1).

$$w_{ij} = w_{ij} + \eta(d_i - y_i)z_j \tag{1}$$

The local error of the winning node s is updated, so the distance between the actual output and target is added.

$$error_s = error_s + dist\,(d, y) \tag{2}$$

The winning node s and its neighbours are moved towards x by fractions e_w and e_n of their distances to the input.

$$\begin{aligned} \overline{w}_s &= \overline{w}_s + e_w * dist(\overline{x}, \overline{w}_s) \\ \overline{w}_i &= \overline{w}_i + e_n * dist(\overline{x}, \overline{w}_i), \forall i \in neighbours(s) \end{aligned} \tag{3}$$

For each node j that was moved by (3), set the width to the mean of the distances between j and all the neighbours of j. Increment the ages of all edges from node s to its topological neighbours. If s and t are connected by an edge, then reset its age. If they are not connected, then create an edge between them with $age = 0$. Remove the edges with an age larger than age_{max}, and then remove the nodes without edges. Recalculate the mean widths of the affected nodes.

If the insertion criterion is met, then insert a new node. The criteria for inserting new nodes could be at a fixed number of iterations. A better criterion is to observe the mean squared error. If the squared error stops decreasing, this can be interpreted as the node movements are not able to adjust sufficiently to lower the error; this in turn means that for the current network size it is as good as it can and a new node should be inserted. We used the first insertion criterion.

The new node r is inserted between the node u with largest error, and its neighbour v with the largest error. Create edges between u and r, and v and r and remove the edge between u and v. Recalculate the mean widths of u, v and r. Decrease the error of u and v and set the error of the node r by (4). Initialise the weights from node r to all output nodes.

$$error_u = error_u / 2 \qquad error_v = error_v / 2 \qquad error_r = (error_u + error_v)/2 \tag{4}$$

Decrease the error variables of all GNG nodes by a factor β, thus recently measured errors have a greater influence than older ones (5).

$$error_j = (1 - \beta) * error_j \tag{5}$$

If the stopping criterion is not met, then repeat the process. The stop criterion could be defined by the maximum size that the network may reach. However, this requires a prior knowledge about the distribution. Another criterion is by defining a maximum error allowed, and learning until some predefined threshold is reached. Also an upper limit on the allowed number of misclassifications can be defined and used as a stop criterion. Since the number of misclassifications converges towards lower values faster than the error, we used the last stop criteria.

2.4. Support Vector Machines (SVM)

In [47], we used the Support Vector Machines (SVM) method in combination with the protein voxel based descriptor. In this chapter, additionally we use the SVM method for protein classification by using the protein ray based descriptor.

The support vector machine is a binary classification method proposed by Vapnik and his colleagues at the Bell laboratories [48], [49]. As a binary problem, it has to find the optimal hyperplane which separates the positive and negative examples.

Examples are presented as data points: $\{x_i, y_i\}$, $i = 1,...,N$, $y_i \in \{-1, 1\}$, $x_i \in R_d$. In our approach, x corresponds to the descriptor of the *i*-th training protein. The points x which lie on the hyperplane satisfy $w \cdot x + b = 0$, where w is normal to the hyperplane, $|b|/||w||$ is the distance from the hyperplane to the origin, and $||w||$ is the Euclidean norm of w. Let suppose that all training examples satisfy the constraints (6), so they can be combined in an inequality (7).

$$\begin{aligned} \mathbf{x}_i * \mathbf{w} + b \geq +1, \quad & \text{for } y_i = +1 \\ \mathbf{x}_i * \mathbf{w} + b \leq -1, \quad & \text{for } y_i = -1 \end{aligned} \tag{6}$$

$$y_i(\mathbf{x}_i * \mathbf{w} + b) - 1 \geq 0, \quad \forall i \tag{7}$$

The points which satisfy the equality (7) lie on a hyperplane. The hyperplanes are parallel and distinguish the positive from negative examples. The goal is to find a pair of hyperplanes which gives the maximum margin by minimizing $||\mathbf{w}||^2/2$. Once the hyperplanes are found, the model is formed from the examples that lie on the separating hyperplanes, and these examples are called support vector machines.

Nonnegative Lagrange multipliers α_i are used for each example. In this way, the primal Lagrangian gets the form (8). Then, we have to minimize L_P with respect to w, and b; and maximize with respect to all α_i at the same time. This is a convex quadratic programming problem, since the function is itself convex, and those points which satisfy the constraints form a convex set. This means that we can equivalently solve the following "dual" problem: maximize L_P, subject to the constraints that the gradient of L_P with respect to w and b vanish, and subject also to the constraints that $\alpha_i \geq 0$. This gives the conditions (9). Then, (9) is substituted into (8), which leads to (10).

$$L_P = \frac{1}{2}\|\mathbf{w}\|^2 - \sum_{i=1}^{N} \alpha_i y_i (\mathbf{x}_i * \mathbf{w} + b) + \sum_{i=1}^{N} \alpha_i \tag{8}$$

$$\mathbf{w} = \sum_i \alpha_i y_i \mathbf{x}_i \,, \qquad \sum_i \alpha_i y_i = 0 \tag{9}$$

$$L_D = \sum_i \alpha_i - \frac{1}{2} \sum_{i,j} \alpha_i \alpha_j y_i y_j \mathbf{x}_i * \mathbf{x}_j \tag{10}$$

L_P and L_D show the Lagrangian which arise from the same objective function but under different constraints. In this way, the problem can be solved by minimizing L_P or by maximizing L_D.

However, this algorithm is not suitable for non-separable problems. Therefore positive slack variables e_i, $i = 1,\ldots,N$ are introduced in (6), thus forming the constraints (11). An extra cost for the error is assigned, so the objective function to be minimized is $\|w\|^2/2 + C(\Sigma_i e_i)$ instead $\|w\|^2/2$. The parameter C is defined by the user. Larger value for C corresponds to a higher penalty to the errors.

$$\begin{aligned} \mathbf{x}_i * \mathbf{w} + b &\geq +1 - e_i, \quad \text{for } y_i = +1 \\ \mathbf{x}_i * \mathbf{w} + b &\leq -1 + e_i, \quad \text{for } y_i = -1 \end{aligned} \qquad e_i \geq 0, \ \forall i \tag{11}$$

In order to generalize the above method to be applicable for non-separable problems, the data are mapped into other feature space using some mapping Φ. If there were a kernel function K such that $K(\mathbf{x}_i, \mathbf{x}_j) = \Phi(\mathbf{x}_i) \cdot \Phi(\mathbf{x}_j)$, we would only need to use K in the training, and would never need to explicitly know what Φ is. The Gaussian kernel given by (12) is an example for this function, where σ stands for the standard deviation.

$$K(\mathbf{x}_i, \mathbf{x}_j) = \exp(-\|\mathbf{x}_i - \mathbf{x}_j\|^2 / 2\sigma^2) \tag{12}$$

Although the SVM method is a binary classifier, there are many approaches that perform multi-class classification [50], but are computationally much more expensive than solving several binary problems. On the other hand, many approaches decompose the multi-class problem into several binary problems thus leading to faster classifier.

One possible approach is one-against-all (OvA), where M separate classifiers are constructed, if there are M classes. In the i-th SVM classifier, samples from the i-th class are taken as positive, while samples from all other classes are taken as negative examples. In the testing phase, the query sample is presented to all M SVMs and is classified according to the outputs from the M classifiers. The main disadvantage of this approach is that all examples are used in building all M classifiers, thus leading to complex models.

On the other hand, the one-against-one (OvO) approach can be used by building a separate classifier for each pair of classes, thus leading to $M(M\text{-}1)/2$ classifiers. Each classifier is trained using the samples of the first class as positive and the samples of the second class as negative examples.

The number of examples used for training of each one of the OvO classifiers is smaller, since only examples from two classes are considered. The main disadvantage of this approach is that every test example has to be presented to $M(M\text{-}1)/2$classifiers, thus leading to slower testing, especially as the number of classes arises.

In this research, our dataset contains protein chains from 150 classes, so we used the OvA approach leading to 150 SVMs, instead of 11175 classifiers in the OvO case.

2.5. Hidden Markov Model (HMM)

In [51] we presented our protein classification approach based on Hidden Markov Model (HMM). Hidden Markov Models (HMMs) [52] are statistical models which are applicable to time series or linear sequences. They are widely used in speech recognition applications [53], and are introduced to bioinformatics in the late 80's [54].

A HMM can be visualized as a finite state machine. Finite state machines move through a series of states and produce some kind of output, either when the machine has reached a particular state or when it is moving from a state to state. Hidden Markov models, which are extensions of the Markov chains, have a finite set of states ($a_1,\ldots,a_n$), including a begin state and an end state. The HMM generates a protein sequence by emitting symbols as it progresses through a series of states. Each state has transition probabilities T_{ij} that a state a_i will transit to another state a_j, and emission probabilities $E(x|j)$ that a state a_j will emit a particular symbol x. Each sequence can be represented by a path through the model. This path follows the Markov assumption, that is, the choice of the next state is only dependent on the choice of the current state (first order Markov Model). However, the state sequence is not known; it is hidden. A HMM is generated for each class. To obtain the probability that a query belongs to the corresponding class, the query sequence is compared to the HMM by aligning it to the model. The most likely path through the model can be computed with the Viterbi [55] or forward algorithm [56]. In this research we have used the Viterbi algorithm where the most probable sequence is determined recursively by backtracking.

In this approach we consider only the C_α atoms of the protein which form its backbone. The main idea of this approach is to model the folding of the protein backbone around its centre of mass by using HMM. In this way we perform protein tertiary structures alignment.

First, the protein backbone is uniformly interpolated. Different number of interpolation points can be used. In this research we interpolate the backbone with 64 points, which has shown as sufficient for extracting the most relevant features of protein tertiary structure [42]. After backbone interpolation, we calculate the Euclidean distances from the interpolated points to the centre of mass. In this way the folding of the protein backbone around its centre of mass is considered. Additionally, distances are quantized in order to obtain discrete values of symbols that can be emitted in each state of the HMM. Different type of quantization can be used. In order to best model the hydrophobic effect, we use a uniform quantization. Experimentally we determined that 20 quantization levels are enough. In this way, by quantizing the distances from the interpolated points to the centre of mass, this approach models the folding of the protein backbone into concentric spheres.

3. Experimental Results

We have implemented a system for protein classification based on the approaches described above. The evaluation of the approaches is made according to the SCOP hierarchy. Our standard of truth data contains 6979 randomly selected protein chains from the 150 most populated SCOP domains from the SCOP 1.73 database [3], [57]. 90% of the data serves as a training data and the other 10% serves as test data. The classification accuracy is used as a precision measure of the approaches.

First, we will present the results obtained by the SGNG approach by using the voxel and ray based descriptors. We trained two separate SGNG networks (one for the voxel and one for the ray based descriptor). 1147 training protein chains from 30 SCOP domains are used for training. The test set contains 127 protein chains from 30 SCOP domains. In Table 1, some classification results are shown.

Table 1. Experimental results of the SGNG approach

	Protein voxel based descriptor	Protein ray based descriptor
Descriptor length	450	64
Number of GNG nodes	163	78
Classification time per test protein (*sec*)	0.024	0,004
Classification accuracy (%)	83.5	98.4

By using the ray based descriptor we get better classification accuracy of 98.4%. This is because proteins that belong to a same domain have very close ray based descriptors, and they are far enough from the descriptors of the proteins from other domains. This result was expected, since the average precision of the voxel based descriptor is 77.8% while the average precision of the ray descriptor is 92.9% [42].

Beside better classification accuracy, by using the ray based descriptor we also achieve faster classification. The classification time is smaller due to the smaller number of GNG nodes in the model and the lower dimensionality of the descriptor. SGNG is faster than the k nearest neighbour method since in the SGNG approach we compare the descriptor of the query protein only with 78/163 vectors, while the query descriptor would be compared with 1147 descriptors in the nearest neighbour's case.

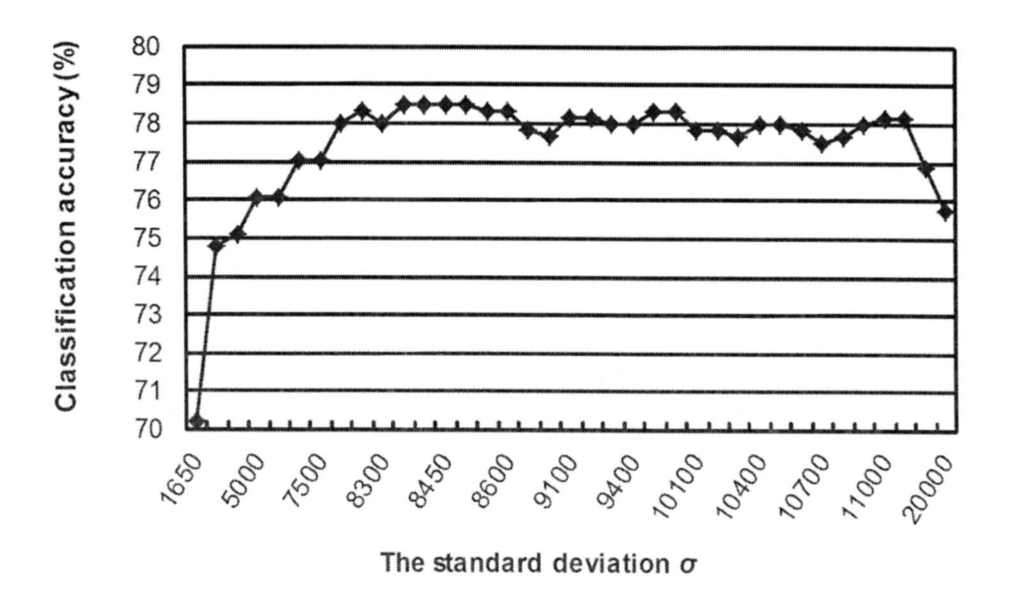

Figure 3. The influence of the standard deviation σ on the classification accuracy of the SVM based approach by using the protein voxel based descriptor.

Next, we will present some results obtained by the SVM based approach by using the voxel and ray based descriptors. First we present some results obtained by mapping the protein tertiary structure in the protein voxel descriptors' feature space. We examined the influence of standard deviation σ of the Gaussian kernel on the classification accuracy. Fig. 3 shows the classification accuracies when different values for σ are used. In this analysis, the entire dataset of 6979 protein chains was used.

The analysis showed that for small value of σ, the training lasts longer, and leads to over-fitting. By decreasing the standard deviation, the classification accuracy is getting worse (70.2% classification accuracy for σ=1650). By increasing σ, the classification accuracy increases. In [47] more detailed results of this analysis are given.

Further, we examined how the penalty given to the errors c influences the classification accuracy. This analysis is performed for the best values of σ (according to Fig. 3) and shows that the error penalty c has a minor influence on the classification accuracy [47]. The highest classification accuracy of 78.83% is obtained for σ =8350 and c=50. The training phase lasts several minutes, while the test phase takes several seconds.

The results obtained by the SGNG approach showed that it is better to map the protein structures in the ray based descriptors' feature space. So, next we examined the performances of the SVM based approach when protein ray based descriptors are used. We examined the influence of the standard deviation σ on the classification accuracy, see Table 2. As can be seen, the highest classification accuracy of 98.7% is achieved for σ=0.7 and σ=1.

Table 2. The influence of the standard deviation σ on the classification accuracy of the SVM based approach by using the protein ray based descriptor

Standard deviation σ	Classification accuracy (%)
0.5	98.37
0.7	98.70
0.8	96.91
0.9	96.74
1	98.70
2	98.37
3	98.37
5	97.56
10	96.74

According to the training and test time, we performed analysis on a smaller dataset. 1875 training and 208 test protein chains were selected from 50 SCOP domains. For σ=1, the training phase took 44 seconds, while the classification time for all test proteins was 3 seconds. For σ=10, the training phase took 25 seconds, while the classification time for all test proteins was 1 second. The training and test phases for σ=1 last longer, but, as can be seen from Table 2, by using σ=1, the precision is higher for 2% than by using σ=10.

In this chapter beside the SGNG and SVM based approach, we also presented a HMM based approach for classifying protein tertiary structures. In this research we interpolated the protein backbone up to 64 interpolation points. First, we examined the influence of number of HMM states (Q) on the classification accuracy [51]. When 20 HMM states are used, the highest precision of 92.51% is achieved. By using 30 HMM states the classification time increases, while the classification accuracy is getting worse.

Additionally, we compared our HMM based approach against an existing approach named 3D HMM [37]. We used models with 20 HMM states. In this analysis, we used a dataset of protein chains from the globins and IgV (V set from immunoglobulin superfamily) families, which are used in [37]. We randomly chose one training protein chain from each domain, while the other protein chains serve as test data. In this way the test set consists of 754 protein chains from the globins and 1326 protein chains from the IgV family. The analysis showed that our approach achieves classification accuracy higher for 1.5% for the globins and 1.7% for the IgV family. More experimental results are given in [51].

However, classifiers which are based on HMM can be used for classification at lower levels, but aren't suitable at upper levels in the SCOP hierarchy. To be precise, this algorithm builds a profile for a given class, so if we use this classifier at upper levels we want to build a profile for dissimilar proteins. So, if we want to classify proteins at upper levels we have to use some other classifier. However, this approach can be incorporated into a hybrid hierarchical classifier, where it can be used at the family and lower levels in the SCOP hierarchy.

Table 3 compares our classification approaches. When the mapping in the feature space is done by using the ray based descriptors, better classification accuracy is achieved. SGNG obtain higher precision than SVM when voxel descriptors are used, but it should be mentioned that the results are obtained on different datasets. Namely, the results for SGNG are obtained on a smaller subset which contains 1147 training and 127 test protein chains. If ray based descriptors are used, the highest precision is obtained using the SVM based approach.

Table 3. Comparison of our protein classification approaches

	SGNG	SVM	HMM
Classification accuracy (%) by using the protein voxel based descriptor	83.5	78.83	/
Classification accuracy (%) by using the protein ray based descriptor	98.4	98.7	92.5
Number of training protein chains	1147	5531	5531
Number of test protein chains	127	614	614

Conclusion

In this chapter we presented three approaches for protein tertiary structures classification. First, we extracted our protein voxel and ray based descriptors for mapping the proteins structures in the feature space. Then, we used three classification methods: Supervised Growing Neural Gas (SGNG), Support Vector Machines (SVM) and Hidden Markov model (HMM). A part of the SCOP 1.73 database was used to evaluate the classification accuracy of our approaches.

The results showed that the SGNG approach achieves more than 83,5% classification accuracy by using the protein voxel based descriptor and 98,4% classification accuracy by using the protein ray based descriptor. In future, our SGNG classifier can be used for hierarchical classification by training separate SGNG network for each parent node in the SCOP hierarchy.

According to the SVM based approach, we investigated the influence of the standard deviation and the error penalty on the classification accuracy. The results showed that the error penalty has a minor influence on the accuracy, while the standard deviation drastically affects the adequacy of the classifier. Further analysis can be made to find the optimal value of the standard deviation. The results showed that the SVM based approach achieves 78,83% classification accuracy by using the protein voxel based descriptor and 98,7% classification accuracy by using the protein ray based descriptor.

The analysis showed that our HMM based approach achieves high precision of 92.5%. Additionally we compared our HMM based approach against 3D HMM [37]. The results showed that our HMM based approach is more accurate than 3D HMM. To be precise, our HMM based approach achieves classification accuracy higher for 1.5% for the globins and 1.7% for the IgV family.

All three approaches showed as much faster (last minutes, seconds) than SCOP, DALI and CATH, therefore we expect that they can significantly decrease the gap between protein holdings of PDB and SCOP databases.

Our future work is concentrated on investigating other protein classifiers in order to obtain higher precision. Also, we want to make a hybrid hierarchical classifier, where the most appropriate classifier would be used at each level in the SCOP hierarchy. Also our future work will be concentrated on increasing the efficiency of the algorithm by investigating new 3D descriptors in order to map the protein structures in the feature space better.

References

[1] Berman, H. M.; Westbrook, J.; Feng, Z.; Gilliland, G.; Bhat, T. N.; Weissig, H.; Shindyalov, I. N.; Bourne, P.E. The Protein Data Bank. *Nucleic Acids Research*, 2000, vol. 28, 235-242.

[2] Protein Data Bank [online]. 2010, Available from: *http://www.rcsb.org.*

[3] Murzin, A. G.; Brenner, S. E.; Hubbard, T.; Chothia, C. Scop: a structural classification of proteins database for the investigation of sequences and structures. *J. Mol. Biol.,* 1995, vol. 247, 536-540.

[4] Andreeva, A.; Howorth, D.; Chandonia, J.-M.; Brenner, S. E.; Hubbard, T. J. P.; Chothia, C.; Murzin, A. G. Data growth and its impact on the SCOP database: new developments. *Nucleic Acids Research*, 2008, vol. 36, 419-425.

[5] Orengo, C. A.; Michie, A. D.; Jones, D. T.; Swindells, M. B.; Thornton, J. M. CATH-- A Hierarchic Classification of Protein Domain Structures. *Structure*, 1997, vol. 5, no. 8, 1093-1108.

[6] Taylor, W. R.; Orengo, C. A. Protein structure alignment. *J. Mol. Biol.*, 1989, vol. 208, 1-22.

[7] Holm, L.; Sander, C. The FSSP Database: Fold Classification Based on Structure-Structure Alignment of Proteins. *Nucleic Acids Research*, 1996, vol. 24, 206-210.

[8] Holm, L.; Sander, C. Protein structure comparison by alignment of distance matrices. *J. Molecular Biology,* 1993, vol. 233, no. 1, 123-138.

[9] Needleman, S. B.; Wunsch, C. D. A general method applicable to the search for similarities in the amino acid sequence of two proteins. *Journal of Molecular Biology,* 1970, vol. 48, no. 3, 443-453.
[10] Altschul, S. F.; Gish, W.; Miller, W.; Myers, E. W.; Lipman, D. J. Basic local alignment search tool. *Journal of Molecular Biology*, 1990, vol. 215, no. 3, 403-410.
[11] Altschul, S. F.; Madden, T. L.; Schaffer, A. A.; Zhang, J.; Zhang, Z.; Miller, W.; Lipman, D. J. Gapped BLAST and PSI-BLAST: a new generation of protein database search programs. *Nucleic Acids Research*, 1997, vol. 25, no. 17, 3389-3402.
[12] Shindyalov, H. N.; Bourne, P. E. Protein structure alignment by incremental combinatorial extension (CE) of the optimal path. *Protein Engineering*, 1998, vol. 9, 739-747.
[13] Ortiz, A. R.; Strauss, C. E.; Olmea, O. Mammoth: An automated method for model comparison. *Protein Science*, 2002, vol. 11, 2606-2621.
[14] Holm, L.; Sander, C. Protein structure comparison by alignment of distance matrices. *Journal of Molecular Biology*, 1993, vol. 233, 123-138.
[15] Cheek, S.; Qi, Y.; Krishna, S. S.; Kinch, L. N.; Grishin, N. V. SCOPmap: Automated assignment of protein structures to evolutionary superfamilies. *BMC Bioinformatics*, 2004, vol. 5, 197-221.
[16] Sadreyev, R.; Grishin, N. COMPASS: a tool for comparison of multiple protein alignments with assessment of statistical significance. *Journal of Molecular Biology*, 2003, vol. 326, 317-336.
[17] Madej, T.; Gibrat, J. F.; Bryant, S. H. Threading a database of protein cores. *Proteins*, 1995, vol. 23, 356-369.
[18] Holm, L.; Sander, C. Dali: a network tool for protein structure comparison. *Trends in Biochemical Science,* 1995, vol. 20, 478-480.
[19] Tung, C. H.; Yang, J. M. fastSCOP: a fast web server for recognizing protein structural domains and SCOP superfamilies. *Nucleic Acids Research*, 2007, vol. 35, W438-W443.
[20] Yang, J. M.; Tung, C. H. Protein structure database search and evolutionary classification. *Nucleic Acids Research*, 2006, vol. 34, no. 13, 3646-3659.
[21] Chi, P. H. *Efficient protein tertiary structure retrievals and classifications using content based comparison algorithms*, PhD thesis, University of Missouri-Columbia, 2007.
[22] Marsolo, K.; Parthasarathy, S.; Ding, C. A Multi-Level Approach to SCOP Fold Recognition. *IEEE Symposium on Bioinformatics and Bioeng.*, 2005, 57-64.
[23] Kim, Y. J.; Patel, J. M. A framework for protein structure classification and identification of novel protein structures. *BMC Bioinformatics*, 2006, vol. 7, no. 456.
[24] Hastie, T.; Tibshirani, R. Discriminant adaptive nearest neighbor classification. *IEEE Transactions on Pattern Analysis and Machine Intelligence,* 1996, vol. 18, no. 6, 607-616.
[25] Cortes, C.; Vapnik, V. Support vector networks. *Machine Learning*, 1995, vol. 20, 273-297.
[26] Ankerst, M.; Kastenmuller, G.; Kriegel, H. P.; Seidl, T. Nearest Neighbor Classification in 3D Protein Databases. *Seventh Int'l Conf. Intelligent Systems for Molecular Biology* (ISMB '99), 1999.
[27] Fritzke, B. A growing neural gas network learns topologies. *Advances in Neural Information Processing Systems,* 1995, vol. 7, 625-632.

[28] Kohonen, T. Self-Organizing Maps. Third ed. *Series in Information Sciences,* Heidelberg: Springer; 2000.

[29] Prudent, Y.; Ennaji, A. A K Nearest Classifier design. *Electronic Letters on Computer Vision and Image Analysis*. 2005, vol. 5, no. 2, 58-71.

[30] Martinetz, T. Competitive *Hebbian learning rule forms perfectly topology preserving maps. ICANN'93*, 1993, Springer, 427-434.

[31] Holmström, J. *Growing Neural Gas, Experiments with GNG*, GNG with Utility and Supervised GNG. 2002, Master Thesis, Uppsala University, Uppsala, Sweden.

[32] Chen, S.; Cowan, C. F.; Grant, P. M. Orthogonal least squares learning algorithms for radial basis function networks. *IEEE Trans. Neural Networks*, 1991, vol. 2, 302-309.

[33] Mirceva, G.; Kulakov, A.; Davcev, D. Protein Classifier by Matching 3D Structures. *Hybrid Artificial Intelligence Systems* (HAIS 2009), 2009, LNCS/LNAI, 5572, Springer-Verlag Berlin Heidelberg, 425-432.

[34] Pooja, K. *Comparative analysis of protein classification methods*. 2004, Master Thesis, University of Nebraska, Lincoln.

[35] Plötz, T. *Advanced Stochastic Protein Sequence Analysis*. 2005, PhD Thesis, Technical Faculty, Bielefeld University.

[36] Plötz, T.; Fink, G. A. Pattern recognition methods for advanced stochastic protein sequence analysis using HMMs. *Pattern Recognition*, 2006, vol. 39, issue 12, 2267-2280.

[37] Alexandrov, V.; Gerstein, M. Using 3D Hidden Markov Models that explicitly represent spatial coordinates to model and compare protein structures. *BMC Bioinformatics,* 2004, vol. 5, no. 2.

[38] Regad, L.; Guyon, F.; Maupetit, J.; Tufféry, P.; Camproux, A. C. A Hidden Markov Model applied to the protein 3D structure analysis. *Computational Statistics & Data Analysis,* 2008, vol. 52, issue 6, 3198-3207.

[39] Kolodny, R.; Koehl, P.; Guibas, L.; Levitt, M. Small libraries of protein fragments model native protein structures accurately. *Journal of Molecular Biology*, 2002, vol. 323, no. 2, 297-307.

[40] Fujita, M.; Toh, H.; Kanehisa, M. Protein sequence-structure alignment using 3D-HMM. *Fourth International Workshop on Bioinformatics and Systems Biology*, 2004, 7-8.

[41] Kalajdziski, S.; Mirceva, G.; Trivodaliev, K.; Davcev, D. Protein Classification by Matching 3D Structures. *IEEE Frontiers in the Convergence of Bioscience and Information Technologies*, 2007, Jeju Island, Korea, 147-152.

[42] Mirceva, G.; Kalajdziski, S.; Trivodaliev, K.; Davcev, D. Comparative analysis of three efficient approaches for retrieving protein 3D structures. *4-th Cairo International Biomedical Engineering Conference* (CIBEC'08), 2008, Cairo, Egypt.

[43] Ephraim, Y.; Merhav, N. Hidden Markov processes. *IEEE Transactions on Information Theory,* 2002, vol. 48, issue 6, 1518-1569.

[44] Rabiner, L. R. A tutorial on hidden Markov models and selected applications in speech recognition. *Proceedings of the IEEE*, 1989, vol. 77, no. 2, 257-286.

[45] Vranic, D. V. 3D Model Retrieval. 2004, Ph.D. Thesis, University of Leipzig.

[46] Daras, P.; Zarpalas, D.; Axenopoulos, A.; Tzovaras, D.; Gerassimos, M. S. Three-Dimensional Shape-Structure Comparison Method for Protein Classification.

IEEE/ACM Transactions on computational biology and bioinformatics, 2006, vol. 3, no. 3, 193-207.

[47] Mirceva, G.; Naumoski, A.; Davcev, D. A Protein Classifier Based on SVM by Using the Voxel Based Descriptor. *Rough Sets and Current Trends in Computing* (RSCTC 2010), 2010, LNCS/LNAI, 6086, Springer-Verlag Berlin Heidelberg, 2010, 640-648.

[48] Vapnik, V. *The Nature of Statistical Learning Theory. 2nd ed.* New York: Springer; 1999.

[49] Burges, C. J. C. A tutorial on support vector machine for pattern recognition. *Data Min. Knowl. Disc.*, 1998, 2, 121.

[50] Weston, J.; Watkins, C. *Multi-class support vector machines*. ESANN99. 1999, M. Verleysen, Ed., Brussels, Belgium.

[51] Mirceva, G.; Davcev, D. HMM based approach for classifying protein structures. *International Journal of Bio- Science and Bio- Technology*, 2009, vol. 1, no. 1, 37-46.

[52] Ephraim, Y.; Merhav, N. Hidden Markov processes. *IEEE Transactions on Information Theory*, vol. 48, 2002, 1518-1569.

[53] Rabiner, L. R. A tutorial on hidden Markov models and selected applications in speech recognition. *Proceedings of IEEE,* 1989, vol. 77, no. 2, 257-285.

[54] Churchill, G. A. Stochastic models for heterogeneous DNA sequences. *Bulletin of Mathematical Biology*, 1998, vol. 51, no. 1, 79-94.

[55] Viterbi, A. J. Error bounds for convolutional codes and an asymptotically optimum decoding algorithm. *IEEE Transactions on Information Theory*, 1967, vol. 13, no. 2, 260-269.

[56] Durbin, R.; Edy, S.; Krogh, A.; Mitchison, G. Biological sequence analysis: Probabilistic models of proteins and nucleic acids. Cambridge University Press; 1998.

[57] SCOP Database [online]. 2010, Available from: *http://scop.mrc-lmb.cam.ac.uk/scop.*

In: Protein Structure
Editor: Lauren M. Haggerty, pp. 71-82

ISBN 978-1-61209-656-8

Chapter 4

Common Structural Characteristics of Fibrous and Globular Proteins

N. Kurochkina*

The School of Theoretical Modeling
Chevy Chase, MD, USA

Abstract

Fibrous proteins form coiled coil structure. Long stretches of α-helices form tightly packed arrangements of coiled coil. Helices contain 3.5 residues per turn. Unwinded at the ends, α-helices bind DNA molecules, or dimerizing subunits. In globular proteins, α-helices are shorter. The average number of residues per turn is 3.6. Similarly to coiled coil structure, α-helices include highly packed helix-helix interface region and unwinded at the ends ligand binding region containing heme, metals, and other ligands. Left-handed and right handed, parallel and antiparallel arrangements of α-helices exist. Core regions with similar structure carry common amino acid combinations. Ligand binding regions can also be recognized by a specific amino acid sequence pattern. Besides, α-helices participate in assemblies of protein motifs and assemblies of protein subunits. Distribution of contact areas along helix surface and their link to helix edges can identify the type of motif or type of assembly. Location of the core regions, ligand binding regions, and assembly-related regions on the surface of the α-helix together with their specific combinations are important for the fold of the molecule.

In this work, specific interactions of α-helices, their ligands, and oligomerization properties of protein molecules are considered. Specific patterns of amino acid combinations together with their location at the surface of the α-helix are associated with particular fold of the molecule.

* Corresponding author: N. Kurochkina The School of Theoretical Modeling P.O.Box 15676 Chevy Chase, MD 20825 Phone: (240) 381-2383 e-mail: info@schtm.org

Introduction

Fibrous proteins and globular proteins compose two large groups each exhititing clearly distinct features (Review: Parry et al., 2008). Fibrous proteins are characterized by elongated shape of molecules. Long stretches of keratin, myosin, and fibrinogen contain coiled coil structure – a pair of α-helices winded around each other along their entire length. Collagen three strands form a triple helix. In globular proteins, length does not deviate significantly from width or height. Secondary structure elements such as α-helices and β-sheets associate in a variety of shapes. Amino acid sequence of collagen contains repeat of three residues Gly-Pro-Hyp. Parallel coiled-coil structure can be recognized by a heptad repeat of leucine residues, 11-residue hendecad, or 15-residue pentadecad repeats. Some proteins that contain antiparallel coiled coil structure also exhibit specific sequence patterns such as heptad repeat of alanines or other small residues named alacoil (Gernert et al., 1997). Not only short sequence motifs but also larger repeating structures can be found in globular proteins as for instance myohemerithrin two pairs of antiparallel α-helices homologous and related to each other by a two-fold symmetry (Sheriff et al., 1987). Position of apolar and polar residues within repeat as well as type of amino acid residue at each position and complementarity of amino acids from the opposing α-helices are important for the formation of structure.

In this chapter, parallel and antiparallel arrangements of α-helices in proteins are described, each of these arrangements having either positive or negative interhelical angles. Small structural units located at the interfaces of the two interacting α-helices are shown to determine to great extent the type of interface. These units are composed mainly of core positions a and d in the leucine zipper nomenclature (Hodges et al., 1972). When two α-helices form an interface, amino acids at these positions come into close contact, form hydrophobic core and contribute significantly to specificity of interaction. Although this type of interaction considered was first designed for coiled coil structures, it was successfully applied for all interactions of α-helices in globular proteins. Identical and homologous sequence combinations at positions a and d exist in fibrous and globular proteins that are responsible for the similar type of fold.

Parallel coiled coil structure is present in many proteins that are involved in transcription regulation, muscle contraction, and blood coagulation. Coiled coil domains of globular proteins are very often used for dimerization of subunits. Structure of a coiled coil homodimeric domain of the cGMP-dependent protein kinase Iα, which functions in the NO-mediated relaxation of the vascular myosin muscle was determined by NMR dipolar couplings (Schnell, 2005). Cardiac phospholamban forms homopentamer shaped as a flower held together by leucine/isoleucine zipper motif along membrane-spanning helices (Oxenoid, 2005).

Heterotrimeric parallel four-helix bundle of the synaptic fusion complex (Sutton et al., 1998) and homo-oligomeric parallel arrangement of α-helical antiparallel bundles of the HIV/SIV gp41 (Caffrey et al., 1998) and influenza virus haemagglutinin (Stevens et al., 2004) show core organization similar to leucine zipper. However, the assembly of a heterotrimer itself leads to the membrane fusion of a synaptic complex, while HIV/SIV gp41 and influenza virus haemagglutinin undergo conformational changes in the fusion protein that leads to viral fusion.

Left-handed supercoiling occurs most frequently. Naturally occurring right-handed supercoiled structures are very rare. Parallel right-handed coiled coil tetramer of surface layer protein tetrabrachion from *Staphylothermus marinus* is characterized by 11-residue repeats and in contrast to left-handed tightly packed structures contains large water-filled cavities (Stetefeld et al., 2000). Right-handed α-helical coiled coil is found in tetramerization domain of the vasodilator-stimulated phosphoprotein (VASP) with 15-residue repeat exhibiting a characteristic pattern of hydrophobic residues combined with salt bridges (Kuhnel et al., 2004).

The bacterial flagellar filament switches between left- and right-handed supercoiled forms when bacteria switches their swimming mode between running and tumbling. Packing interactions of a protein called flagellin are responsible for supercoiling of a filament. Pairs of R-type straight protofilaments studied by x-ray crystallography exhibit long stretches of antiparallel coiled coil (Samatey et al., 2001).

Many proteins contain parallel arrangement of α-helices.

Eight parallel α-helices surround eight-stranded β-barrel in 8α/8β fold of triose phosphate isomerase. This enzyme gave TIM-barrel name to 8α/8β fold that is found in a variety of proteins. In contrast to long supercoiled structures, all parallel α-helices form short interfaces along 1-2 turns of the helix. Each helix contacts two other helices of the barrel, one helix preceding and one helix following in amino acid sequence. In spite of low sequence homology all triose phosphate isomerase enzymes show highly similar structure of the 8α/8β domain. This is demonstrated by crystallographic structures of triose phoshate isomerase at high resolution determined for *T. brucei* triose phosphate isomerase in complex with glycerol-3-phosphate and sulfate (Wierenga, 1991), *L. mexicana* triose phosphate isomerase in complex with 2-phosphoglycolate (Kursula, 2003), *S. cereviciae* triose phosphate isomerase (Lolis, 1990) in complex with phosphoglycolohydroxamate (Davenport, 1991) and dihydroxyacetone phosphate (Jogl, 2003), *P. falciparum* triose phosphate isomerase in complex with 2-phosphoglycerate (Parthasarathy, 2003), *G. gallus* triose phosphate isomerase in complex with phosphoglycolohydroxamate. TIM-barrel fold domain is also present in pyruvate kinase from *Orictolagus cuniculus* and *Homo sapiens* (Smith, 2003), malate synthase from *E. coli* (Anstrom, 2003) and *M. tuberculosis* (Smith, 2003), methylmalonyl CoA mutase (Mancia, 1996), fructose-1,6-biphospate aldolase from *Thermus aquaticus* (Izard, 2004), *Drosophila melanogaster* (Hester, 1988), *Homo sapiens* (Gamblin, 1990), *Orictolagus cuniculus* (St Jean, 2005), KDPG aldolase from *E. coli* and other proteins. When 8α/8β fold also includes antiparallel arrangement, one helix and one strand reverses direction at the C-terminus of the barrel as it is found in enolase from *S. cereviciae* (Lebioda, 1991) *H. sapiens* (Chai, 2004), *T. Brucei* (Da Silva Giotto, 2003), and epimerase from *B. subtilis* (Klenchin, 2004).

Long antiparallel helices occur as coiled coil of SIN Nombre virus nucleocapsid protein, serine t-RNA synthetase, DNA binding transcription factor STAT-4, activator of transcription STAT-1, two-stranded homodimer PROP, DNA-binding protein SSO 10, colicin e3 and other proteins. Alacoil, which carries alanines or other small residues in every seventh position (Gernert et al., 1995), is found in several proteins including repressor of primer and ferritin. One of the frequently observed motifs is an antiparallel arrangement of helices named four-α-helix bundle (Review: Harris et al., 1994, Kohn et al., 1997). The four-α-helical motif is found in hemerythrins (Sheriff et al., 1987), ferritins (Andrews et al., 1989), tobacco mosaic virus coat protein (Namba and Stubbs, 1986), cytochrome b562, cytochrome c' (Mathews et al., 1985), transcription factors (Banner et al., 1987), membrane M2 proton channel of influenza A virus (Schnell et al., 2008; Pielak et al., 2009) and other proteins. The four-α-helical structure also represents a unit of higher assemblies (Chou et al., 1988). Bacteriorhodopsin (Chou et al., 1992) and annexin, for instance, combine four-α-helix bundle substructures as a part of 5-, 6-, 7-, and 8-helical assemblies. The subunit of a tobacco mosaic virus coat protein, which is a four-α-helix bundle, assembles in a helical rod wrapped around viral RNA. Three turns of the helical rod contain 49 subunits (Namba et al., 1989, Bhyravbhatla et al., 1998). Helices of the four-α-helix bundle in these assemblies are arranged so that all interhelical angles are positive. Arrangement of α-helices, which exhibits negative interhelical angles exists in the six-helix motif of the proteins that are components of oligomeric signaling complexes MYDDosome, PIDDosome (Lin et al., 2010), apoptosome (Yuan et al., 2010), inflammosome (de Alba, 2009), and death inducing signaling complex (DISC) (Scott et al., 2009; Wang et al., 2010). Domain carrying six-helix bundle fold with a Greek key topology (Lasker et al., 2005) is the major unit involved in assembly and belongs to apoptosis (DD) superfamily (Weber and Vincenz, 2001), which includes caspase recruitment domain (CARD), pyrin and death effector domain (DED) . The helical assembly of subunits MYDDosome forms a tower shaped structure 110Å in height and 70 Å in diameter comprising 6 myeloid differentiation primary response proteins (MYD88), 4 interleukin-1 receptor-associated kinases 4 (IRAK4), and 4 IRAK2, each interacting within the complex via their apoptosis domains (DD) (Lin et al., 2010). Similar domain-domain interfaces are observed in PIDDosome PIDD-RAIDD complex (Park et al., 2007) although mode of assembly resembles double-stranded helical oligomer (Lin et al., 2010).

Parallel helices A, B, and C of alcohol dehygrogenase from *Equus caballus* (Li, 1994) and parallel and antiparallel interfaces of beta-lactamase from *Bacillus licheniformis* (Knox, 1991), *E. coli* (Thomas, 2005), and *Mycobacterium fortuitum* are arranged as if they form an arc of a TIM-barrel. Both contain α/β fold.

Glutathione S-transferase (GST) is a detoxifying enzyme, which catalyses conjugation of glutathione to xenobiotic compounds so that they become more soluble and are therefore more easily excreted This enzyme is present virtually in all organisms. A classification of GSTs based on substrate/inhibitor specificity, amino acid sequence homology, structure similarity, and immunological identity divides all members of this family into alpha, mu, pi, theta, and omega classes (Sheehan et al., 2001). Crystallographic structures of GSTs exist for some members of each class.

GST structural family includes soluble chloride channels *Homo sapiens* CLIC1 (Dulhunty et al., 2001), CLIC4 (Li et al., 2006), *Caenorhabditis elegans* excretory canal abnormal protein 4, and *Drosophila melanogaster* CLIC (Littler et al., 2008), which are similar to omega class GSTs. This group of chloride channels exists in soluble and membrane-bound forms as other intracellular ion channels such as annexin, BCL-XL, and some bacterial toxins (Harrop et al., 2001). Microsomal GSTs are membrane-bound; their tertiary structure differs from a thioredoxin-like fold of cytosolic GSTs. Microsomal glutathione S-transferases I, II, and III are members of a common gene family together with 5-Lipoxygenase-activating protein (FLAP) and leukotriene-C4 (LTC4) synthase (Iacobsson et al, 1997). However, microsomal GSTs and cytosolic GSTs possess a common four-helical structure, which is present in a transmembrane region of LTC4 (Ago et al., 2007) and FLAP (Ferguson et al, 2007).

GST is folded as a molecule with two domains, the N-terminal thioredoxin-like domain of $\alpha_1\beta\alpha_2\beta\beta\alpha_3$ fold, which is also a part of glutaredoxin, glutathione peroxidase, bacterial Dsba enzyme, and glutathione reductase, and the C-terminal α-helical domain. A GST-like domain is also reported in a part of elongation factor 1B from *Saccharomyces cerevisiae* (Jeppesen et al., 2003). The N-terminal (α_1) and C-terminal (α_3) helices of the N-terminal GST domain are located on the same side of the β-sheet and interact with each other and with C-terminal α-helical domain. The middle helix (α_2) of the N-terminal GST domain is on the opposite side of the β-sheet. GST contains a conserved glutathione binding G-site within domain 1 and a xenobiotic substrate binding H-site in the domain 2 that is not conserved and accommodates various ligands (Sheehan et al., 2001). The interface between N-terminal and C-terminal helices of the N-terminal GST domain exhibits both positive and negative interhelical angles in various species. Glutathione S-transferase from *Rattus norvegicus,* pi and theta class *Mus musculus,* pi and mu class *Homo sapiens, Gallus gallus, Caenorhabditis elegans,* and *Drosophila melanogaster* exhibits positive interhelical angles similar to annexin. Interfaces with negative interhelical angle were found to be present in glutathione S-transferase from *Ochrobactrum anthropi*, *Anopheles dirus*, *Anopheles gambiae, Aegilops tauschii, Oryza sativa, Burkholderia xenovorans, Arabidopsis thaliana,* theta and omega class *Homo sapiens,* and zeta class *Mus musculus.*

Annexins contain four homologous domains (domain I, II, III and IV). Each domain consists of five α-helices as a combination of a four-helix bundle with an extra helix C connecting helix-turn-helix substructures AB and DE (Huber et al., 1990). The annexin ion channel is formed by the interior residues of AB substructures of domain II and domain IV (Hofmann et al., 2000). The different annexin domains exhibit different ligand binding properties. Annexin binding of phospholipids is Ca^{2+}-dependent. Annexin V binding to sulfatide regulates coagulability in the blood stream (Ida et al., 2004). Annexins IV, V, and VI specifically bind glycosaminoglycans and may function as recognition elements for these molecules (Ishitsuka et al., 1998). Heparin binds to annexin A2 (Shao et al., 2006).

Parallel and antiparallel arrangements of α-helices carrying positive and negative interhelical angles are illustrated in Figure 1. The rule to determine sign of an angle assumes that angle is positive if helix distant from the viewer is rotated clockwise relative to proximal helix and negative if it is rotated counterclockwise. Helix–helix preferred orientations, which occur at angles of 20° and -70°, correspond to ''knobs into holes'' packing proposed by Crick (1953). Ridges into grooves packing gives basis for -50° interhelical angle (Chothia et al., 1981).

Repeating sequence combinations occur at a and d positions of parallel helix-helix interfaces with similar values of interhelical angles not only in homologous proteins but also within the same protein and in nonhomologous proteins. Within each group of observed combinations, correlation exists between the size of amino acid and magnitude of the interhelical angle (Kurochkina, 2007). Distinct sequence combinations are present in leucine zipper and TIM-barrel proteins. Therefore, it is possible to distinguish between parallel interfaces with positive and negative interhelical angles (Kurochkina, 2008).

Parallel arrangement of α-helices

Distant helix

Proximal helix

α negative α positive

Triose phosphate isomerase GCN4 leucine zipper

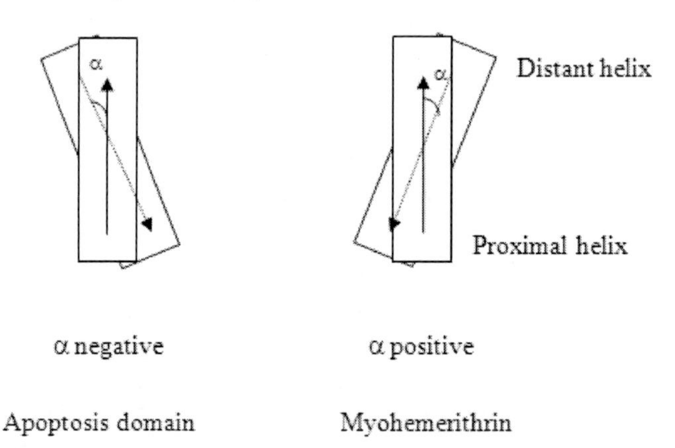

Figure 1. Positive and negative interhelical angles in parallel and antiparallel arrangements of α-helices.

More evidence is provided by the prediction of interhelical angles for glutathione S-transferase, intracellular chloride channel and annexin molecules from various sources. Correct results were achieved in 58 out of 62 examined proteins (Data not published; manuscript under review). Correlation between specific position of methyl groups and interhelical angle is found for parallel and antiparallel types of packing. The basis for preference of a particular helix-helix arrangement appears to be determined by position of the side chain methyl groups. A good correlation was observed between the position of side chain methyl groups and the interhelical angle in a set of 200 proteins (Kurochkina, 2008). Mutations at core positions of helix-helix interfaces can significantly modify interhelical angles and fold of the protein and can lead to misfolded structure and impaired function (Kurochkina et al., 2010). Helix-helix interfaces are key determinants of protein structure and protein assemblies (Kurochkina, 2010).

Fold of proteins and assemblies of protein subunits can be seen in Figure 2. Crystallographic and NMR three-dimensional structures were used (Bernstein et al., 1977) from the Brookhaven Protein Data Bank.

Parallel arrangement of helices

GCN4 transcription factor dimer; coiled coil /2zta/

Cardiac phospholamban pentamer /2zll/

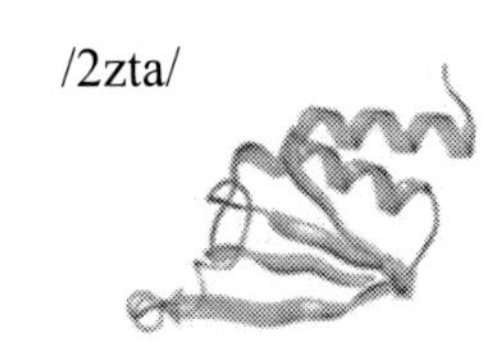

Glutathione s-transferase N-terminal domain /11gs/

Triose phosphate isomerase TIM-barrel $8\alpha/8\beta$ /1n55/

Antiparallel arrangement of helices

Repressor of Primer; coiled coil dimer /1rop/

Tobacco mozaic virus coat protein subunit – a four-α-helix bundle /2tmv/

Tobacco mozaic virus coat protein assembly of 49 subunits; each subunit is a four-α-helix bundle /2tmv/

Figure 2. (Continued).

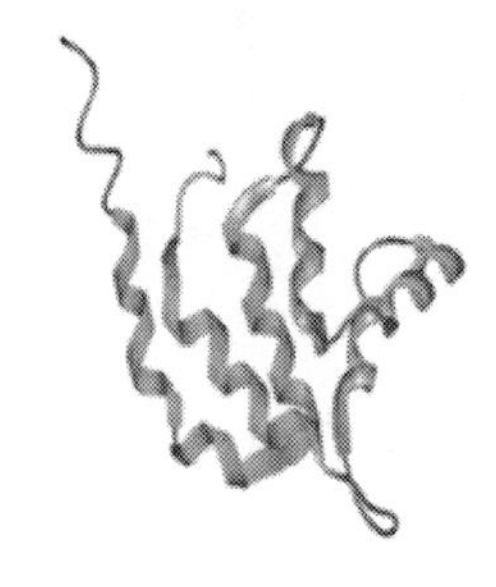

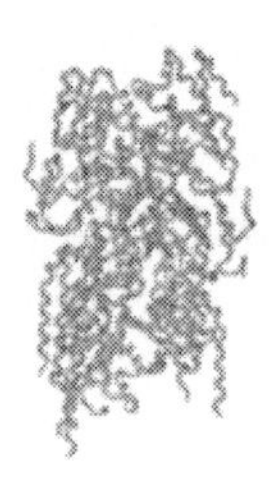

Six-α-helix bundle /3mop/

Apoptosome - helix of 14 subunits; each subunit is a six-α-helix bundle /3mop/

Figure 2. Fold of proteins carrying parallel and antiparallel helix-helix interfaces; PDB accession number is given as /xxxx/.

References

Ago. H, Kanaoka. A.H., Irikura D, et al. (2007) Crystal structure of a human membrane protein involved in cysteinyl leukotriene biosynthesis. *Nature 448,* 609-612.

De Alba, E. (2009) Structure and interdomain dynamics of apoptosis-associated speck-like protein containing a CARD (ASC). *J Biol Chem. 284*, 32932-32941.

Anstrom, D.M., Kallio, K., Remington, S.J. (2003) Structure of the Escherichia coli malate synthase G: pyruvate:acetyl-coenzyme A abortive ternary complex at 1.95 Å resolution. *Protein Science 12*, 1822-1832.

Bernstein, F.C., Koetzle, T.F., Williams, G.J.B., Meyer, E.F., Jr, Brice, M.D., Rodgers, J.R., Kennard, O., Shimanouchi, T., Tasumi, M., (1977) The protein data bank: a computer-based archival file for macromolecular structures. *J Mol Biol. 112*, 535-542.

Bhyravbhatla, B., Watowich, S. J. and Caspar, D. L. (1998) Refined atomic model of the four-layer aggregate of the tobacco mosaic virus coat protein at 2.4-Å resolution. *Biophys J. 74*, 604-615.

Caffrey, M., Cai, M., Kaufman, J., Stahl, S. J., Wingfield, P. T., Covell D. G., Gronenborn, A. M., Clore, G. M. (1998) Three-dimensional solution structure of the 44 kDa ectodomain of SIV gp41. *Embo J. 17*, 4572-4584.

Chai, G., Brewer, J., Lovelace, L., Aoki, T., Minor, W., Lebioda, L. (2004) Expression, purification and the 1.8 Å resolution crystal structure of human neuron specific enolase. *J Mol Biol 341*, 1015-1021.

Chothia, C., Levitt, M., Richardson, D. (1981) Helix to helix packing in proteins. *J Mol Biol. 145*, 215-250.

Chou, K.C., Maggiora, G.M., Némethy, G., Scheraga, H.A. (1988) Energetic approach to 4-alpha-helix packing. *Proc Natl Acad Sci. 85*, 4295–4299.

Chou, K.C., Carlacci, L., Maggiora, G.M., Parodi, L.A., Schultz, M.W. (1992) An energy-based approach to packing the 7-helix bundle of bacteriorhodopsin. *Protein Sci. 1*, 810–827.

Crick, F. (1953) *Acta Crystallogr. 6*, 689.

Dulhunty A., Gage P., Curtis S., et al. (2001) The Glutathione Transferase Structural Family Includes a Nuclear Chloride Channel and a Ryanodine Receptor Calcium Release Channel Modulator. *J Biol Chem. 276*, 3319–3323.

Da Silva Giotto, M. T., Hannaert, V., Vertommen, D., Navarro, M., Rider, M. H., Michels, A. M., Garratt, R. C., Rigden, D. J. (2003) The crystal structure of trypanosome brucei enolase: visualization of the inhibitory metal binding site III and potential as target for selective, irreversible inhibition. *J Mol Biol. 331*, 653-665.

Davenport, R. C., Bash, P. A., Seaton, B. A., Karplus, M., Petsko, G. A., Ringe, D. (1991) Structure of triose phosphate isomerase phosphoglycolohydroxamate complex: an analogue of the intermediate on the reaction pathway. *Biochemistry 30*, 5821-5826.

Ferguson A. D., Mckeever B. M , Wisniewski D. et al. (2007) Crystal structure of inhibitor-bound human 5-lipoxygenase-activating protein. *Science 317*, 510-512.

Gamblin, S. J., Davies, G. J., Grimes, J. M., Jackson, R. M., Littlechild, J. A., Watson, H. C. (1991) Activity and specificity of human aldolases. *J Mol Biol. 219*, 573-576.

Gernert, K. M., Surles, M. C., Labean, T. H., Richardson, J. S., Richardson, D. C. (1995) Alacoil: a very tight, antiparallel coiled-coil of helices. *Protein Science 4*, 2252-2260.

Harris, N. L., Presnell, S. R., Cohen, F. E. (1994) Four helix bundle diversity in proteins. *J. Mol. Biol. 236*, 1356-1368.

Harrop, S. J., DeMaere, M, Z., Fairlie, W. D., et al. (2001) Crystal structure of a soluble form of the intracellular chloride ion channel CLIC1 (NCC27) at 1.4-A resolution. *J Biol Chem. 276*, 44993–45000.

Hester, G., Brenner-Holzach, O., Rossi, F. A., Struck-Donatz, M., WinterHalter, K. H., Smit, J., Piontek, K. (1991) The crystal structure of fructose-1,6-biphosphate aldolase from *Drosophila Melanogaster* at 2.5 Å resolution. *Febs Letters 292*, 237-242.

Hodges, R. S., Sodek. J., Smillie, L. B. et al. (1972) Amino-acid sequence of rabbit skeletal tropomyosin and its coiled-coil structure. *Proc Natl Acad Sci 69,* 3800-3804.

Hofmann, A, Proust, J., Dorowski, A. et al. (2000) Annexin 24 from *Capsicum annuum* . *J Biol Chem. 275*, 8072-8082.

Huber, R., Romish, J., Paques, E.P. (1990) The crystal and molecular structure of human annexin V, an anticoagulant protein that binds to calcium and membranes. *Embo J. 9*, 3867-3874.

Ida, M., Sato, A., Matsumoto, I. et al. (2004) Human annexin V binds to sulfatide: contribution to regulation of blood coagulation. *J Mol Biol. 135*, 583-588.

Ishitsuka, R., Kojima, K., Utsumi, H. et al. (1998) Glycosaminoglycan Binding Properties of Annexin IV, V, and VI. *J Biol Chem. 273*, 9935-9941.

Izard, T., Sygusch, J. (2004) Induced Fit Movements and Metal Cofactor Selectivity of Class II Aldolases. Structure of Thermus aquaticus fructose-1,6-biphospate aldolase. *J Biol Chem. 279*, 11825-11833.

Jacobsson, P.J., Mansini, J.A., Riendeau, D. et al. (1997) Identification and characterization of a novel microsomal enzyme with glutathione-dependent transferase and peroxidase activities. *J Biol Chem. 272*, 22934-22939.

Jeppesen, M.G., Ortiz, P., Shepard, W, et al. (2003) The Crystal Structure of the Glutathione *S*-Transferase-like Domain of Elongation Factor 1B from *Saccharomyces cerevisiae. J Biol Chem. 278*, 47190–47198.

Jog, J., Rozovsky, S., MsDermott, A.E., Tong, L. (2003) Optimal alignment for enzymatic proton transfer: Structure of the Michaelis complex of triosephosphate isomerase at 1.2-Å resolution. *Proc Natl Acad Sci 100*, 50-55.

Klenchin, V. A., Schmidt, D. M., Gerlt, J. A., Rayment, I. (2004) Evolution of enzymatic activities in the enolase superfamily: structure of a substrate-liganded complex of the L-ALA-D/L-GLU epimerase from *Bacillus subtilis*. *Biochemistry 43*, 10370-10378.

Knox, J. R., Moews, P. C. (1991) Beta-lactamase of Bacillus licheniformis 749-C. Refinement at 2 Angstroms resolution and analysis of hydration. *J Mol Biol 220*, 435-455.

Kohn, W. D., Mant, C. T., Hodges, R. S. (1977) α-helical protein assembly motifs. *J Biol Chem. 272*, 2583-2586.

Kühnel, K., Jarchau, T., Wolf, E., Schlichting, I., Walter, U., Wittinghofer, A., Strelkov, S. V. (2004) The VASP tetramerization domain is a right-handed coiled coil based on a 15-residue repeat. *Proc Natl Acad Sci 101*, 17027-17032.

Kurochkina N. (2007) Amino acid composition of parallel helix-helix interfaces. *J Theor Biol 247*, 110-121.

Kurochkina N. (2008) Specific sequence combinations at parallel and antiparallel helix-helix interfaces. *J Theor Biol 255*, 188-198.

Kurochkina, N. (2010) Helix-helix interactions and their impact on protein motifs and assemblies. *J Theor Biol 264*, 585-592.

Kurochkina, N., Yardeni, T., Huizing, M. (2010) Molecular modeling of the bifunctional enzyme UDP-GlcNAc 2-epimerase/ManNAc kinase and predictions of structural effects of mutations associated with HIBM and sialuria. *Glycobiology 20*, 322-337.

Kursula, I., Wierenga, R. (2003) Crystal Structure of Triosephosphate Isomerase Complexed with 2-Phosphoglycolate at 0.83-Å Resolution. *J Biol Chem 278*, 9544-9551.

Lasker, M. V., Gajjar, M. M., Nair, S. K. (2005) Molecular Structure of the IL-1R-Associated Kinase-4 Death Domain and Its Implications for TLR signaling. *J Immun 175*, 4175-4179.

Lebioda, L., Stec, B., Brewer, J. M. (1989) The structure of yeast enolase at 2.25-Å resolution. *J Biol Chem 264*, 3685-2693.

Li, Y., Li, D., Wang, D. (2006) Trimeric structure of the wild soluble chloride intracellular ion channel CLIC4 observed in crystals. *Biophys Biochem Research Comm 343*, 1272-1278.

Li, H., Hallows, A., Punzi, S., Marquez, E., Carrell, H. L., Pankiewicz, K.W., Watanabe, K. A., Goldstein, B. M. (1994) Crystallographic studies of two alcohol dehydrogenase-bound analogs of thiazole-4-carboxamide adenine dinucleotide (TAD), the active anabolite of the antitumor agent tiazofurin. *Biochemistry 33*, 23-32.

Lin, S-C, Lo, Y-C, Wu, H. (2010) Helical assembly in the MyD88–IRAK4–IRAK2 complex in TLR/IL-1R signaling. *Nature 465*, 885-890.

Littler, D. R., Harrop, S. J., Brown, L., J. et al. (2008) Comparison of vertebrate and invertebrate CLIC proteins: the crystal structures of Caenorhabditis elegans EXC-4 and Drosophila melanogaster DmCLIC. *Proteins 71*, 364-378.

Lolis, T., Alber, T., Davenport, R.C., Rose, D., Hartman, F. C., Petsko, G. A. (1990) Structure of yeast triosephosphate isomerase at 1.9 Angstrom resolution. *Biochemistry 29*, 6609-6618.

Mancia, F., Keep, N. H., Nakagawa, A., Leadlay, P. F., Mcsweeney, S., Rasmussen, B., Bosecke, P., Diat, O., Evans, P. R. (1996) How coenzyme B12 Radicals are generated: the crystal structure of methylmalonyl-CoA mutase at 2 Angstrom resolution. *Structure (London) 4*, 339-350.

Mathews, F. S. (1985) The structure, function, and evolution of cytochromes. *Prog Biophys Mol Biol 45*, 1-56.

Namba, K., Stubbs, G. (1986) Structure of tobacco mosaic virus at 3.6 Angstroms resolution. Implications for assembly. *Science 231*, 1401-1406.

Namba, K., Pattanayek, R., Stubbs, G. (1989) Visualization of protein-nucleic acid interactions in a virus. Structure of intact tobacco mosaic virus at 2.9 Angstrom resolution by X-ray fiber diffraction. *J Mol Biol 208*, 307-325.

Oxenoid, K., Chou, J. J. (2005) The structure of phospholamban pentamer reveals a channel-like architecture in membranes. *Proc Natl Acad Sci 102*, 10870-10875.

Park, H. H., Logette, E., Raunser, S., Cuenin, S., Walz, T., Tschopp, J., Wu1, H. (2007) Death Domain Assembly Mechanism Revealed by Crystal Structure of the Oligomeric PIDDosome Core Complex. *Cell 128*, 533-546.

Parry, D. A. D., Fraser, R. D., Squire, J. M. (2008) Fifty years of coiled-coils and α-helical bundles: A close relationship between sequence and structure. *J Struct Biol 163*, 258-269.

Parthasarathy, S., Eaazhisai, K., Balaram, H., Balaram, P. (2003) Structure of *Plasmodium Falciparum* triose-phosphate isomerase-2-phosphoglycerate complex at 1.1-Å resolution. *J Biol Chem 278*, 52461-52470.

Pielak, R. M., Schnell, J.R., and Chou, J. J. (2009) Mechanism of drug inhibition and drug resistance of influenza A M2 channel. *Proc Natl Acad Sci 106*, 7379-7384.

Samatey, F. A., Imada, K., Nagashima, S., Vonderviszt, F., Kumasaka, T., Yamamoto, M., Namba, K. (2001) Structure of the bacterial flagellar protofilament and implications for a switch for supercoiling. *Nature 410*, 331-337.

Schnell, J. R., Zhou, G-P., Zweckstetter, M., Rigby, A. C., Chou, J. J. (2005). Rapid and accurate determination of coiled-coil domains using NMR dipolar couplings: application to cGMP-dependent protein kinase Ia. *Protein Science 14*, 2421-2428.

Schnell, J. R., and Chou, J. J. (2008) Structure and mechanism of the M2 proton channel of influenza A virus. *Nature 451*, 591-595.

Scott, F. L., Stec, B., Pop, C., Dobaczewska, M. K., Lee, J. J., Monosov, E., Robinson, H., Salvesen, G. S., Schwarzenbacher, R., Riedl, S. J. (2009) The Fas/FADD death domain complex structure unravels signaling by receptor clustering. *Nature 457*, 1019-1022.

Shao, C, Zhang F, Kemp, MM et al. (2006) Crystallographic Analysis of Calcium-dependent Heparin Binding to Annexin A2. *J Biol Chem 281*, 31689–31695.

Sheehan, D., Meade, G., Foley, V. M., Dowd, C., A. (2001) Structure, function and evolution of glutathione transferases : implications for classification of non-mammalian members of an ancient enzyme superfamily. *Biochem J 360*, 1-16.

Smith, C.V., Huang, C.-C., Miczak, A., Russell, D. G., Sacchettini, J. C., and Honer zu Bentrup, K. (2003) Biochemical and structural studies of malate synthase from *Mycobacterium tuberculosis. J Biol Chem 278*, 1735-1743.

Sheriff, S., Hendrickson, W. A., Smith, J. L. (1987) Structure of myohemerythrin in the azidomet state at 1.7/1.3 angstroms resolution. *J Mol Biol 197*, 273-296.

Stetefeld, J., Jenny, M., Schulthess, T., Landwehr, R., Engel, J., Kammerer, R. A. (2000) Crystal structure of a naturally occurring parallel right-handed coiled coil tetramer. *Nat Str Biol 7*, 772-776.

Stevens, J., Corper, A. L., Basler, C. F., Taubenberger, J. K., Palese, P., Wilson, I. A. (2004) Structure of the uncleaved human H1 hemagglutinin from the extinct 1918 influenza virus. *Science 303*, 1866-1870.

Sutton, R. B., Fasshauer, F., Jahn, R., Brunger, A. T. (1998) Crystal structure of a SNARE complex involved in synaptic exocytosis at 2.4 Å resolution. *Nature 395*, 347-353.

Thomas, V. L., Golemi-Kotra, D., Kim, C., Vakulenko, S. B., Mobashery, S., Shoichet, B. K. (2005) Structural consequences of the inhibitor-resistant Ser130Gly substitution in TEM beta-lactamase. *Biochemistry 44*, 9330-9338.

Wang, L., Yang. J. K., Kabaleeswaran. V., Rice, A. J., Cruz, A. C., Park, A. Y., Yin, Q. (2010) The Fas-FADD death domain complex structure reveals the basis of DISC assembly and disease mutations. *Nat Struct Mol Biol 17*, 1324-1329.

Weber, C. H., Vincenz, C. (2001) The death domain superfamily: a tale of two interfaces? *TIBS 26*, 475-481.

Wierenga, R. K., Noble M. E. M., Vriend, G., Naughe, S., Hol, W. G. J. (1991). Refined 1.83 Angstrom structure of trypanosomal triosephospate isomerase, crystallized in presence of 2.4M- Ammonium sulphate. A comparison with the structure of the trypanosomal triosephosphate isomerase-glycerol-3-phosphate complex. *J Mol Biol. 220*, 995-1015.

Yuan, S., Yu, X., Topf, M., Ludtke, S. J., Wang, X., Akey, C. W. (2010) Structure of an Apoptosome-Procaspase-9 CARD Complex. *Structure 18*, 571-583.

In: Protein Structure
Editor: Lauren M. Haggerty, pp. 83-92

ISBN 978-1-61209-656-8

Chapter 5

Structural Analysis of the Transcriptional Activation Domain of the Human c-Myc Oncogene

Shagufta H. Khan and Raj Kumar*
Department of Basic Sciences, The Commonwealth Medical College,
Scranton, PA, USA

Abstract

The c-Myc gene is an important regulator of normal cell growth, differentiation, and programmed cell death, and plays a central role in tumorigenesis. The c-Myc protein contains a potential transactivation domain within its N-terminal region, which contains short acidic, proline-, and glutamine- rich clusters similar to the activation domains of many transcription factors, and is required for the biological activities of c-Myc. However, like the activation domains of most transcription factors, the structural basis for the activity of the c-Myc AD region is not completely understood. In this study, using various structural analytical tools, we determined the structural characteristics of the activation domain of c-Myc protein. Our analyses show that the c-Myc activation domain possesses all the characteristics of an intrinsically disordered protein, a common phenomenon in many cancer related proteins. In recent years, there have been several indications that relative to prokaryotes, eukaryotic genomes are highly enriched in ID proteins, and it has been predicted that more than two-thirds of proteins associated with cancer-related genes possess long stretches of ID regions. Our results suggest that c-myc activation domain is capable of making multiple interactions with its target binding partners, the specificity of which may result in regulation of transcription.

Keywords: c-Myc, activation domain, intrinsically disordered, transcription factor.

* Address for Correspondence: Raj Kumar, Ph.D.; Professor of Biochemistry; Department of Basic Sciences; The Commonwealth Medical College; 501 Madison Avenue, Suite 205; Scranton, PA-18510 Phone: 570-504-9675; Fax: 570-504-9660; E-mail: rkumar@tcmedc.org

Introduction

The c-Myc gene has emerged as a centerpiece and key in cancer biology. The expression of c-myc has been reported to play a central role in tumorigenesis (1). It has been reported that c-Myc is a key molecular integrator of cell cycle machinery and cellular metabolism (Kevin et al., 1996). The c-*myc* gene is located on human chromosome 8, and comprises of three exons (Battey et al., 1983).

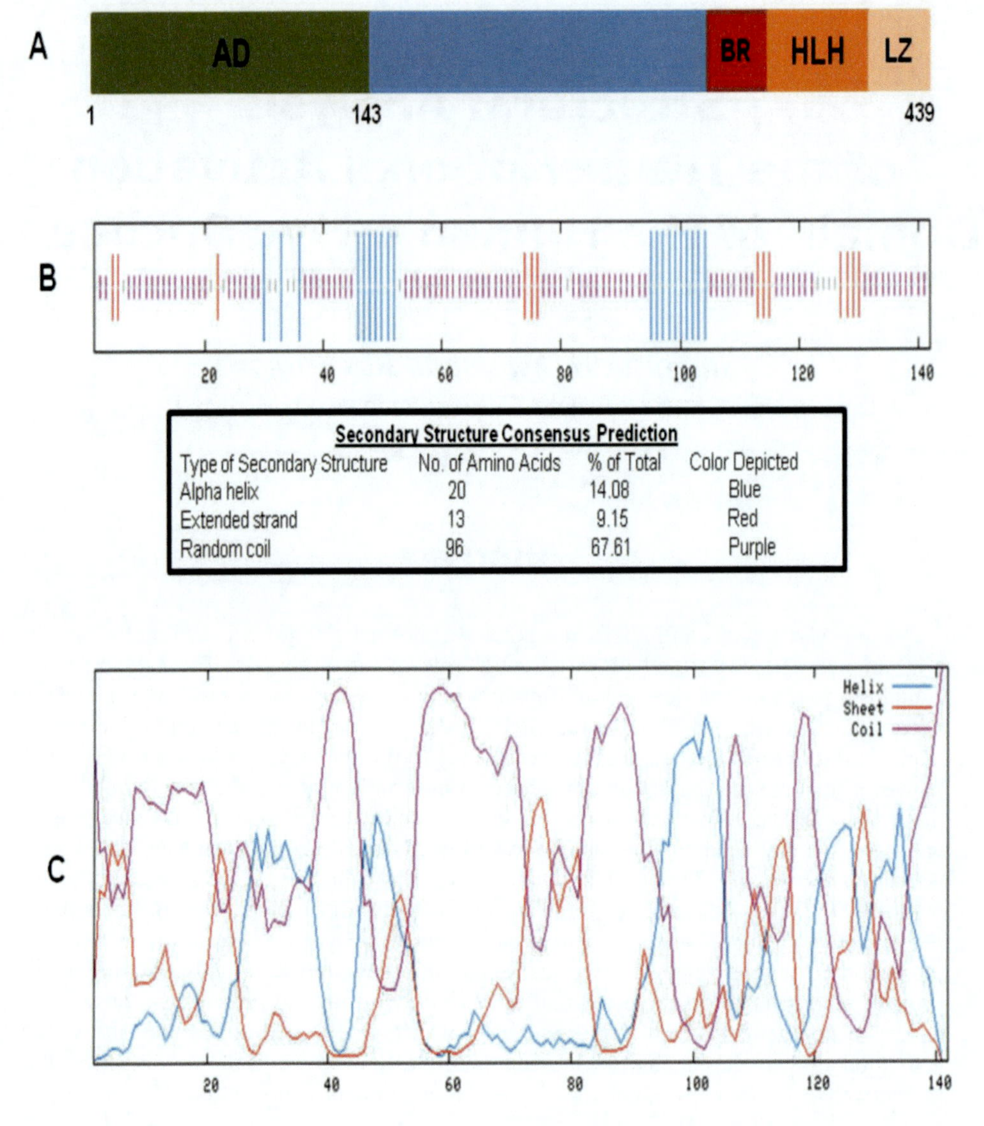

Secondary Structure Consensus Prediction

Type of Secondary Structure	No. of Amino Acids	% of Total	Color Depicted
Alpha helix	20	14.08	Blue
Extended strand	13	9.15	Red
Random coil	96	67.61	Purple

Figure 1. A) A topological diagram of human c-Myc protein showing different functional domains. AD, activation domain; BR, basic region; HLH, helix-loop-helix; and LZ, leucine zipper. B) Consensus Secondary structural elements predictions of c-Myc AD using NPS (Network Protein Sequence Analysis based on SOPM, HNNC, MLRC, DPM, DSC, GOR I, GOR III, and PHD Predictors) consensus method as described (Combet et al., 2000). A summary of secondary structural elements of c-Myc AD is shown in the box. C) Secondary structural elements predictions of c-Myc AD using HNN method as described (Combet et al., 2000). Blue, red, and purple colors indicate helix, β-sheet, and random coil configurations, respectively.

The *c-myc* is a nuclear protein, and is an important regulator of normal cell growth, differentiation, and programmed cell death (Marcu et al., 1992). The amino acid sequence of c-Myc protein suggests that it contains a potential transactivation domain within its N-terminal, and a dimerization interface consisting of a helix-loop-helix leucine zipper (HLH/LZ) domain at its C-terminal end (Figure 1A) (Dang, 1999). Towards that end also lies the DNA binding domain (DBD), which is rich in basic amino acids, and contacts specific DNA sequences within the DNA major groove (Dang et al., 1992). The c-Myc protein binds to and transactivates through consensus 5′-CACGTG-3′ nucleotide sequences (Amati et al., 1992). The AD of c-Myc is localized to its first 143 amino acids (Kato et al., 1990), which contains short acidic, proline-, and glutamine- rich clusters similar to the ADs of many transcription factors (Kato et al., 1990). Like the DBD, the AD is also required for the biological activities of c-Myc. However, like the ADs of most transcription factors, the structural basis for the activity of the c-Myc AD region is not completely understood. Secondary structural analyses of these ADs (when expressed independently as recombinant peptides) show that they are largely unstructured in aqueous solution at neutral pH (Garza et al., 2010; Garza et al., 2009; Kumar and Thompson, 2003; Combet et al., 2000). In other words, they exist as an intrinsically disordered (ID) protein. However, ADs have a propensity to form ordered structure under specific physiological conditions (Garza et al., 2009). In this study, we determined the structural characteristics of the c-Myc AD protein. We used various parameters to determine the ID characteristics of the c-Myc's AD, which consists of first 143 amino acids of the molecule. Our analyses show that the c-Myc AD possesses many characteristics of an ID protein.

Secondary Structural Predictions of the c-Myc AD

Transcription factors possess a modular structure that typically consists of two functionally essential yet separable modules consisting of the DBD and AD (Garza et al., 2009). The DBDs of many transcription factors have been well characterized structurally whereas structurally, ADs are less well-defined. This in part is due to the fact the ADs of a number of TFs possess ID regions (Garza et al., 2009; Uversky et al., 2000; Dyson and Wright, 2005). Therefore, it is important to find out their existence and structural basis for the function in transcriptional regulations. In spite of the importance of the ID nature of ADs in gene regulation by the TFs, scientific community has still not been able to fully understand how the transcriptional signal to specific gene(s) is transmitted. The c-M*yc* gene is involved in the establishment of many types of human malignancies (Battey et al., 1983; Marcu et al., 1992; Kevin et al., 1996). The members of this family have been shown to function as transcription factors, and through a designated target sequence bring about continued cell-cycle progression, cellular immortalization and blockages to differentiation in many lineages (Battey et al., 1983; Marcu et al., 1992; Kevin et al., 1996). We performed secondary structural analysis of the AD of the c-Myc (amino acids 1-143; Figure 1A) using Network Protein Sequence (NPS) analysis, which predicts the secondary structural elements of a protein based on amino acid sequences. The prediction for secondary structural elements in c-Myc AD was performed using NPS consensus and HNN methods available at http://npsa-pbil.ibcp.fr as described (Combet et al., 2000). This method gives an average consensus value from several independent methods (Combet et al., 2000). The results obtained for c-Myc AD from this method are shown in Figure 1B and summarized in the box. It is evident from the secondary structural elements that more than 67% of c-Myc contains random coli conformation with only a small proportion as helix (14%) and sheet (9%). Similar results were obtained using HNN secondary structure prediction method (Figure 1C) using Network protein Sequence Analysis (Combet et al., 2000). These results clearly support the notion that the c-Myc AD is mostly unstructured.

C-Myc AD Possesses Characteristics of an ID Protein

In recent years, there have been several indications that relative to prokaryotes, eukaryotic genomes are highly enriched in ID proteins. ID proteins are biologically functional under physiological conditions; however, unlike globular proteins they lack well-defined three-dimensional structures (Uversky et al., 2000; Dyson and Wright, 2005). It has been predicted that more than two-thirds of proteins associated with cancer-related genes possess long stretches of ID regions. Since ID proteins can exist in an ensemble of conformers to perform biological functions under physiological conditions, they can play critical role in a variety of biological processes related to cell-signaling pathways including cell cycle control, and signal transduction (Garza et al., 2009; Uversky et al., 2000; Dyson and Wright, 2005). However, due to highly dynamic nature of ID molecules, characterization of such proteins has become a major challenge. Since, structural uniqueness of most proteins determines their biological functions under physiological conditions, in recent years, determining the structural basis for the functional activity of ID domains/regions of signaling proteins, e.g. transcription factors, involved in cancer biology has become a central issue. To determine whether this AD exists as an ID protein, we first performed PONDR VL3 Analysis for ID prediction. Predictions for the degree of disordered profile (VL3) was performed using PONDR prediction from the website www.pondr.com, as described (Romero et., 1997; Li et al., 1999; Uversky et al., 2000; Romero et al., 2001), The VL3 predictor is a feed forward neural network trained on long regions of disordered conformations as characterized by various methods (Romero et al., 2001; Romero et al., 1997; Li et al., 1999). The results are shown in Figure 2A. It is evident from the PONDR Score that most of the peptide sequence consists of ID region (based on PONDR Score of more than 0.5). To confirm these findings, we used FoldIndex method of disorder prediction, which predicts the probability of a protein/peptide to fold (Prilusky et al., 2005). Fold Index was performed at http://bip.weizmann.ac.il/fldbin/findex as described (Prilusky et al., 2005). A large red area (unfolded) compared to small green area (folded) suggests that a large fraction of the AD is unfolded. Further the unfolded area correlates well with the low hydrophobicity (blue line) and high charge (pink line) (Figure 2B). It is well known that the disorder promoting residues also have large exposed surfaces, as is the case with many ID proteins (Dyson and Wright, 2005; Jones and Thornton, 1997; Oldfield et al., 2005). These results suggest that c-Myc AD region is similar to the ID proteins in terms of composition of amino acids and their propensity to form large surface area, which can be utilized for protein:protein interactions.

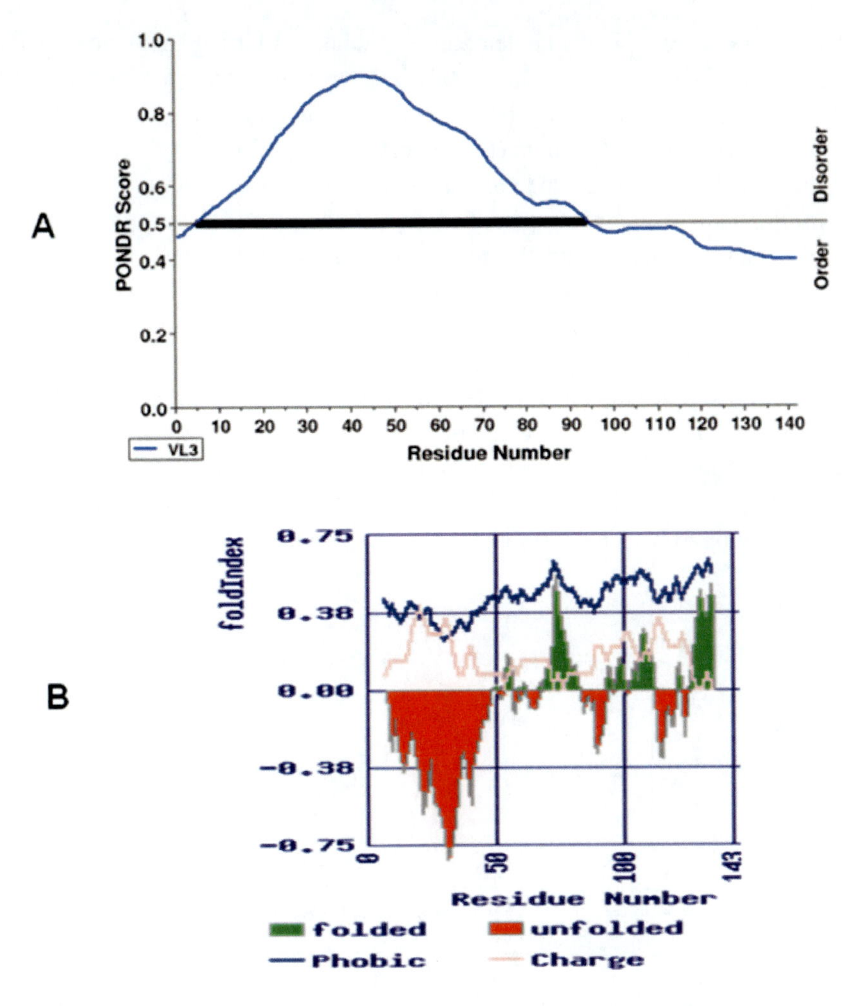

Figure 2. A) PONDR plot for c-Myc AD protein disorder prediction the based on VL3 analysis. Score above 0.5 are considered to be disordered sequences where as score below 0.5 indicates ordered conformation (Romero et., 1997; Li et al., 1999). X axis shows amino acid numbers in the c-Myc AD sequences, and Y axis shows probability score. B) Fold Index showing the probability of c-Myc AD sequences for the propensity to fold (Prilusky et al., 2005). The area in red represents the propensity for being unfolded and green area for folded amino acids. The blue plot indicates the hydrophobicity index, and pink plot shows the charged amino acids.

Charge/Hydropathy Analysis

Like many other transcription factors, the AD of c-Myc is also predicted to exist as an ID domain (McEwan et al., 1996). However, not much is known about its structural characteristics. It is well known that the basis of predicting ID nature of a protein depends upon the differences in sequence characteristics between folded and unfolded proteins. Various factors have been suggested to be important in terms of protein disorder, including flexibility, aromatic content, secondary structure preferences, and hydrophobicity (Uversky et al., 2000). Among other factors, low mean hydrophobicity and high net charge are suggested to significantly contribute to disordered conformations (Uversky et al., 2000). It is predicted that the ID sequences are typically depleted in the number of buried amino acids and enriched in exposed amino acids (Dunker et al., 2002). Due to these characteristics, the ID regions have potential to form a sufficiently large number of favorable interactions that can allow globular proteins to interact with them (Dyson and Wright, 2005). Uversky et al. (Uversky et al., 2000) have introduced a method for the analysis to distinguish ordered and disordered protein conformations based only on net charge and hydropathy. The charge/hydropathy plots compare the mean net charge and the mean hydropathy. A linear boundary separates the ordered and disordered proteins when plotted in the charge/hydropathy space. The boundary between the ordered and disorder proteins is determined using a linear discriminate function. Predictions for the Charge/Hydropathy analysis was performed using PONDR prediction from the website www.pondr.com, as described (Romero et., 1997; Li et al., 1999; Uversky et al., 2000; Romero et al., 2001). We applied this method to the c-Myc AD sequences to determine whether this peptide qualifies as an ID region based on these parameters. The plots are shown in Figure 3A. It is evident from the results that this peptide falls within the ID proteins; further supporting the notion that c-Myc AD qualifies as an ID protein.

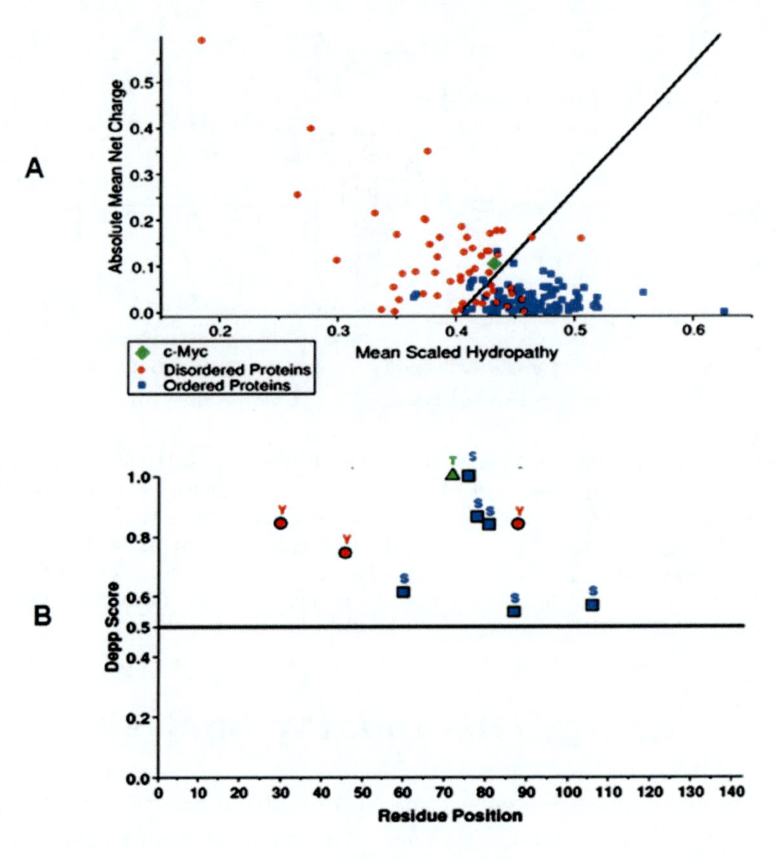

Figure 3. A) Charge-hydropathy analysis using Uversky plot (Uversky et al., 2000; Romero et al., 2001). The plot of the mean hydrophobicity vs. mean net charge of 54 completely disordered proteins (red circles), and 105 completely ordered proteins (blue squares). The solid line represents the border between ordered and disordered proteins. The cyan square corresponds to the c-Myc AD. B) Disorder Enhanced Phosphorylation Predictor plot showing the propensity of specific amino acids for phosphorylation under physiological conditions (Romero et., 1997; Li et al., 1999; Uversky et al., 2000; Romero et al., 2001). Six Serine (S) residues (blue square), two tyrosine (Y) residues (red circle), and one Threonine (T) residue is found to be phosphorylable. Under cellular conditions, their phosphorylation may be kinase- and cell- specific.

Disorder Enhanced Phosphorylation of Specific Amino Acids

It has been predicted that ID regions have a much higher frequency of known phosphorylation sites than ordered regions and that signaling via phosphorylation-regulated protein-protein interactions often involves ID regions (Kyriakis, 2000; Zor et al., 2005; Dunker and Uversky, 2008; Iakoucheva et al., 2004). Site-specific phosphorylation represents an important regulatory mechanism in the activities of signaling proteins. It has been hypothesized that protein phosphorylation of Ser residue predominantly occurs within ID regions of signaling molecules rather than merely on surface residues. Therefore, we determined whether the ID c-Myc AD has propensity to phosphorylation sites. Disorder Enhanced Phosphorylation Predictor was performed using PONDR prediction from the website www.pondr.com as described (Romero et., 1997; Li et al., 1999; Uversky et al., 2000; Romero et al., 2001). Our results clearly supports that the AD of c-Myc is rich in phosphorylation sites (Figure 3B). Relatively few ID regions have been structurally characterized, and yet a significant fraction of them contain phosphorylation sites (Iakoucheva et al., 2004). One of the main reasons for such propensity is to facilitate extensive formation of hydrogen binding between the backbones and/or side chains that can occur through disorder-order transition within the ID region. The formation of these hydrogen bonds would be difficult if the sites of phosphorylation were located within ordered regions. These results further support the functional importance of ID nature of the c-Myc AD.

Conclusion

Our analyses in this study clearly show that the c-Myc AD possesses the characteristics of an ID protein as judged by various parameters. A structural analysis of c-myc AD using various methods reveals that this polypeptide is relatively unstructured which conforms to the current picture of AD domains being relatively flexible and unstructured, until they interact with a specific target protein within the general transcriptional machinery. Protein-protein interaction sites are a common phenomenon of ID ADs of transcription factors. Many of these hetero-complexes are nonobligatory, associating and disassociating upon the cellular factors (Jones and Thornton, 1996). The c-Myc AD is also known to interact with specific proteins from the basal transcription machinery, which have been reported to bind/fold ID regions in many transcription factors (McEwan et al., 1996). These results indicate that one of the reasons for the existence of c-Myc AD as an ID protein may be due in part to allow its flexible surfaces for such protein:protein interactions. Our results suggest that c-myc AD can make multiple interactions with the target proteins, the specificity of which may result in either positive or negative regulation of transcription. Based on these observations, we hypothesize that an induced conformation or limited set of conformations may occur in the c-Myc AD in order for it to carry out its transcriptional activation functions. This could happen either by an induced fit mechanism, in which the c-Myc AD may fold into functionally active conformation due to its direct interaction with its binding partner protein(s), by shifting the equilibrium between a large proportion of the unstructured conformers and a small proportion of relatively structured conformers, which can make physically interact with binding partners (Figure 4).

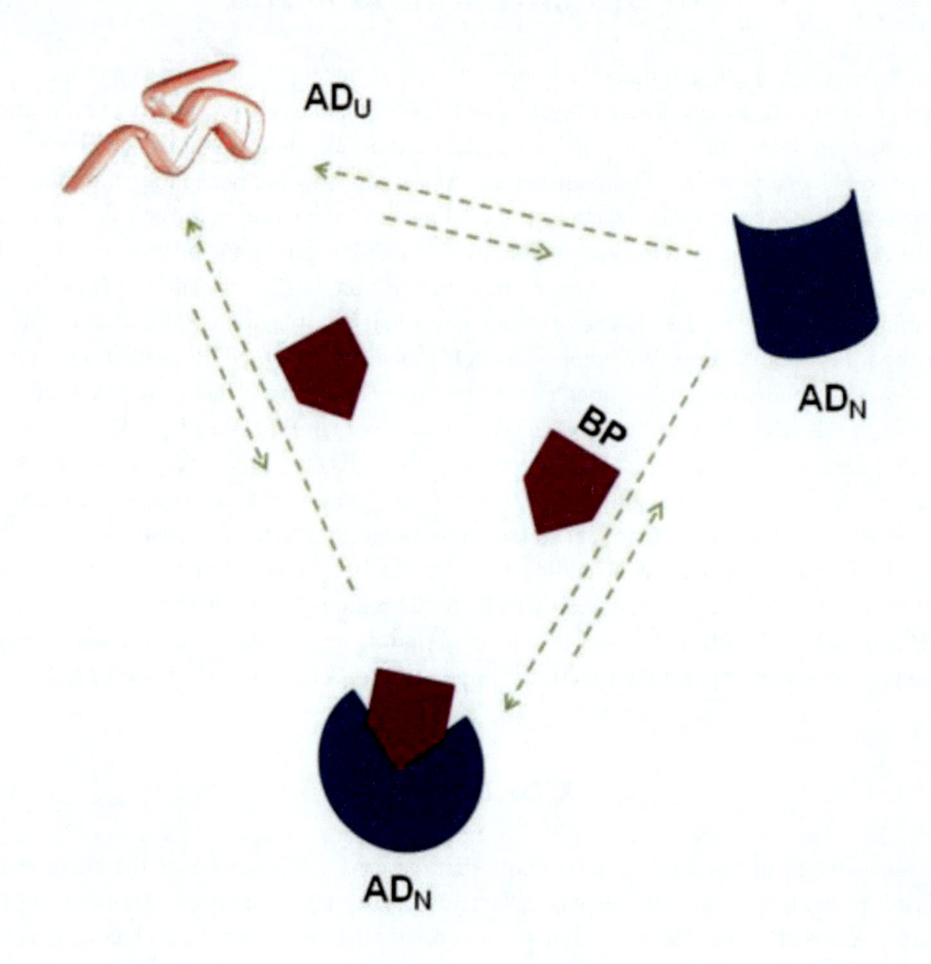

Figure 4. A model for the folding of the c-Myc AD (based on Kumar et al., 2001). ADU represents the assembly of ID conformers of the c-Myc AD in the absence of its proper binding partner (BP). In this circumstance, the c-Myc AD may exist only in its unfolded state(s) or in equilibrium with a folded state, ADN. BP could induce folding directly to AF1N by binding, shifting the c-Myc AD directly to the AD_N·BP complex. Alternatively, AD_U could be in equilibrium with its natively folded state AD_N. Without BP, $[AD_N]$ is very small relative to $[AD_U]$; however, BP could bind AD_N and shift AD to the complex, by the law of mass action.

References

Amati, B., Dalton, S., Brooks, M.W., Littlewood, T.D., Evan, G.I., and Land, H. (1992). Transcriptional activation by the human c-Myc oncoprotein in yeast requires interaction with Max. *Nature*. 359, 423-426.

Bairoch, A., Apweiler, R., Wu, C.H., Barker, W.C., Boeckmann, B., Ferro, S., Gasteiger, E., Huang, H., Lopez, R., Magrane, M., Martin, M.J., Natale, D.A., O'Donovan, C., Radaschi, N., and Yeh, L.S. (2005). The universal protein resource (UniProt). *Nucleic Acids res*. 33, D154-159.

Battey, J., Moulding, C., Taub, R., Murphy, W., Stewart ,T., Potter, H., Lenoir, G., and Leder, P. (1983). The human c-myc oncogene: structural consequences of translocation into the IgH locus in Burkitt lymphoma. *Cell*, 34, 779-787.

Combet, C., Blanchet, C., Geourjon, C. and Deléage, G. (2000). NPS@: Network Protein Sequence Analysis. *Trends Biochem. Sci.* 25, 147-150.

Dang, C.V., Dolde, C., Gillison, M.L., and Kato, G.J. (1992). Discrimination between related DNA sites by a single amino acid residue of Myc-related basic-helix-loop-helix proteins. *Proc. Natl. Acad. Sci. USA*. 89, 599-602.

Dang, C.V. (1999). c-Myc Target Genes Involved in Cell Growth, Apoptosis, and Metabolism. Mol. Cell. Biol. 19, 1–11.

Dunker, A.K., Brown, C.J., Lawson, J.D. Iakoucheva, L.M. and Obradovic, Z. (2002). Intrinsic disorder and protein function. *Biochemistry*, 41, 6573–6582.

Dunker, A.K., and Uversky, V.N. (2008). Signal transduction via unstructured protein conduits. *Nat. Chem. Biol.* 4, 229–230.

Dyson, H.J., and Wright, P.E. (2005). Intrinsically unstructured proteins and their functions. Nat. rev. Mol. Cell. Biol. 6, 197-208.

Garza, A.S., Ahmad, N., and Kumar R. (2009). Role of Intrinsically disordered Protein Regions/Domains in Transcriptional Regulation. *Life Sciences*, 84, 189-193.

Garza, A.S., Khan, S.H., and Kumar, R. (2010). Site-specific phosphorylation induces functionally active conformation in the intrinsically disordered N-terminal activation function domain (AF1) of the glucocorticoid receptor. *Mol. Cell. Biol.* 30, 220-230.

Iakoucheva, L.M., Radivojac, P., Brown, C.J., O'Connor, T.R., Sikes, J.G., Obradovic, Z., and Dunker, A.K. (2004). The importance of intrinsic disorder for protein phosphorylation. *Nucleic Acids Res*. 32, 1037–1049.

Janin, J., Miller, S., and Chothia, C. (1988). Surface, subunit interfaces and interior of oligomeric proteins. *J. Mol. Biol.* 204, 155-164.

Jones, S., and Thornton, J. (1996). Principles of protein-protein interactions. *Proc. Natl. Acad. Sci. USA*. 93, 13-20.

Jones, S., and Thornton, J.M. (1997). Analysis of protein-protein interaction sites using surface patches. *J. Mol. Biol.* 272, 121-132.

Kato, G.J., Barrett, J., Villa-Garcia, M., and Dang, C.V. (1990). An amino-terminal c-myc domain required for neoplastic transformation activates transcription. *Mol. Cell. Biol.* 10, 5914-5920.

Kevin M. RYAN, K.M., and BIRNIE, G.D. (1996). *Myc* oncogenes: the enigmatic family. *Biochem. J.* 314, 713-721.

Kumar, R., Lee, J.C., Bolen, D.W., and Thompson, E.B. (2001). The Conformation of the Glucocorticoid Receptor AF1/tau1 Domain Induced by Osmolyte Binds Co-regulatory Proteins. *J. Biol. Chem.* 276, 18146-18152.

Kumar, R., and Thompson, E.B. (2003). Transactivation functions of the N-terminal domains of nuclear hormone receptors: protein folding and coactivator interactions. *Mol. Endocrinol. 17*, 1-10.

Kyriakis, J.M. (2000). MAP kinases and the regulation of nuclear receptors. *Sci.* STKE PE1.

Li, X., Romero, P., Rani, M., Dunker A.K., and Obradovic, Z. (1999). Predicting protein N- and C-, and internal regions. *Ser. Workshop Genome Inform. 10*, 30-40.

Marcu, K.B., Bossone, S.A., and Patel, A.J. (1992). myc function and regulation. *Annu. Rev. Biochem. 61*, 809-860.

McEwan, I.J., Dahlman-Wright, K., Ford, J., Wright, A.P. (1996). Functional interaction of the c-Myc transactivation domain with the TATA binding protein: evidence for an induced fit model of transactivation domain folding. *Biochemistry. 35*, 9584-9593.

Oldfield, C.J., Cheng, Y., Cortese, M.S., Romero, P., Uversky, V.N., and Dunker, A.K. (2005). Coupled folding and binding with alpha-helix-forming molecular recognition elements. *Biochemistry*, 44, 12454-12470.

Prilusky, J., Felder, C.E., Zeev-Ben-Mordehai, T., Rydberg, E.H., Man, O., Beckman, J.S., Silman, I., and Sussman J.L. (2005). FoldIndex: a simple tool to predict whether a given protein sequence is intrinsically unfolded. *Bioinform. 21*, 3435-3438.

Romero, P., Obradovic, Z., and Dunker, A.K. (1997). Sequence data analysis for long disordered regions prediction in the calcineurin family. *Genome Inform. Ser. Workshop Genome Inform. 8*, 110-124.

Romero, P., Obradovic, Z., Li, X., Garner, E.C., Brown, C.J., and Dunker, A.K. (2001). Sequence complexity of disordered protein. *Proteins*, *42*, 38-48.

Uversky, V.N., Gillespie, J.R., Fink, A.L. (2000). Why are "natively unfolded" proteins unstructured under physiologic conditions? *Proteins*, *41*, 415-427.

Vehinen, M., Torkkila, E., and Riikonen, P. (1994). Accuracy of protein flexibility predictions. *Proteins*, *19*, 141-149.

Zor, T., Mayr, B.M., Dyson, H.J., Montminy, M.R. and Wright, P.E. (2002). Roles of phosphorylation and helix propensity in the binding of the KIX domain of CREB-binding protein by constitutive (c-Myb) and inducible (CREB) activators. *J. Biol. Chem. 277*, 42241–42248.

In: Protein Structure
Editor: Lauren M. Haggerty, pp. 93-103
ISBN 978-1-61209-656-8

Chapter 6

From Coding Variant to Structure and Function Insight

***Aaron J. Friedman*[1], *Ali Torkamani*[2], *Gennady Verkhivker*[3], *and Nicholas J. Schork*[2*]**

[1] Graduate program in Biomedical Sciences, University of California, San Diego, La Jolla, CA (AF), USA

[2] The Scripps Translational Science Institute and Department of Molecular and Experimental Medicine, The Scripps Research Institute, La Jolla, CA (AT, NJS), USA

[3] The University of Kansas, Kansas, MO and Department of Pharmacology, University of California, San Diego (GV), USA

Abstract

The availability of cost-effective DNA sequencing technologies has led to the identification and cataloguing of millions of naturally occurring inherited and somatic human genomic variations. As a result, questions concerning the ultimate phenotypic and functional significance of these variations have been raised. Although some large-scale initiatives have been launched for this purpose, including the Encyclopedia of DNA Elements (ENCODE) initiative, there is a need for the DNA sequencing and genomics communities to reach out and foster greater collaborative opportunities with the functional genomics community. One area that is ripe for this kind of activity is the functional characterization of coding variations, as structural proteomics and crystallography researchers may both benefit from, and contribute to, an understanding of the molecular and phenotypic influence of such variations. We briefly review the motivation for this kind of interaction and draw on publicly available data to showcase its need. We also consider how relevant research could be pursued.

[*] Address correspondence to: Nicholas J. Schork, Ph.D. The Scripps Translational Science Institute Department of Molecular and Experimental Medicine The Scripps Research Institute 3344 North Torrey Pines Court, Suite 300 La Jolla, CA 92037 nschork@scripps.edu 858-554-5705 (admin) 858-546-9284 (fax)

Introduction

Contemporary large-scale human DNA sequencing and polymorphism discovery initiatives, such as the Human Genome Project [Lander et al. 2001; Venter et al. 2001], the International HapMap Project [International HapMap Consortium 2005; 2007; International HapMap3 Consortium 2010], the '1000 Genomes Project' [1000 Genomes Project Consortium 2010] and the International Cancer Genome Consortium (ICGC) [ICGC 2010], have led to the identification and cataloguing of millions of naturally occurring inherited and somatic human genomic variations. As a result, the ultimate phenotypic and physiologic (i.e., 'functional') significance of these variations has begun to occupy the attention of many researchers in the biomedical sciences. In fact, many initiatives have been launched for this purpose, including the NIH-coordinated Genetic Association Information Network (GAIN), which promotes genome-wide association studies (GWAS) [Manolio et al. 2008], and the 'Encyclopedia of DNA Elements (ENCODE) initiative' [The ENCODE Project Consortium 2007], which seeks to characterize potential regulatory features of DNA sequence and the variants that reside within them. Notably, the large-scale assessment of the functional significance of identified human DNA sequence variations requires interdisciplinary approaches, with epidemiologists, bioinformaticians, statisticians, and molecular biologists all playing important roles.

As useful as these collaborative functional genomic initiatives are, however, they are limited and could thus be easily complemented by others. For example, although there has been some historical debate about the contribution of common non-synonymous coding variations (i.e., protein polymorphisms) to common complex congenital diseases as well as somatic perturbations in cancer, it is now becoming clear that multiple rare coding variations within genes of relevance can, as a collection, contribute to both congenital disease susceptibility and somatically-mediated tumorigenesis in epidemiologically significant ways (see, e.g., Bodmer and Bonilla [2008], Schork et al. [2009] and Torkamani et al. [2008] for a review). In fact many recent studies have borne this out by identifying multiple rare coding variations in biologically-based candidate genes that collectively distinguish individuals with and without a particular phenotype [Schork et al. 2009]. This suggests that the ability to reliably assess the functional effect of novel coding variants will be crucial for future disease-oriented sequencing studies.

Further Attempts to Catalog Sequencing Variants

In addition to historical studies that have identified coding variants whose functional significance is unknown, DNA sequencing methods have advanced to the point where sequencing the entire genomes of individuals and/or tumors is possible. The publication of the complete genomes of Craig Venter [Levy et al. 2007], James Watson [Wheeler et al. 2008], an entire family [Roach et al. 2010] and paired tumor/normal tissue samples [Mardis and Wilson 2009] are cases-in-point. These studies identified an enormous number of novel non-synonymous coding variations that are likely to have deleterious effects on relevant encoded proteins, but few have been subjected to laboratory assay-based functional verification. Proposed large-scale initiatives to sequence the entire genomes of thousands of individuals, such as the National Human Genome Research Institute's led '1000 Genomes Project' [1000 Genomes Project Consortium 2010] will undoubtedly identify even more coding variations, especially since much of the initial focus of this effort will be on the 'exome' or coding portion of the genome. These initiatives are complemented by equally ambitious DNA sequence variation discovery programs within the cancer research community, such as the International Cancer Genome Consortium (ICGC) [ICGC 2010]. As with the 1000 genomes project, initial focus of the work for the ICGC has been on the identification of coding variations that alter proteins fundamental to cell proliferation and other relevant tumorigenic processes. In this light, three extraordinary papers by Vogelstein and Stratton and colleagues suggest, in fact, that many cancers exhibit a variety of common and rare non-synonymous coding variations that are likely to contribute to tumorigenesis in some way or another [Torkamani et al. 2008].

Functional Assessment of Coding Variations

The transition from the mere identification of disease-associated sequence variant toan understanding of its molecular and physiologic effects that might lead, e.g., to the design of therapeutics is not trivial. For coding variants it is essential to leverage protein structural information in this process. The identification of naturally occurring human DNA sequence variants, including non-synonymous coding variations, has indeed increased rapidly and steadily over the years due to the development of large-scale genomics initiatives and technologies (Figure 1).

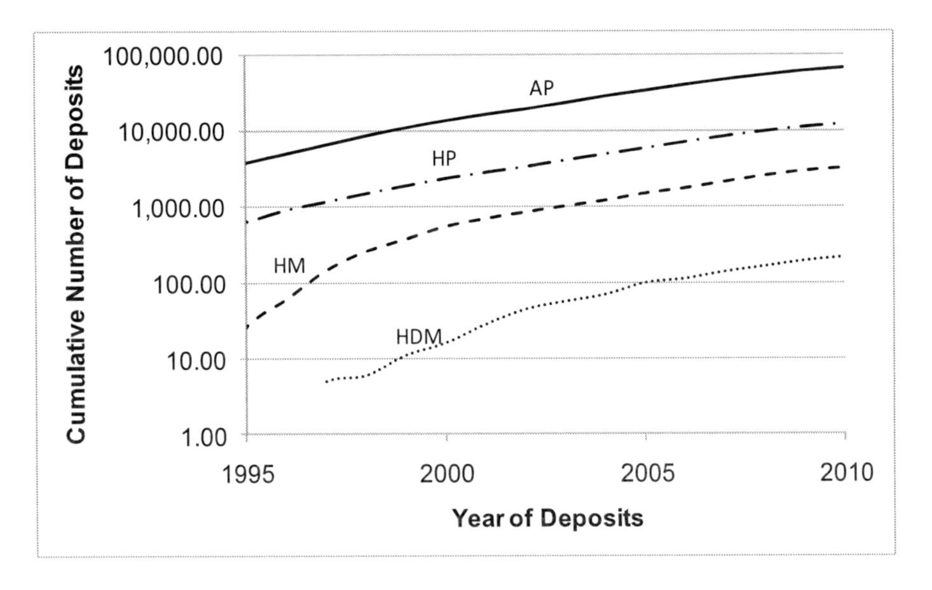

Figure 1.The relative increase in the number of deposited protein structures in the Protein Data Back (PDB) over the last 15 years, including all structures (AP; solid line), specifically human protein structures (HP; dashed-dotted line), all human protein structures accommodating or assessing a known mutation in the amino acid sequence (HM; dashed line), and all human protein structures accommodating or assessing a known disease-causing mutation in the amino acid sequence (HDM; dotted line). The number of deposits is on the log based 10 scale on the y-axis.

Table 1. Mutant forms of previously crystallized human protein structures in the PDB

Protein	Disease	PDB ID(s)
Abl tyrosine kinase	Cancer	2F4J
AL-09	Amyloidogenesis	3CDC, 3CDF
AL-12	Amyloidosis	3DVF
Alcohol dehydrogenase 2 variant	Alcohol-induced flushing syndrome	3INL
Alpha-1-antichymotrypsin	Chronic obstructive pulmonary disease	1QMN
Alpha-1-antitrypsin	Pittsburgh Mutant	1QMB, 1OO8, 1OPH
Alpha-actinin-4	Focal segmental glomerulosclerosis	2R0O
Androgen Receptor	Androgen-independent Prostate Cancer	1GS4, 1Z95
Apolipoprotein A	Lysine Binding Deficiency	4KIV
Argininosuccinate lyase	Argininosuccinic aciduria	1K62
Aurora A	Cancer	2WQE
B-RAF	Cancer	1UWJ, 3IDP
BRCA1	Breast Cancer	1N5O, 1T2U, 2ING
Bruton's tyrosine kinase	X-linked agammaglobulinaemia	1BTK, 1BWN
Cell division protein 42 homolog	Cancers	2ASE
CFTCR	Cystic Fibrosis	1XMJ, 2PZF
Complement Factor H	Age-related macular degeneration	2JGX
Copper-transporting ATPase 1	Menkes Disease	1YJR, 1YJT
CRALBP	Bothnia dystrophy	3HX3
Cytochrome p450	Drug metabolism	3IBD
Dihydrofolate reductase	Cancers	1U71
EGFR	Cancer	2ITN, 2ITO, 2ITP, 2ITQ, 2ITT, 2ITU, 2ITV, 2ITZ, 2JIT, 2JIU, 2JIV, 3IKA
Erythrocyte Isozyme	Nonspherocytic hemolytic anemia	2VGF, 2VGG, 2VGI
Ferritin	Hereditary ferritinopathy	3HX2, 3HX5, 3HX7
FGFR2	Apert Syndrome	1IIL, 1II4

Protein	Disease	PDB ID(s)
FGFR2	Cancer and skeletal disorders	2PVY, 2PWL, 2PY3, 2PZ5, 2PZP, 2PZR, 2Q0B
Fibrinogen gamma chain	Thrombosis	3HUS
Filamin A	Otopalatodigital syndrome 2	3HOC
Filamin-B	Dominant skeletal malformations	2WA6, 2WA7
Fructose-bisphosphate aldolase B	Hereditary fructose intolerance	1XDL, 1XDM
Gamma-crystallin D	Cataracts	2KFB
GPC	Osteoporosis	2AFS
Glutathione S-transferase P	Leukemia	3CSI
Glycyl-tRNA synthetase	Motor nerve degeneration	2Q5I
Glycyl-tRNA synthetase	Charcot-Marie-Tooth Disease	2PMF
Human male SRY	46X,Y sex reversal	1J47
Human Serum Albumin	Familial dysalbuminemic hyperthyroxinemia	1HK2, 1HK3, 1HK5
Intestinal fatty acid binding protein	Insulin resistance/Altered Lipid Metabolism	1KZX
Isocitrate dehydrogenase	Glioma	3INM
Lamin-A/C	familial partial lipodystrophy	3GEF
LDL Receptor	Familial hypercholesterolemia	2W2N, 2W2O, 2W2Q, 3GCW
Lysozyme	Amyloid Fibrillogenesis	1LOZ, 1LYY, 1IOC
Mast cell growth factor receptor	Gastrointestinal stromal tissues	3G0F
Mineralocorticoid receptor	Hypertension	1Y9R, 1YA3
MtDNA aldehyde dehydrogenase	Ethanol Metabolism Deficiency	1ZUM, 2ONM
mtDNA SCO1 protein homolog	Cytochrome c oxidase assembly defects	2HRF, 2HRN
Myosin heavy chain beta isoform	Familial hypertrophic cardiomyopathy	2FXO
Nucleoside diphosphate kinase A	Neuroblastoma	2HVE

Table 1. (Continued)

Protein	Disease	PDB ID(s)
p53	Cancers	2BIM, 2BIN, 2BIO, 2BIP, 2BIQ, 2J1W, 2JIX, 2JIY, 2J1Z, 2J20, 2J21, 2PCX, 2VUK, 1UOL, 2X0U, 2X0V, 2X0W
p97	Paget's Disease, Frototemporal dementia	3HU1, 3HU3
Phenylalanine Hydroxylase	Phenylketonuria	1TDW, 1TG2
PI3Ka	Cancer	3HHM, 3HIZ
Porphobilinogen deaminase	Acute intermittent porphyria	3EQ1
Porphobilinogen synthase	Tyrosinemia	1PV8
Prion Protein	Creutzfeldt-Jakob disease	1FKC, 1FO7, 3HEQ, 3HER, 3HES, 3HJX, 2KUN
Protein DJ-1	Parkinson's Disease	2RK3, 2RK4, 2RK6, 3B36, 3B38, 3B3A, 3BWY
Pyridoxine 5'-phosphate oxidase	Neonatal epileptic encephalopathy	3HY8
Ras	Cancers	1CLU, 1RVD
Sorcin	Hypertrophic cardiomyopathy	2JC2
Superoxide Dismutase	Amyotrophic Lateral Sclerosis	1VAR, 1N19, 1OEZ, 1OZT, 1OZU, 1PTZ, 1UXL, 1UXM, 2NNX, 2VR6, 2VR7, 2VR8, 3GQF, 2WKO, 3GZO, 3GZP, 3GZQ
Thrombospondin-1	Coronary Artery Disease	2RHP

Protein	Disease	PDB ID(s)
Topoisomerase I	Cancers	1RR8, 1RRJ, 1SEU
Transthyretin	Early-onset amyloidotic polyneuropathy	5TTR, 1F86, 1III, 1IIK, 1TTR, 1X7S, 1X7T, 1IJN, 2B14, 2B16, 2G3X, 2NOY, 3KGS, 3KGT, 3NG5
Triosephosphate isomerase	Human triosephosphate deficiency	2VOM
Ubiquitin carboxyl isozyme L1	Parkinson's Disease	3IFW, 3IRT, 3KVF
UDP-galactose 4-epimerase	Galactosemia	1I3K, 1I3L, 1I3M, 1I3N
Uroporphyrinogen decarboxylase	Familial porphyria cutanea tarda	1JPH, 1JPI, 1PJK
Uroporphyrinogen decarboxylase	Hepatoerythropoietic porphyria	2Q6Z, 2Q71
Vitamin D nuclear receptor	Rickets	3M7R
von Willebrand Factor	von Willebrand disease	1IJB, 1IJK

In fact, the number of non-coding and coding variations whose biological significance might need functional deciphering as a result of these initiatives now total over 10,000,000 and 100,000, respectively (Figure 1)! However, the growth of empirically derived structures in the Protein Data Bank [Berman et al. 2000] has progressed much more slowly (Figure 1) due to the obvious complexities and costs involved in structure determination. Relevant structures of mutant forms of what were thought to be the 'wild-type' or 'normal' human (or preexisting) protein structures in the PDB currently number approximately 150 and cover only 59 proteins (Table 1), which amounts to less than 10% of all human protein structures in the PDB that involve a mutation of any sort (including induced mutations). This is many orders of magnitude less than the 100,000+ naturally occurring human amino acid variations in hundres of proteins identified thus far in sequencing efforts. Greater strides need to be made in structurally characterizing these variants, which would be facilitated by increased collaboration between genomic scientists and structural biologists/crystallographers.

Structure Determination Initiatives

The large-scale, flagship project among crystallographers and structural biologists, the Protein Structure Initiative (PSI), was, at one time, under great scrutiny owing, in part, to the belief that the scientific strategy behind the PSI of achieving a breadth of structural insights by focusing on distantly related proteins does not lend itself to immediate biomedical application [Service 2008a,b; Chazin 2008]. Proponents of the initial motivation for the PSI have argued that sampling diverse structures from across the set of all proteins will produce a more complete sampling of the three-dimensional fold space. While homology modeling leverages this structural breadth to predict novel structures of proteins, its accuracy is limited by the ability of the existing algorithms, and, as is well known, the predicted structures may not be as accurate as those derived from actual structure determination studies. Inferring structure through homology modeling exposes the theoretical structures to noise and hence the resulting structures are typically less accurate than their empirical counterparts. Combining the empirical structures with conformational sampling approaches, such as molecular dynamics simulations, has improved computational predictions of potential variant functional characterization and drug lead identification [Lin et al. 2002].

In diseases where mutations are common, such as cancer, it is important to leverage as much protein structural information that might be relevant to any structural perturbations in those proteins as possible in order to, e.g., understand functional effects or design potentially therapeutic compounds that target those proteins that are effective generally (i.e., and not specific to individual mutations). Given that the National Institutes of Health, the National Science Foundation, and other funding agencies are becoming more focused on the practical and immediate applications of scientific research, it makes sense for the structural biology community to consider characterizing and solving structures of variant or mutant forms of individual proteins within the human species – especially those associated with disease – rather than those evolutionay protein orthologous variants that manifest merely across protein families and species.

Research Synergy and Potential

There is increasing evidence that an in-depth examination of the structural effects of naturally occurring human coding variants can lead to insights into the function of the affected residues as well as the molecular pathologies associated with the diseases they might induce. At least one very recent study examining the structural consequences of individual amino acid substitutions in a protein whose crystal structure was solved that are associated with a number of inherited diseases has shown how the different substitutions could lead to different diseases (Fan et al. 2008). In a similar manner, a recent study solved the crystal structure of EGFR kinase mutations in order to show how these mutations are not only likely to drive tumorigenesis, but also likely to influence binding activities of specific anti-cancer compounds such as gefitinib which are designed to target the EGFR protein [Yun et al. 2007, 2008].

This is not to say, however, that every disease variant needs to have the protein structure harboring that variant determined by experimental (i.e., crystallography) methods, as many disease variants do not influence the structure significantly. For example, mutations of catalytic residues in enzymes can reduce the efficacy of the protein without changing the overall structure of the protein. Several characteristics therefore need to be considered when ranking the need and usefulness of crystallizing a mutant form of a protein. For example, mutations that are determined by substitution matrices to be more severe (e.g. R→D) will likely produce more significant changes in protein structure than a more conserved mutation (e.g. R→K). Additionally, it is important to consider the location of the nucleotide and amino acid substitutions on the structure of the protein. Mutations of residues that are far away from active sites, involved in secondary structure, or participate in major stabilizing interactions, such as disulfide bonds, are more likely to cause the protein to adopt a different fold. Combining these characteristics with other factors, such as the viability of the protein as a drug target, the frequency of the variant, and predicted amenability to crystallization may provide a starting point for ranking targets for crystallization.

Conclusion

The rapidly increasing number of human coding variants identified through DNA sequencing and genotyping initiatives should motivate the mapping of these variants onto the corresponding protein structures to elucidate functional changes induced by those variants. In this light, given the contemporary emphasis on translational research, integrative science, interdisciplinary training, and, ultimately, cost-sharing within the biomedical science community, it seems that a stronger relationship between human DNA sequencers and protein structural biologists is simply as timely and practical as it is logical.

Acknowledgments

This work was supported in part by the following research grants: U19 AG023122-01, R01 MH078151-01A1, N01 MH22005, U01 DA024417-01, P50 MH081755-01, UL1 RR025774, the Price Foundation and Scripps Genomic Medicine. This work is the authors' sole responsibility and does not necessarily represent funding agencies' views.

References

1000 Genomes Project Consortium, Durbin RM, Abecasis GR, Altshuler DL, Auton A, Brooks LD, Durbin RM, Gibbs RA, Hurles ME, McVean GA. A map of human genome variation from population-scale sequencing. *Nature*. 2010 Oct 28;467(7319):1061-73. PMID: 20981092

Berman, H.M., Westbrook, J., Feng, Z., Gilliland, G., Bhat, T.N., Weissig, H., Shindyalov, I.N., and Bourne, P.E. (2000). The Protein Data Bank. *Nucleic Acids Res*. 28, 235-242.

Bodmer W, Bonilla C. Common and rare variants in multifactorial susceptibility to common diseases. Nat Genet. 2008 Jun;40(6):695-701. Review. PMID: 18509313

Chazin WJ. Evolution of the NIGMS Protein Structure Initiative. *Structure*. 2008 Jan;16(1):12-14. PMID: 18184576

Cooper GM, Goode DL, Ng SB, Sidow A, Bamshad MJ, Shendure J, Nickerson DA. Single-nucleotide evolutionary constraint scores highlight disease-causing mutations. *Nature Methods*, 7, 250–251 (1 April 2010)

ENCODE Project Consortium, Birney E, Stamatoyannopoulos JA, Dutta A, Guigó R, Gingeras TR, et al. Identification and analysis of functional elements in 1% of the human genome by the ENCODE pilot project. *Nature*. 2007 Jun 14;447(7146):799-816. PMID: 17571346.

Fan L, Fuss JO, Cheng QJ, Arvai AS, Hammel M, Roberts VA, Cooper PK, Tainer JA. XPD helicase structures and activities: insights into the cancer and aging phenotypes from XPD mutations. *Cell*. 2008 May 30;133(5):789-800. PMID: 18510924.

International Cancer Genome Consortium, Hudson TJ, Anderson W, Artez A, Barker AD, Bell C, et al.. International network of cancer genome projects. *Nature*. 2010 Apr 15;464(7291):993-8. PMID: 20393554.

International HapMap Consortium. A haplotype map of the human genome. *Nature*. 2005 Oct 27;437(7063):1299-320. PMID: 16255080.

International HapMap Consortium, Frazer KA, Ballinger DG, Cox DR, Hinds DA, et al. A second generation human haplotype map of over 3.1 million SNPs. *Nature*. 2007 Oct 18;449(7164):851-61. PMID: 17943122.

International HapMap 3 Consortium, Altshuler DM, Gibbs RA, Peltonen L, Altshuler DM, Gibbs RA, Integrating common and rare genetic variation in diverse human populations. *Nature*. 2010 Sep 2;467(7311):52-8. PMID: 20811451.

Lin, J.H., Perryman, A.L., Schames, J.R., and McCammon, J.A. (2002). Computational drug design accommodating receptor flexibility: the relaxed complex scheme. *J Am Chem Soc*. 124, 5632-5633.

Lander ES, Linton LM, Birren B, Nusbaum C, Zody MC, et al. Initial sequencing and analysis of the human genome. *Nature*. 2001 Feb 15;409(6822):860-921. PMID: 11237011.

Levy S, Sutton G, Ng PC, Feuk L, Halpern AL, Walenz BP, Axelrod N, Huang J, Kirkness EF, Denisov G, Lin Y, MacDonald JR, Pang AW, Shago M, Stockwell TB, Tsiamouri A, Bafna V, Bansal V, Kravitz SA, Busam DA, Beeson KY, McIntosh TC, Remington KA, Abril JF, Gill J, Borman J, Rogers YH, Frazier ME, Scherer SW, Strausberg RL, Venter JC. The diploid genome sequence of an individual human. *PLoS Biol*. 2007 Sep 4;5(10):e254. PMID: 17803354.

Manolio TA, Brooks LD, Collins FS. A HapMap harvest of insights into the genetics of common disease. *J Clin Invest*. 2008 May;118(5):1590-605. Review. PMID: 18451988.

Mardis ER, Wilson RK. Cancer genome sequencing: a review. *Hum Mol Genet*. 2009 Oct 15;18(R2):R163-8. Review. PMID: 19808792.

Roach JC, Glusman G, Smit AF, Huff CD, Hubley R, Shannon PT, Rowen L, Pant KP, Goodman N, Bamshad M, Shendure J, Drmanac R, Jorde LB, Hood L, Galas DJ. Analysis of genetic inheritance in a family quartet by whole-genome sequencing. *Science*. 2010 Apr 30;328(5978):636-9. Epub 2010 Mar 10. PMID: 20220176.

Schork NJ, Murray SS, Frazer KA, Topol EJ. Common vs. rare allele hypotheses for complex diseases. *Curr Opin Genet Dev*. 2009 Jun;19(3):212-9. Epub 2009 May 28. Review. PMID: 19481926.

Service RFa Structural biology. Researchers hone their homology tools. *Science*. 2008a Mar 21;319(5870):1612. PMID: 18356502

Service RFb. Structural biology. Protein structure initiative: phase 3 or phase out. *Science*. 2008b Mar 21;319(5870):1610-3. PMID: 18356501.

Torkamani A, Verkhivker G, Schork NJ. Cancer driver mutations in protein kinase genes. *Cancer Lett*. 2009 Aug 28;281(2):117-27. Epub 2008 Dec 10. Review. PMID: 19081671.

Venter JC, Adams MD, Myers EW, Li PW, Mural RJ, et al. The sequence of the human genome. Science. 2001 Feb 16;291(5507):1304-51. Erratum in: *Science*, 2001 Jun 5;292(5523):1838. PMID: 11181995.

Wheeler DA, Srinivasan M, Egholm M, Shen Y, Chen L, McGuire A, He W, Chen YJ, Makhijani V, Roth GT, Gomes X, Tartaro K, Niazi F, Turcotte CL, Irzyk GP, Lupski JR, Chinault C, Song XZ, Liu Y, Yuan Y, Nazareth L, Qin X, Muzny DM, Margulies M, Weinstock GM, Gibbs RA, Rothberg JM. The complete genome of an individual by massively parallel DNA sequencing. *Nature*. 2008 Apr 17;452(7189):872-6. PMID: 18421352.

Yun CH, Boggon TJ, Li Y, Woo MS, Greulich H, Meyerson M, Eck MJ. Structures of lung cancer-derived EGFR mutants and inhibitor complexes: mechanism of activation and insights into differential inhibitor sensitivity. Cancer Cell. 2007 Mar;11(3):217-27. PMID: 17349580.

Yun CH, Mengwasser KE, Toms AV, Woo MS, Greulich H, Wong KK, Meyerson M, Eck MJ. The T790M mutation in EGFR kinase causes drug resistance by increasing the affinity for ATP. Proc Natl Acad Sci U S A. 2008 Feb 12;105(6):2070-5. Epub 2008 Jan 28. PMID: 18227510.

In: Protein Structure
Editor: Lauren M. Haggerty, pp. 105-121
ISBN 978-1-61209-656-8

Chapter 7

MATHEMATICAL MODELLING OF HELICAL PROTEIN STRUCTURE

James McCoy*
Institute of Mathematics and its Applications
School of Mathematics and Applied Statistics
University of Wollongong, Wollongong, NSW 2522
Australia

Abstract

We review recent work on modelling helical parts of protein structure mathematically. We use classical calculus of variations and consider mathematical energies dependent on the curvature and torsion of the protein backbone curve. We demonstrate classes of mathematical energies for which a given helix, or all helices, are extremisers. A future goal is to find within these energies some which permit less regular protein backbone curves, as also observed in nature. As a step in this direction we incorporate a new function into the model, dependent on position along the backbone curve, which permits some allowance for variability in structure due to different amino acid components, side chains, or obstacles.

Keywords: proteins, polymers, calculus of variations, curvature, torsion, helices

AMS classification scheme numbers: 49K15, 49K20, 49N30, 14H50, 92C05.

1. Introduction

Modelling the strikingly regular cylindrical and conical helices which appear in parts of protein backbone structure has captured the attention of the broad scientifc community and generated much research using a wide variety of techniques and experiments by physicists, biologists, chemists, engineers and mathematicians. In particular, mathematical modelling such as in [1], [5] and [9] overlooks chemical structure and focuses on geometric properties of the backbone curve. Biological reasons to support doing this are that it is known that

*E-mail address: jamesm@uow.edu.au

proteins with the same biological task in different animal species have very similar geometric shapes but may have only 25% of their constituent amino acids in common [8]. It is also known that many different sequences fold into the same native state structure [2, 4, 6].

As in classical physics, we assume the protein backbone curve seeks to minimise some energy, but here, as in [5], [10], [9] and [7] we assume the energy depends on the shape of the curve. A more advanced mathematical model including the assembly of various helical pieces and interaction energies appears in [3]. A smooth curve in three space is uniquely determined up to orientation and translation by its curvature and torsion, so these geometric quantities are obvious candidates for the associated energy to depend upon. Likewise derivatives of these are reasonable however dependence on orientation or position in space is undesirable unless we wish to account for nonuniformity in the ambient environment.

Here we adopt an approach along the lines of [7], seeking general conditions an energy functional must satisfy such that the helices observed in protein structure do minimise, or at least extremise, this functional. Later work could refine this approach to build in different observed protein backbone shapes also as extremisers, or consider adding in additional features to accommodate side-chains, inhomogeneity of the ambient fluid and other factors. We give a step in this direction later in the chapter where we add a function of the arc length along the backbone which might allow for varying bending behaviour due to side-chains or obstacles nearby the backbone.

The structure of this chapter is as follows. In Section 2. we describe the geometry of curves in three-dimensional Euclidean space which is necessary to model the protein backbone curve in our approach. In Section 3. we briefly describe the classical calculus of variations branch of mathematics in a simplified setting. Then in Section 4. we apply the theory of Sections 2. and 3. to our particular problem of using the calculus of variations to extremise energy functionals define on curves in three-dimensional space. In Section 5. we provide the all important resulting Euler-Lagrange equations, which must be satisfied in order for a given curve to be an extremiser of the energy. In the subsequent sections we analyse when cylindrical helices are solutions of the Euler-Lagrange equations and also when conical helices may appear. We complete the chapter with a density extension to the energy functional and some concluding remarks.

2. Geometry of Space Curves

We model the protein backbone as a smooth curve C in ambient three-dimensional Euclidean space $\mathbb{R}^3$. The principal directions in the ambient space we label x, y and z. Unit vectors in each of these directions are labelled $\mathbf{i}$, $\mathbf{j}$ and $\mathbf{k}$ respectively and they obey the usual 'right-hand' rule, in the sense that their vector cross products satisfy

$$\mathbf{i} \times \mathbf{j} = \mathbf{k}, \mathbf{j} \times \mathbf{k} = \mathbf{i} \text{ and } \mathbf{k} \times \mathbf{i} = \mathbf{j}.$$

Taking some particular point in the ambient space as the origin, O, points on the curve C are traced out by the position vector

$$\mathbf{r}\,(s) = (x\,(s)\,, y\,(s)\,, z\,(s)) = x\,(s)\,\mathbf{i} + y\,(s)\,\mathbf{j} + z\,(s)\,\mathbf{k},$$

as the parameter s varies in some interval $[a, b]$. The points with position vectors $\mathbf{r}\,(a)$ and $\mathbf{r}\,(b)$ correspond to the endpoints of the curve. As the curve C is smooth, it makes sense to differentiate $\mathbf{r}(s)$ componentwise with respect to the parameter s; we write

$$\dot{\mathbf{r}}(s) := (\dot{x}(s), \dot{y}(s), \dot{z}(s)) = \left(\frac{d}{ds}x(s), \frac{d}{ds}y(s), \frac{d}{ds}z(s)\right)$$

and call this vector the *velocity vector* of the curve C. Throughout this work we will use the dot to denote differentiation with respect to the parameter s. We will assume that our curve C is *regular*, meaning means that $\dot{\mathbf{r}}(s) \neq 0$ for any $s \in [a, b]$. This means that the vector $\dot{\mathbf{r}}(s)$ defines at each point $\mathbf{r}(s)$ the tangent direction to C and also implies that C may be parametrised by *unit speed.* In the unit-speed parametrisation the magnitude of the velocity vector is identically equal to 1 everywhere on the curve, $|\dot{\mathbf{r}}(s)| \equiv 1$. In this case we write

$$\mathbf{T}(s) = \dot{\mathbf{r}}(s)$$

for a unit tangent vector to C (the other unit tangent vector being $-\mathbf{T}$). We will further assume that our curve C has no self-intersections, that is, there are no $s_1, s_2 \in [a, b]$, $s_1 \neq s_2$ with $\mathbf{r}(s_1) = \mathbf{r}(s_2)$.

Throughout this work we will assume s denotes the unit-speed parameter for curve C. From calculus, the arc length function of a curve given by some parametrisation $\mathbf{v}(t)$, not necessarily of unit speed, beginning at point $\mathbf{v}(a)$ is

$$\ell(t) = \int_a^t \left|\dot{\mathbf{v}}(\tilde{t})\right| d\tilde{t}.$$

Note that for the vector $\mathbf{v}(t)$, the dot represents differentiation with respect to its parameter t, rather than with respect to the unit-speed parameter. We will use this notation when there is no chance of confusion. Observe that, for a regular curve, if t is in fact the unit-speed parameter s, then the fact that $|\dot{\mathbf{r}}(s)| \equiv 1$ implies, by simple integration, that

$$\ell(s) = s - a,$$

that is, the arc length is simply s, up to translation by $-a$. For this reason the unit-speed parametrisation of C is also called the parametrisation of C by arc length.

The tangent space to the regular curve C at a point with position vector $\mathbf{r}(s)$ is the set of all scalar multiples of $\dot{\mathbf{r}}(s)$. That $\dot{\mathbf{r}}(s)$ is tangent to C at $\mathbf{r}(s)$ can be seen formally by considering the definition of the derivative $\mathbf{r}(s)$, together with vector addition in $\mathbb{R}^3$. Now, vectors perpendicular to $\dot{\mathbf{r}}(s)$ are normal to the curve C. If C were in two-dimensional Euclidean space, the normal direction at $\mathbf{r}(s)$ could be written down explicitly: in this case $\dot{\mathbf{r}}(s) = (\dot{x}(s), \dot{y}(s))$ so the vector $(-\dot{y}(s), \dot{x}(s))$ is clearly perpendicular as the dot product of the two vectors is equal to zero. However, if C is in three-dimensional Euclidean space, that is, C is a *space-curve*, then the set of all vectors perpendicular to $\dot{\mathbf{r}}(s)$ has two dimensions. One could generalise the approach we used above, however we will define a specific 'normal' vector to C and a mutually perpendicular 'binormal' vector. To do this we first define the *curvature* of C at the point $\mathbf{r}(s)$ as

$$\kappa(s) := |\ddot{\mathbf{r}}(s)|,$$

in the case that C is parametrised by unit speed. In other words, in the unit-speed parametrisation, $\kappa(s)$ is precisely the magnitude of the second derivative of $\mathbf{r}$, that is, the magnitude of the *acceleration vector*. Since the parametrisation is constant speed, the acceleration serves purely to change the shape of the path C, that is, to cause it to curve in the ambient space. It is worth noting that if $\kappa(s) \neq 0$ then $\frac{1}{\kappa(s)}$ is the radius of the 'circle of best fit' which touches C at $\mathbf{r}(s)$ and bends in the direction of C. If $\kappa(s) = 0$ then C is 'locally' flat.

It is a straightforward calculation to check that if a regular curve is parametrised by some other vector valued function $\mathbf{v}(t)$, not necessarily with unit speed, then the curvature of C is given in terms of the derivatives of $\mathbf{v}$ (derivatives with respect to its parameter t) by

$$\tilde{\kappa}(t) := \frac{|\ddot{\mathbf{v}}(t) \times \dot{\mathbf{v}}(t)|}{|\dot{\mathbf{v}}(t)|^3}, \tag{1}$$

where here again dot denotes derivative with respect to t.

Returning to the arc length parametrisation, in view of unit speed we have the identity

$$|\mathbf{T}(s)| = |\dot{\mathbf{r}}(s)|^2 = 1$$

and differentiating this expression with respect to s yields

$$2\,\dot{\mathbf{r}}(s) \cdot \ddot{\mathbf{r}}(s) = 0.$$

This means that at each point $\mathbf{r}(s)$ on C, the vector $\ddot{\mathbf{r}}(s)$ is either equal to zero or perpendicular to the tangent vector, $\dot{\mathbf{r}}(s)$. In the case that $\kappa(s) \neq 0$, we define the *principal normal* vector to C at $\mathbf{r}(s)$ as the unit vector

$$\mathbf{N}(s) := \frac{1}{\kappa(s)}\dot{\mathbf{T}}(s) = \frac{1}{\kappa(s)}\ddot{\mathbf{r}}(s).$$

For a smooth curve C, even if κ is equal to zero somewhere, it is possible with care to define the unit principal normal vector, however, we will not pursue this here. From our two perpendicular unit vectors $\mathbf{T}(s)$ and $\mathbf{N}(s)$ at the point $\mathbf{r}(s)$ we now find a mutually perpendicular direction using the vector cross product and define the *binormal* to C at $\mathbf{r}(s)$ as the unit vector

$$\mathbf{B}(s) := \mathbf{T}(s) \times \mathbf{N}(s),$$

Like the basis vectors $\mathbf{i}$, $\mathbf{j}$ and $\mathbf{k}$ of the ambient space, the vectors $\mathbf{T}$, $\mathbf{N}$ and $\mathbf{B}$ obey the right hand rule at each point along C, but $\mathbf{T}$ is always the unit tangent vector to C at $\mathbf{r}(s)$ and $\mathbf{N}$ and $\mathbf{B}$ are mutually perpendicular unit normal vectors. The basis of the ambient three-dimensional Euclidean space given by $\{\mathbf{T}, \mathbf{N}, \mathbf{B}\}$ is called the *Frenet frame* for C.

Using the properties of $\mathbf{T}$, $\mathbf{N}$ and $\mathbf{B}$ we can show that $\dot{\mathbf{B}}(s)$ is proportional to $\mathbf{N}$ at each point along C; we write

$$\dot{\mathbf{B}}(s) = -\tau(s)\,\mathbf{N}(s)$$

for some scalar function $\tau(s)$, called the *torsion* of C. The torsion measures the deviation of C from being locally planar. Indeed, it $\tau \equiv 0$, then $\mathbf{B}$ is a constant vector so the curve C lies in the two-dimensional plane with normal $\mathbf{B}$.

The *Frenet-Serret* formulas for a regular curve C in the unit-speed parametrisation are obtained from the above:

$$\begin{aligned}\dot{\mathbf{T}}(s) &= \kappa(s)\,\mathbf{N}(s)\\ \dot{\mathbf{N}}(s) &= -\kappa(s)\,\mathbf{T}(s) + \tau(s)\,\mathbf{B}(s)\\ \dot{\mathbf{B}}(s) &= -\tau(s)\,\mathbf{N}(s).\end{aligned} \tag{2}$$

In a similar way that we had above an expression for $\tilde{\kappa}(t)$ in a general parametrisation $\mathbf{v}(t)$ not necessarily of unit speed, we could also rewrite the Frenet-Serret formulas in a more general form for the general parametrisation. We will not do this here, but we will note that the expression for torsion of space-curves in the more general setting is:

$$\tilde{\tau}(t) = \frac{(\dot{\mathbf{v}}(t) \times \ddot{\mathbf{v}}(t)) \cdot \dddot{\mathbf{v}}(t)}{|\dot{\mathbf{v}}(t) \times \ddot{\mathbf{v}}(t)|^2} = \frac{\det(\dot{\mathbf{v}}(t), \ddot{\mathbf{v}}(t), \dddot{\mathbf{v}}(t))}{|\dot{\mathbf{v}}(t) \times \ddot{\mathbf{v}}(t)|^2}. \tag{3}$$

The numerator of the third expression here is the determinant of the matrix formed by placing the vectors $\dot{\mathbf{v}}$, $\ddot{\mathbf{v}}$ and $\dddot{\mathbf{v}}$ as rows (or as columns, since $\det A^T = \det A$ for a square matrix A). The expression on the numerator above is also called the vector triple product; both expressions above are equivalent. Notice that we are not dividing by zero in this expression since the formula is only valid where $\tilde{\kappa}(t) \neq 0$.

That a smooth space curve C is determined uniquely up to position and orientation in space, once we have its curvature and torsion functions, $\kappa(s)$ and $\tau(s)$, corresponds to a result on existence and uniqueness of solutions to systems of ordinary differential equations. This result is obviously important to us, however we will not discuss its proof. Usually it is not possible to integrate the Frenet-Serret equations to reconstruct $\mathbf{r}(s)$ from given functions $\kappa(s)$ and $\tau(s)$. Integration is possible for plane curves, for which the Frenet-Serret equations are simpler but we are not so interested in plane curves in our application. In a case more relevant for us, the circular helix with position vector

$$\mathbf{r}(s) = (\alpha\cos t, \alpha\sin t, \beta t)$$

with $\alpha, \beta \neq 0$ has constant curvature and constant torsion, given by

$$\kappa = \frac{\alpha}{\alpha^2 + \beta^2}, \quad \tau = \frac{\beta}{\alpha^2 + \beta^2}.$$

Generalised helices are those space-curves for which there is a fixed vector $\mathbf{a}$ in space, the direction of the *axis* of the helix, for which the angle between the tangent vector $\dot{\mathbf{r}}(s)$ and $\mathbf{a}$ is constant. It can be shown that a curve is a generalised helix if and only if the ratio $\frac{\tau}{\kappa}$ is identically constant, where $\kappa \neq 0$, or $\tau = 0$ where $\kappa = 0$. Generalised helices are important for us as cylindrical helices fit into this class of space-curves and so do conical helices, which are also observed in protein structure. In the case where $\tau = \tau_0 \neq 0$ is constant, then a generalised helix must necessarily have $\kappa = \kappa_0$ constant and so is a circular helix.

A conical helix lies on the surface of a cone and if its axis is the z-axis with tip at the origin, it can be parametrised by

$$\mathbf{v}(t) = (\alpha\, t\cos(\beta\log t), \alpha\, t\sin(\beta\log t), \gamma\, t)$$

for some constants $\alpha, \beta, \gamma \neq 0$. The curvature and torsion functions of the conical helix can be calculated using the previous formulas and are respectively

$$\kappa(s) = \frac{\kappa_0}{s}, \quad \tau = \frac{\tau_0}{s},$$

where κ_0 and τ_0 are constants given by

$$\kappa_0 = \frac{\alpha\,\beta\sqrt{1+\beta^2}}{\delta} \text{ and } \tau = \frac{\gamma\,\beta}{\delta}$$

and $\delta = \sqrt{\alpha^2\left(1+\beta^2\right)+\gamma^2}$. The parameter t and the unit-speed parameter s are related by $s = \delta\, t$. For more details of this we refer the reader to [9].

3. Background on the Calculus of Variations

The central idea of the calculus of variations is, in the simplest setting, to find a function $y(x)$ which minimises (or maximises) some integral such as

$$I[y] := \int_a^b f\left(x, y, y'\right) dx,$$

given some conditions at the endpoints, for example, $y(a) = A$ and $y(b) = B$, for some known constants A and B. The integrand f is a suitably smooth function of the independent variable x, the unknown function y and in this case the first derivative of y, namely $y'(x)$. More generally higher derivatives of y can be included. In practical problems, the integral will often be some physical energy, so it is necessarily positive and obtains a nonzero minimum. Usually the precise form of the function f is known although this will not be the case in our later sections.

Suppose $u(x)$ is a suitably smooth function which minimises I and set $y(x) = u(x) + \varepsilon\, v(x)$, where ε is a parameter and v is a smooth function defined on $[a, b]$ with *compact support*. This means that v and all its derivatives go smoothly to zero at the endpoints of the interval a and b. For any ε, the function y is a valid 'trial function' in the integral $I[y]$ since it has the correct behaviour at the endpoints of the interval $[a, b]$. We may now consider the function i of one variable, ε, defined by

$$i(\varepsilon) = I[y] = I[u + \varepsilon\, v]$$

which, written out in full is simply

$$i(\varepsilon) = \int_a^b f\left(x, u(x) + \varepsilon\, v(x), u'(x) + \varepsilon\, v'(x)\right) dx. \tag{4}$$

Since $u(x)$ is a minimiser of I, then assuming $i(\varepsilon)$ is differentiable, it must be the case that

$$i'(0) = 0 \tag{5}$$

that is, the function $i(\varepsilon)$ has a 'critical point' at $\varepsilon = 0$. Differentiating (4) with respect to ε using the chain rule we have

$$i'(\varepsilon) = \int_a^b \left[\frac{\partial f}{\partial y} \left(x, u(x) + \varepsilon v(x), u'(x) + \varepsilon v'(x)\right) v(x) \right.$$
$$\left. + \frac{\partial f}{\partial y'} \left(x, u(x) + \varepsilon v(x), u'(x) + \varepsilon v'(x)\right) v'(x) \right] dx, \quad (6)$$

where $\frac{\partial f}{\partial y}$ and $\frac{\partial f}{\partial y'}$ denote respectively differentiation of f with respect to its second and its third argument. Now setting $\varepsilon = 0$, equation (6) becomes

$$i'(0) = \int_a^b \left[\frac{\partial f}{\partial y} \left(x, u(x), u'(x)\right) v(x) + \frac{\partial f}{\partial y'} \left(x, u(x), u'(x)\right) v'(x) \right] dx. \quad (7)$$

This is called the *first variation* of the energy. Integrating the second term by parts and using the compact support of v, (7) becomes, in view of (5)

$$0 = i'(0) = \int_a^b \left[\frac{\partial f}{\partial y} \left(x, u(x), u'(x)\right) - \frac{d}{dx} \frac{\partial f}{\partial y'} \left(x, u(x), u'(x)\right) \right] v(x)\, dx.$$

Since the above holds for any v with compact support, the *fundamental lemma of the calculus of variations* gives that we must in fact have

$$\frac{\partial f}{\partial y} \left(x, u(x), u'(x)\right) - \frac{d}{dx} \frac{\partial f}{\partial y'} \left(x, u(x), u'(x)\right) = 0.$$

This is an ordinary differential equation that the function u must satisfy in order that (5) is satisfied. This differential equation for u is called the *Euler-Lagrange* equation for local extrema of the energy functional I.

As we are using calculus of one variable, note that a solution function $u(x)$ of the differential equation (5) simply corresponds to a critical point of $i(\varepsilon)$; to determine whether it achieves a local minimum, local maximum, or a point of inflection of the function $i(\varepsilon)$ we could check the second derivative of $i(\varepsilon)$ at $\varepsilon = 0$, if i is twice differentiable. If we found that

$$i''(0) > 0 \quad (8)$$

held, we would be sure that i had a local minimum at $\varepsilon = 0$. If we found instead

$$i''(0) < 0$$

then i would have a local maximum at $\varepsilon = 0$, while if $i''(0) = 0$ we would not be able to determine the nature of the critical point of i from this information alone and we would need to look at higher derivatives of i or use other information about the energy. The calculation of $i''(\varepsilon)$ is called the *second variation* of i. In practice, often this calculation is not necessary if we know a physical energy is necessarily being minimised.

It is worth noting that there are many more general settings in which the calculus of variations can be successfully applied. One which is relevant in our context is integrands which depend on more functions and also their derivatives, for example

$$I[y_1, y_2] = \int_a^b F\left(x, y_1, y_2, y_1', y_2'\right) dx.$$

We will not derive Euler-Lagrange equations for this general problem here, but we will specialise further to our setting in the subsequent section.

4. Calculus of Variations of Space Curves

In this section we will take the background from the previous two sections to apply calculus of variations to finding extremal curves in three-dimensional Euclidean space. Suppose we have a curve C, represented by position vector $\mathbf{r}(s)$ for $s \in [a, b]$, which minimises some energy

$$I[\mathbf{r}] = \int_C \mathcal{F}(\kappa, \tau)\, d\mathcal{L}.$$

Observe that the energy is a line integral along the curve C which has arc length element $d\mathcal{L}$ and we have an energy density $\mathcal{F}(\kappa, \tau)$ which depends upon the curvature and torsion functions of the curve C. These functions are given in terms of the derivatives of $\mathbf{r}$ by the formulas (1) and (3). The choice that $\mathcal{F}$ depends on curvature and torsion is because, as described in Section 2., once these functions are determined, the shape of C is known, up to translations and rotations in space. Later we will include additional dependence in the integrand.

To obtain Euler-Lagrange equations for minimising $I[\mathbf{r}]$, we use a similar process as in the previous section, however we must take care with the extra geometry involved. Firstly observe that we may parametrise the line integral as

$$I[\mathbf{r}] = \int_a^b \mathcal{F}(\kappa(s), \tau(s))\, |\dot{\mathbf{r}}(s)|\, ds,$$

since the arc length element is $d\mathcal{L} = |\dot{\mathbf{r}}(s)|\, ds$. Now suppose $\mathbf{r}(s)$ is a suitably smooth curve which minimises $I[\mathbf{f}]$. This means that, as in Section 3., we have a corresponding local minimum of a function i, but how do we set up this function in this situation? Noting that $\mathbf{r}$ has at each $s \in (a, b)$ a Frenet frame $\{\mathbf{T}(s), \mathbf{N}(s), \mathbf{B}(s)\}$, we define a varied curve

$$\underline{\mathbf{r}}(s) = \mathbf{r}(s) + \varepsilon_1\, \psi_1(s)\, \mathbf{N}(s) + \varepsilon_2\, \psi_2(s)\, \mathbf{B}(s)\,. \tag{9}$$

Here ψ_1 and ψ_2 are smooth real-valued functions on $[a, b]$ which are equal to zero at the endpoints of the interval, so the varied curves are all fixed at the endpoints. The ε_1 and ε_2 are parameters. This is called a *normal variation* of the curve given by $\mathbf{r}(s)$. We do not include a term in the direction of the tangent vector $\mathbf{T}(s)$ as it is well known that such variations do not lead to a nontrivial Euler-Lagrange equation. Observe that the two parameters arise from the fact that the normal space to the curve at each point $\mathbf{r}(s)$ has two dimensions; an orthonormal basis of this space is $\{\mathbf{N}(s), \mathbf{B}(s)\}$.

We may now compute the energy I of the varied curve with position vector $\underline{\mathbf{r}}$:

$$I[\underline{\mathbf{r}}] = \int_a^b \mathcal{F}(\underline{\kappa}(s), \underline{\tau}(s))\, |\dot{\underline{\mathbf{r}}}(s)|\, ds,$$

where $\underline{\kappa}$ and $\underline{\tau}$ are the curvature and torsion functions of the varied curve.

Under the assumption that the curve C with position vector $\mathbf{r}(s)$ is a minimiser, the function $i(\varepsilon_1, \varepsilon_2)$ given by

$$i(\varepsilon_1, \varepsilon_2) = I[\underline{\mathbf{r}}]$$

has a critical point at the point $\varepsilon_1 = \varepsilon_2 = 0$, which we will write using vector notation as $\underline{\varepsilon} = \underline{0}$. That i has a critical point here means

$$\left.\frac{\partial}{\partial \varepsilon_1} i(\varepsilon_1, \varepsilon_2)\right|_{\underline{\varepsilon}=\underline{0}} = 0 \tag{10}$$

and

$$\left.\frac{\partial}{\partial \varepsilon_2} i(\varepsilon_1, \varepsilon_2)\right|_{\underline{\varepsilon}=\underline{0}} = 0 \tag{11}$$

must both be satisfied by the minimising curve. These two equations give rise to the two Euler-Lagrange equations which must be simultaneously satisfied by the curvature and torsion functions of a 'critical' curve. These equations will relate curvature, torsion and their derivatives to $\mathcal{F}$ and its derivatives. Given a particular $\mathcal{F}$, one would hope to solve for curvature and torsion. However, asking the question around the other way, we determine which energy densities $\mathcal{F}$ allow helices as solutions of the resulting Euler-Lagrange equations in several scenarios.

It is worth noting that while the calculations involved in establishing (10) and (11) follow a similar pattern as in Section 3., there are additional steps involved. To obtain the relevant expression for $i(\varepsilon_1, \varepsilon_2) = I[\underline{\mathbf{r}}]$ in terms of ε_1 and ε_2, one must compute $\dot{\underline{\mathbf{r}}}$ to obtain $|\dot{\underline{\mathbf{r}}}|$ and compute $\underline{\kappa}(s)$ and $\underline{\tau}(s)$ for the varied curve using the formulas (1) and (3). It is important to note that the varied curve $\underline{\mathbf{r}}(s)$ is not necessarily parametrised by unit-speed. However $\underline{\mathbf{r}}(s)$ is written using the Frenet frame $\{\mathbf{T}, \mathbf{N}, \mathbf{B}\}$ for the minimising curve, given by $\mathbf{r}(s)$, where s is its unit-speed parameter, so one can use the Frenet-Serret formulae with respect to this parametrisation to simplify derivatives of $\mathbf{T}$, $\mathbf{N}$ and $\mathbf{B}$.

In particular, from (9), we have after simplification using (2)

$$\dot{\underline{\mathbf{r}}}(s) = [1 - \varepsilon_1 \kappa \psi_1]\,\mathbf{T} + [\varepsilon_1 \psi_1' - \varepsilon_2 \tau \psi_2]\,\mathbf{N} + [\varepsilon_1 \psi_1 \tau + \varepsilon_2 \psi_2']\,\mathbf{B}.$$

Since the frame $\{\mathbf{T}, \mathbf{N}, \mathbf{B}\}$ it follows that

$$|\dot{\underline{\mathbf{r}}}|^2 = [1 - \varepsilon_1 \kappa \psi_1]^2 + [\varepsilon_1 \psi_1' - \varepsilon_2 \tau \psi_2]^2 + [\varepsilon_1 \psi_1 \tau + \varepsilon_2 \psi_2'],$$

so this means that

$$\left.\frac{\partial}{\partial \varepsilon_1} |\dot{\underline{\mathbf{r}}}|^2\right|_{\underline{\varepsilon}=\underline{0}} = -2\,\kappa\,\psi_1 \text{ and } \left.\frac{\partial}{\partial \varepsilon_2} |\dot{\underline{\mathbf{r}}}|^2\right|_{\underline{\varepsilon}=\underline{0}} = 0.$$

To calculate the variations of curvature and torsion we have to use (9) and its simplified derivative expressions in (1) and (3) and if we want the variations of the derivatives of these quantities further lengthy computations are required. We refer the interested reader to the appendix of [9] for details of these calculations.

We should also note that the first order conditions on i above, (10) and (11) guarantee nothing more than a critical point of this function of two variables ε_1 and ε_2. We could be sure such a point was a local minimum if we could show for all vectors $\xi = (\xi_1, \xi_2) \in \mathbb{R}^2$,

$$\sum_{k,\ell=1}^{2} \frac{\partial^2 i}{\partial \varepsilon_k \partial \varepsilon_\ell}\bigg|_{\underline{\varepsilon}=\underline{0}} \xi_k \, \xi_\ell > 0.$$

This is equivalent to saying that the matrix of second derivatives of i is positive definite when $\varepsilon_1 = \varepsilon_2 = 0$, that is, that the matrix has all positive eigenvalues. However, we will not be concerned with this condition here as we will assume that by analogy with physical energies that our energies are naturally being minimised. The computation of second variations such as above could well be interesting in certain situations and could lead to additional clarification of which energies are relevant within our classes, however, such computations are extremely lengthy and we will not include them.

5. The Euler-Lagrange Equations

Derivations are done using the procedure outlined in Section 4.. We will not include all the lengthy calculations, but we refer the reader to [9] and [7] for details of most of the cases listed below. We will however, in Section 8. include some details of new aspects of an extended situation.

In the case that the energy density has the form $\mathcal{F}(\kappa, \tau, \dot{\kappa}, \dot{\tau})$, that is, we allow the energy density to depend upon the curvature and torsion functions of the backbone curve C and their first derivatives, the Euler-Lagrange equations corresponding to (10) and (11) are

$$\begin{aligned} &\frac{d^2}{ds^2}\left[\frac{\partial \mathcal{F}}{\partial \kappa} - \frac{d}{ds}\left(\frac{\partial \mathcal{F}}{\partial \dot{\kappa}}\right)\right] + \frac{2\,\tau}{\kappa}\frac{d^2}{ds^2}\left[\frac{\partial \mathcal{F}}{\partial \tau} - \frac{d}{ds}\left(\frac{\partial \mathcal{F}}{\partial \dot{\tau}}\right)\right] \\ &+\left(\frac{\dot{\tau}}{\kappa} - \frac{2\,\dot{\kappa}\,\tau}{\kappa^2}\right)\frac{d}{ds}\left[\frac{\partial \mathcal{F}}{\partial \tau} - \frac{d}{ds}\left(\frac{\partial \mathcal{F}}{\partial \dot{\tau}}\right)\right] + \left(\kappa^2 - \tau^2\right)\left[\frac{\partial \mathcal{F}}{\partial \kappa} - \frac{d}{ds}\left(\frac{\partial \mathcal{F}}{\partial \dot{\kappa}}\right)\right] \\ &+2\,\kappa\,\tau\left[\frac{\partial \mathcal{F}}{\partial \tau} - \frac{d}{ds}\left(\frac{\partial \mathcal{F}}{\partial \dot{\tau}}\right)\right] + \kappa\left[\dot{\kappa}\frac{\partial \mathcal{F}}{\partial \dot{\kappa}} + \dot{\tau}\frac{\partial \mathcal{F}}{\partial \dot{\tau}} - \mathcal{F}\right] = 0 \end{aligned} \tag{12}$$

and

$$\begin{aligned} &-\frac{1}{\kappa}\frac{d^3}{ds^3}\left[\frac{\partial \mathcal{F}}{\partial \tau} - \frac{d}{ds}\left(\frac{\partial \mathcal{F}}{\partial \dot{\tau}}\right)\right] + \frac{2\,\dot{\kappa}}{\kappa^2}\frac{d^2}{ds^2}\left[\frac{\partial \mathcal{F}}{\partial \tau} - \frac{d}{ds}\left(\frac{\partial \mathcal{F}}{\partial \dot{\tau}}\right)\right] \\ &+2\,\tau\frac{d}{ds}\left[\frac{\partial \mathcal{F}}{\partial \kappa} - \frac{d}{ds}\left(\frac{\partial \mathcal{F}}{\partial \dot{\kappa}}\right)\right] \\ &+\left(\frac{\tau^2}{\kappa} + \frac{\ddot{\kappa}}{\kappa^2} - \frac{2\,\dot{\kappa}^2}{\kappa^3} - \kappa\right)\frac{d}{ds}\left[\frac{\partial \mathcal{F}}{\partial \tau} - \frac{d}{ds}\left(\frac{\partial \mathcal{F}}{\partial \dot{\tau}}\right)\right] \\ &+\dot{\tau}\left[\frac{\partial \mathcal{F}}{\partial \kappa} - \frac{d}{ds}\left(\frac{\partial \mathcal{F}}{\partial \dot{\kappa}}\right)\right] - \dot{\kappa}\left[\frac{\partial \mathcal{F}}{\partial \tau} - \frac{d}{ds}\left(\frac{\partial \mathcal{F}}{\partial \dot{\tau}}\right)\right] = 0. \end{aligned} \tag{13}$$

These equations can generalise in a natural way if $\mathcal{F}$ depends also on higher derivatives of curvature and torsion. Instead simplifying the situation, if the energy density depends only on curvature and torsion, that is, $\mathcal{F} = \mathcal{F}(\kappa, \tau)$, then the above equations reduce to

$$\frac{d^2}{ds^2}\frac{\partial \mathcal{F}}{\partial \kappa} + \frac{2\,\tau}{\kappa}\frac{d^2}{ds^2}\frac{\partial \mathcal{F}}{\partial \tau} + \left(\frac{\dot{\tau}}{\kappa} - \frac{2\,\dot{\kappa}\,\tau}{\kappa^2}\right)\frac{d}{ds}\frac{\partial \mathcal{F}}{\partial \tau} + \left(\kappa^2 - \tau^2\right)\frac{\partial \mathcal{F}}{\partial \kappa} + 2\,\kappa\,\tau\frac{\partial \mathcal{F}}{\partial \tau} - \kappa\,\mathcal{F} = 0 \quad (14)$$

and

$$-\frac{1}{\kappa}\frac{d^3}{ds^3}\frac{\partial \mathcal{F}}{\partial \tau} + \frac{2\,\dot{\kappa}}{\kappa^2}\frac{d^2}{ds^2}\frac{\partial \mathcal{F}}{\partial \tau} + 2\,\tau\frac{d}{ds}\frac{\partial \mathcal{F}}{\partial \kappa} + \left(\frac{\tau^2}{\kappa} + \frac{\ddot{\kappa}}{\kappa^2} - \frac{2\,\dot{\kappa}^2}{\kappa^3} - \kappa\right)\frac{d}{ds}\frac{\partial \mathcal{F}}{\partial \tau} + \dot{\tau}\frac{\partial \mathcal{F}}{\partial \kappa} - \dot{\kappa}\frac{\partial \mathcal{F}}{\partial \tau} = 0. \quad (15)$$

If the energy density depends only on curvature, $\mathcal{F} = \mathcal{F}(\kappa)$, then the equations reduce further to

$$\frac{d^2}{ds^2}\mathcal{F}' + \left(\kappa^2 - \tau^2\right)\mathcal{F}' - \kappa\,\mathcal{F} = 0 \quad (16)$$

and

$$\dot{\tau}\,\mathcal{F}' + 2\,\tau\frac{d}{ds}\mathcal{F}' = 0, \quad (17)$$

where now $\mathcal{F}'$ is the derivative of $\mathcal{F}$ with respect to its only argument, κ.

In [7] we also considered the subcase of (12) and (13) where $\mathcal{F} = \mathcal{F}(\kappa, \dot{\kappa})$.

6. Results: Very General Energies which Permit a Particular Cylindrical Helix

If one looks directly only for circular helix solutions to (12) and (13), that is $\kappa \equiv \kappa_0$ and $\tau \equiv \tau_0$ for constants κ_0 and τ_0, the equations simplify dramatically. Indeed, $\mathcal{F}$ and its derivatives are all constant along the backbone curve, when evaluated at (κ_0, τ_0) so all the $\frac{d}{ds}$ terms disappear. Equation (13) disappears completely, while (12) becomes

$$\left(\kappa_0^2 - \tau_0^2\right)\frac{\partial \mathcal{F}}{\partial \kappa}(\kappa_0, \tau_0) + 2\,\kappa_0\,\tau_0\frac{\partial \mathcal{F}}{\partial \tau}(\kappa_0, \tau_0) - \kappa_0\,\mathcal{F}(\kappa_0, \tau_0) = 0. \quad (18)$$

Note that we have used the notation $\mathcal{F}(\kappa, \tau)$ above because although $\mathcal{F} = \mathcal{F}(\kappa, \tau, \dot{\kappa}, \dot{\tau})$ for (12) and (13), that $\kappa \equiv \kappa_0$ and $\tau \equiv \tau_0$ means that we are really only considering the more general $\mathcal{F}$ evaluated at $(\kappa_0, \tau_0, 0, 0)$, because the derivatives of curvature and torsion along the curve are each identically equal to zero.

If we write the left hand side of (18) as

$$P[\mathcal{F}](\kappa_0, \tau_0) = \left(\kappa_0^2 - \tau_0^2\right)\frac{\partial \mathcal{F}}{\partial \kappa}(\kappa_0, \tau_0) + 2\,\kappa_0\,\tau_0\frac{\partial \mathcal{F}}{\partial \tau}(\kappa_0, \tau_0) - \kappa_0\,\mathcal{F}(\kappa_0, \tau_0),$$

it might, or more likely might not, be the case that $P[\mathcal{F}](\kappa_0, \tau_0) = 0$. We also need to decide if we want just one circular helix with some values of κ_0 and τ_0 to satisfy this condition or, as a stronger requirement, that *all* cylindrical helices satisfy the condition. In the former case, (18) is a weak condition at a single value of the argument of $\mathcal{F}$. Many $\mathcal{F}(\kappa, \tau)$ will satisfy this condition. For example, if $\mathcal{F}$ is continuous in both arguments and there are points (κ_1, τ_1) and (κ_2, τ_2) such that $P[\mathcal{F}](\kappa_1, \tau_1) < 0$ and $P[\mathcal{F}](\kappa_2, \tau_2) > 0$, then the intermediate value theorem of single variable calculus gives that there is a (κ_0, τ_0) where (18) is satisfied. Indeed, one such value lies along the line in (κ, τ) space given by

$$\ell(\lambda) = \lambda(\kappa_1, \tau_1) + (1-\lambda)(\kappa_2, \tau_2) = (\lambda\kappa_1 + (1-\lambda)\kappa_2, \lambda\tau_1 + (1-\lambda)\tau_2),$$

for some $\lambda = \lambda_0 \in (0,1)$. The function of one variable on which to apply the intermediate value theorem is

$$p(\lambda) = P[\mathcal{F}](\lambda\kappa_1 + (1-\lambda)\kappa_2, \lambda\tau_1 + (1-\lambda)\tau_2).$$

Other suitable (κ, τ) may be constructed using different curves between (κ_1, τ_1) and (κ_2, τ_2).

If instead we have particular κ_* and τ_* in mind, that is, a particular circular helix which we want to satisfy (12) and (13), then we can modify any particular energy density $\mathcal{F}(\kappa, \tau)$ (or more generally $\mathcal{F}(\kappa, \tau, \dot{\kappa}, \dot{\tau})$) rather simply to achieve this. As in [7], we will have, for fixed numbers A, B and C which we can evaluate if we know $\mathcal{F}$,

$$\mathcal{F}(\kappa_*, \tau_*) = A, \quad \frac{\partial\mathcal{F}}{\partial\kappa}(\kappa_*, \tau_*) = B, \quad \frac{\partial\mathcal{F}}{\partial\tau}(\kappa_*, \tau_*) = C.$$

The energy density

$$\mathcal{G}(\kappa, \tau) = \mathcal{F}(\kappa, \tau) + \frac{1}{\kappa_*}\left[\left(\kappa_*^2 - \tau_*^2\right)B + 2\kappa_*\tau_* C - \kappa_* A\right]$$

satisfies $P[\mathcal{G}](\kappa_*, \tau_*) = 0$, that is, (18) is satisfied with $\mathcal{G}$ replacing $\mathcal{F}$ and all evaluations at (κ_*, τ_*). Notice that $\mathcal{G}$ is just translation of $\mathcal{F}$ by a constant, so it corresponds to adding an appropriate multiple of the arc length to the energy.

We remark that in [7] we also found some general classes of energies for which specific helices were extremal by 'freezing coefficients' in the Euler-Lagrange equations, that is, evaluating those coefficients at (κ_*, τ_*) and then solving the resulting linear differential equations.

7. Results: More Specific Energies Permitting All Cylindrical Helices

We will now begin with the simplest energy densities and work through to more complicated situations. If $\mathcal{F}$ is identically constant, then we are seeking to minimise nothing more than the arc length of the backbone curve and the Euler-Lagrange equations should reduce right down to give only straight lines as solutions. Indeed, all derivatives of $\mathcal{F}$ disappear from (16) and (17); (17) is identically satisfied while (16) becomes $\kappa = 0$. Consequently, the straight line between the endpoints $\mathbf{r}(a)$ and $\mathbf{r}(b)$ is indeed the only solution.

The case of which $\mathcal{F}$ of the form $\mathcal{F}(\kappa)$ permit helical solutions of (16) and (17) was examined in detail in [7]. We neglected the possibility of $\tau \equiv 0$ as this would lead only to planar curves, not helices, but we did consider $\tau \equiv \tau_0$, a nonzero constant and $\kappa = \kappa_0$. Equation (17) becomes

$$\frac{d}{ds}\mathcal{F}' = \mathcal{F}''(\kappa)\dot{\kappa} = 0$$

which is definitely satisfied since $\kappa = \kappa_0$ is constant. The other solution of this equation, that we instead prescribe $\mathcal{F}(\kappa) = \alpha + \beta\,\kappa$, was considered in [5]. However, here we reiterate that since we are already considering cylindrical helices, (17) is necessarily satisfied. We also require that (16) is satisfied and for cylindrical helices this reads

$$\left(\kappa_0^2 - \tau_0^2\right)\mathcal{F}'(\kappa_0) - \kappa_0\,\mathcal{F}(\kappa_0) = 0. \tag{19}$$

This is a condition just at one point κ_0 of the argument of $\mathcal{F}$. If, as in Section 6. we wanted (19) satisfied just for a particular κ_0, we could add an additional arc length term to $\mathcal{F}$ to obtain a similar energy for which (19) is satisfied. However, in this section we are more interested in which energies will permit *all* cylindrical helices as solutions to the Euler-Lagrange equations, so we need to consider (19) as an ordinary differential equation for unknown function $\mathcal{F} = \mathcal{F}(\kappa)$, with parameter τ_0. This differential equation has solution

$$\mathcal{F}(\kappa) = C_0\sqrt{\kappa^2 - \tau_0^2}$$

for arbitrary constant C_0 and this corresponds to the energy

$$E[\mathbf{r}] = C_0 \int \sqrt{\kappa^2 - \tau_0^2}\,d\mathcal{L}.$$

In summary, choosing a constant torsion τ_0, for example, due to other modelling requirements, the above energy has as 'critical curves' all cylindrical helices with torsion τ_0.

In [7] we also considered the possibility of generalised helices, not necessarily cylindrical, as solutions to (16) and (17). In other words, for which energies $\mathcal{F} = \mathcal{F}(\kappa)$ are generalised helices, with $\frac{\tau}{\kappa}$ identically constant, solutions of (16) and (17)? In view of (17), the energy density must necessarily have the form

$$\mathcal{F}(\kappa) = \pm\sqrt{\kappa} + C_0,$$

and then (16) simplifies as in [7]. By direct substitution of $\kappa \equiv \frac{\kappa_0}{s}$ and $\tau \equiv \frac{\tau_0}{s}$ one can check then that the resulting Euler-Lagrange equations are never satisfied by conical helices. However some more 'exotic' generalised helices do satisfy the Euler-Lagrange equations and in the particular case

$$\mathcal{F}(\kappa) = \sqrt{\kappa}$$

as described in [7], the projection of such a helix onto a plane perpendicular to its axis is a catenary.

Now we consider the more general setting of allowing $\mathcal{F}$ to depend also on τ, that is, $\mathcal{F} = \mathcal{F}(\kappa, \tau)$. We considered the case of one particular circular helix satisfying the Euler-Lagrange equations (14) and (15) at the beginning of Section 6.. Now instead we suppose all cylindrical helices satisfy (14) and (15). Again (14) is identically satisfied since a circular helix has constant curvature κ_0 and constant torsion τ_0. The other equation, (15) reduces again to (18), but now instead we want this equation to be satisfied for any choice of κ_0 and τ_0, so we must solve this equation as a first order linear partial differential equation. Its solution is

$$\mathcal{F}(\kappa, \tau) = \psi\left(\frac{\kappa^2 + \tau^2}{\tau}\right)\sqrt{1 \pm \sqrt{1 - \frac{4\,\kappa\,\tau}{\left(\kappa^2 + \tau^2\right)^2}}},$$

for arbitrary differentiable function ψ of one variable. This corresponds to the energy functional

$$E[\mathbf{r}] = \int_C \psi\left(\frac{\kappa^2+\tau^2}{\tau}\right)\sqrt{1 \pm \sqrt{1 - \frac{4\,\kappa\,\tau}{(\kappa^2+\tau^2)^2}}}\,d\mathcal{L}.$$

Another class of energies for which all cylindrical helices were solutions of the corresponding Euler-Lagrange equations was found in [7] where we considered $\mathcal{F} = \mathcal{F}(\kappa, \dot{\kappa})$. For the Euler-Lagrange equations to be satisfied for all cylindrical helices required, with $\tau = \tau_0$ a parameter,

$$\frac{\partial \mathcal{F}}{\partial \kappa}(\kappa, \dot{\kappa}) = \frac{\kappa}{\kappa^2 - \tau_0^2}\,\mathcal{F}(\kappa, \dot{\kappa}).$$

This very simple first order partial differential equation has general solution

$$\mathcal{F}(\kappa, \dot{\kappa}) = \psi(\dot{\kappa})\sqrt{\kappa^2 - \tau_0^2},$$

for arbitrary differentiable function ψ of one variable. The corresponding class of energies is

$$E[\mathbf{r}] = \int_C \psi(\dot{\kappa})\sqrt{\kappa^2 - \tau_0^2}\,d\mathcal{L}.$$

In [9], we considered the simple linear energy density $\mathcal{F}(\kappa, \tau) = \alpha + \beta\kappa + \gamma\tau$, for arbitrary constants α, β and γ. This is a generalisation of the energy density $\alpha + \beta\kappa$ considered in [5]. The Euler-Lagrange equations (14) and (15) reduce to

$$-\alpha\,\kappa + \gamma\,\kappa\,\tau - \tau^2\beta = 0 \text{ and } \beta\,\dot{\tau} - \gamma\,\dot{\kappa} = 0.$$

All solutions to these equations have the form

$$\kappa = \frac{-C_0^2\,\beta}{\alpha + C_0\,\gamma}, \tau = \frac{\alpha\,C_0}{\alpha + C_0\,\gamma},$$

for arbitrary constant C_0. Since both curvature and torsion are constant, we recognise these solutions as cylindrical helices. This means that, taking energy density $\mathcal{F}(\kappa, \tau) = \alpha + \beta\,\kappa + \gamma\,\tau$ for fixed constants α, β and γ, we obtain a family of cylindrical helices as solutions of the Euler-Lagrange equations, indexed by parameter C_0.

In [9] we additionally constructed energies which yield conical helices as extremal curves. Such energies take the form

$$\mathcal{F}(\kappa, \tau) = f_0(\xi) + f_1(\xi)\sin(\beta \ln \kappa) + f_2(\xi)\cos(\beta \ln \kappa) + \kappa f_3(\xi),$$

for arbitrary differentiable functions f_i, $i = 0, \ldots, 3$ of the variable $\xi = \frac{\tau}{\kappa}$ with parameter β. Observe that if we take $f_0 \equiv \alpha$, $f_3 \equiv \beta$ and $f_1 \equiv 0$, $f_2 \equiv 0$, we recover the energy density $\mathcal{F} = \alpha + \beta\,\kappa$ which also gave rise to cylindrical helices, as in Section 7. and in [5].

8. An Extension

There are many ways we might wish to generalise the energy density function, to incorporate variability in the ambient space (fluid) and/or obstacles, side-chains off the protein backbone, behaviour of different backbone constituents, and so on. As a first approximation to variable behaviour along the backbone, we may wish to consider adding in a function of the arc length of the backbone curve.[1] We will consider

$$I[\mathbf{r}] = \int_a^b \rho(s)\,\mathcal{F}(\kappa,\tau)\,d\mathcal{L}.$$

The function ρ could be considered analogous to a density along a rod. This could be a smooth function, or it could be discrete valued, for example, a rod which changes composition from uniform plastic to uniform steel at some point. However, in our context ρ might instead represent a penalty restricting the tightness of helical weaving due, for example, to complex side-chains in certain locations. It is reasonable to assume in this setting that ρ is a smooth function of s.

It is important to note that since s is the arc length parameter of an extremal curve, the varied energy has the form

$$I[\underline{\mathbf{r}}] = \int_a^b \rho(\underline{s})\,\mathcal{F}(\underline{\kappa},\underline{\tau})\,|\underline{\mathbf{r}}'(s)|\,ds,$$

where

$$\underline{s}(s) = \int_a^s |\underline{\mathbf{r}}'(t)|\,dt \tag{20}$$

since $\underline{s}$ is the arc length along the varied curves given by position vector $\underline{\mathbf{r}}(s)$. The presence of the integral in the argument of ρ causes additional terms to arise in derivation of the corresponding Euler-Lagrange equations. In fact, the equation corresponding to variation in the binormal direction, analogous to (15) is identical, but, an extra term does arise in the equation corresponding to the normal direction, analogous to (14). This becomes an integro-differential equation. Using all the variations as derived, for example, in the appendix of [9], together with the variation of (20), and an additional integration by parts, we find upon application of the fundamental lemma of the calculus of variations the Euler-Lagrange equation

$$\begin{aligned} 0 = \kappa \int_b^s \rho'(t)\,\mathcal{F}(\kappa(t),\tau(t))\,dt + \frac{d^2}{ds^2}\left[\rho\left(\frac{\partial \mathcal{F}}{\partial \kappa} + \frac{2\,\tau}{\kappa}\frac{\partial \mathcal{F}}{\partial \tau}\right)\right] \\ -\frac{d}{ds}\left[\rho\left(\frac{3\,\dot{\tau}}{\kappa} - \frac{2\,\tau\,\dot{\kappa}}{\kappa^2}\right)\frac{\partial \mathcal{F}}{\partial \tau}\right] + \rho\left(2\,\kappa\,\tau + \frac{\ddot{\tau}}{\kappa} - \frac{\dot{\kappa}\,\dot{\tau}}{\kappa^2}\right)\frac{\partial \mathcal{F}}{\partial \tau} + \rho\left(\kappa^2 - \tau^2\right)\frac{\partial \mathcal{F}}{\partial \kappa} - \rho\,\kappa\,\mathcal{F} \end{aligned}$$

Observe that if we set $\rho \equiv 1$ above, the equation looks slightly different to (14), but this is just because we have not expanded out the $\frac{d}{ds}$ factors.

[1] The author would like to thank Professor Steven Roberts for this suggestion.

If we now, for example, seek circular helix solutions $\kappa \equiv \kappa_0$, $\tau \equiv \tau_0$, the unchanged Euler-Lagrange equation is again identically satisfied, while the new one becomes

$$\left[\frac{\partial \mathcal{F}}{\partial \kappa}(\kappa_0, \tau_0) + \frac{2\,\tau_0}{\kappa_0}\frac{\partial \mathcal{F}}{\partial \tau}(\kappa_0, \tau_0)\right]\rho''(s) + \left[2\,\tau_0\,\kappa_0\frac{\partial \mathcal{F}}{\partial \tau}(\kappa_0, \tau_0) + \left(\kappa_0^2 - \tau_0^2\right)\frac{\partial \mathcal{F}}{\partial \kappa}(\kappa_0, \tau_0)\right]\rho(s) = \kappa_0\,\mathcal{F}(\kappa_0, \tau_0)\,\rho(b)\,.$$

This standard second-order, constant-coefficient ordinary differential equation is a generalisation of (18) (to which it reduces if $\rho \equiv 1$). Solutions to this equation are trigonometric, hyperbolic, linear and constant functions, depending on the coefficients.

An outcome of this is that such densities $\rho(s)$ permit circular helices but could better approximate non-uniform ambient fluid and side chains than $\rho \equiv$ constant. As examples, an exponential decay density (hyperbolic type solution) could model the affect of a local obstacle, while periodic densities (trigonometric) might be more relevant for a nonuniformity parallel to the axis of the helix.

9. Conclusion

In this chapter we have introduced the idea of modelling protein structure, particularly regular helical components of the backbone, using mathematical energies which depend on geometric properties of the backbone curve. To model the protein backbone in this way, we introduced the necessary mathematical concepts and tools from geometry of space-curves. We further outlined the classical applied mathematics topic of the calculus of variations which is the fundamental approach for seeking minimising conditions on our model energies. Next we combined the ideas of geometry of space-curves and the calculus of variations to seek minimising curves amongst families of curves. As a result we presented Euler-Lagrange equations which extremals of our energies must satisfy. By investigating the simplification of these equations in several settings we wrote down general conditions for which specific, or all, cylindrical helices were solutions. We also described some energies which permit conical helices that are also observed protein structures. As an extension and a first step toward incorporating more complex protein behaviour, we added a 'density' term to our energy functional, presenting the extended Euler-Lagrange equation in a special case and exploring the requirements on such energies for circular helix solutions.

References

[1] J. R. Banavar, A. Maritan, Colloquium: Geometrical approach to protein folding: a tube picture, *Rev. Mod. Phys.* **75** (2003), 23–34.

[2] C. Branden, J. Tooze, *Introduction to Protein Structure,* 2nd ed., Garland, New York, 1999.

[3] P. Bruscolini, A coarse-grained, "realistic" model for protein folding, *J. Chem. Phys.* **107** (1997) no. 18, 7512–7529.

[4] T. E. Creighton, Proteins, *Structure and Molecular Properties*, 2nd ed., Freeman, New York, 1993.

[5] A. Feoli, V. V. Nesterenko and G. Scarpetta, Functionals linear in curvature and statistics of helical proteins, *Nucl. Phys. B* **705** (2005), 577–592.

[6] A. R. Fersht, *Structure and mechanism in protein science: A guide to enzyme catalysis and protein folding*, Freeman, New York, 1988.

[7] J. A. McCoy, Helices for mathematical modelling of proteins, nucleic acids and polymers, *J. Math. Anal. Appl.* **347** (2008), 255–265.

[8] C. Micheletti, F. Seno, A. Maritan, J. R. Banavar, Strategies for protein folding and design, *Annals of Combinatorics* **3** (1999), 431–450.

[9] N. Thamwattana, J. A. McCoy, J. M. Hill, Energy density functions for protein structures, *Q. Jl Mech. Appl. Math.* **61** no. 3 (2008), 431–451.

[10] S. Zhang, X. Zuo, M. Xia, S. Zhao, E. Zhang, General equilibrium shape equations of polymer chains, *Phys. Rev. E* **70** (2004), 051902.

In: Protein Structure
Editors: Lauren M. Haggerty, pp. 123-153

ISBN: 978-1-61209-656-8

Chapter 8

Structural Changes in Blood Plasma Lipoproteins in the Physiological Temperature Range and Their Interaction with Steroid Hormones

L. E. Panin, V. G. Kunitsyn and L. M. Polyakov
Scientific Research Institute of Biochemistry, Siberian Branch of Russian Academy of Medical Sciences, Novosibirsk, Russia

Abstract

Temperature dependencies of viscosity (η) and conductivity (σ) of blood lipoproteins and A-I apolipoprotein was studied. It has revealed the presence of the anomalous region at temperature 35–38 ± 0.5°C (T_c). Transition heat is very low in human HDL, VLDL and apoA-I, and in LDL it is higher by a factor of 4-5. Some mechanisms of the cortisol interaction with HDL and apoA-I have been studied by infrared (IR) spectroscopy and conductometry. It was shown that the hormone destroy the initial structure of lipoproteins due to formation of new hydrogen bonds and hidrofobic interaction. The hormone has been found to strengthen the tangle → α-helixes and tangle → β-structures transitions and increase the ordering of lipids. Therewith, ΔH_{trans}. rises markedly (13 and more times), and at the same time the anomalous region is shifted by 1-2°C in apoA-I. The anomalous changes of viscosity and conductivity in the physiological temperature range for all lipoprotein fractions and apoA-I seems to be due to the structural phase transition in both proteins and lipids. The structural transition into apoA-I is likely to contain the elements of phase transition of the first and second types. In human and rat VLDL, the smectic → cholesteric phase transition seems to occur. Enthalpy of viscous flow and structural transition in VLDL is higher for rats than for human.

1. Introduction

There are numerous physicochemical factors able to affect the structures of simple and conjugate proteins, temperature being one of them. Of special interest is the appearance of structural transitions in lipoproteins in the organism temperature range [1].

Structural phase transition in human low density lipoproteins (LDL) in the physiological temperature range has been described previously. It is related to the change not only in the lipid phase but in the secondary structure of apolipoprotein B as well [2]. In rabbit intact VLDL, the transition in the region of 40-41°C has been also detected by means of differential scanning calorimetry [3]. *T*-transition has been recorded in LDL by employing nuclear magnetic resonance (NMR) spectroscopy, but at a lower temperature and in a very wide temperature range (ΔT) [4]. The listed studies considered mainly the temperature transitions in the lipid moiety of the lipoprotein.

Works reporting on changes in the LDL and apoB secondary structure in the physiological temperature range are very few [5, 6]. We suggest that phase transition does occur also in lipoproteins in both lipids and proteins [7]. Of interest is the effect of steroid hormones on T-transition in lipoproteins. This is due to the fact that the complex of glucocorticoid tetrahydroderivatives apolipoprotein A-I is involved in regulation of gene expression [8, 9].

Lipoproteins, like biological membranes, exhibit the properties of liquid crystals, to which viscosimetry and conductometry are widely applied. We were the first to study the structural transitions in serum lipoproteins by viscosimetry and conductometry. Advantages of these methods are their high sensitivity to structural transformations, retention of nativity of the object under study, and their accessibility. EPR-spectroscopy and fluorimetry require to insert into the object the lipophilic probes and spin labels in rather high concentrations (10^{-4} - 10^{-5} M). These actions disrupt the structure of the object under study [3, 10].

Taking into consideration these circumstances, we set three tasks in our study:

1. to examine the changes of fluorescence and infrared (IR) spectra of human and rat lipoproteins after their incubation with cortisol;
2. to compare the temperature dependence of viscosity and conductivity of blood lipoproteins and apoA-I in human and rat;
3. to analyze theoretically the nature of anomalous changes of these parameters in the physiological temperature range.

2. Materials and Methods

Preparative Techniques

Lipoproteins were isolated from blood serum of human and Vistar line male rat by the method of preparative ultracentrifuging in KBr solution [11]. Centrifuging was carried out in a "Optima L-90K" Ultracentrifuge (Beckman-Coulter, USA), 75-Ti rotor, for 18-20 h at 140,000g. The first fraction comprised VLDL (d < 1.006 g/cm^3), the second one, LDL (d = 1.006 - 1.063 g/cm^3), and the third, HDL (d = 1.063 - 1.21 g/cm^3). Dialysis of lipoproteins after centrifuging was run at 4°C for 48 h against the 0.015 M K^+,Na^+-phosphate buffer, pH 7.35. Delipidation was conducted with a cooled chloroform-methanol mixture (1:1); 20 ml of the mixture were taken per 1 ml of lipoprotein solution, with further repeated washing of the protein precipitate with ethyl ester [11].

Chromatographic Methods

A mixture of apolipoproteins was deposited over the column (1.6 × 100 cm) with CL-6B Sepharose (Pharmacia, Sweden) and eluted with 0.01 M Tris/HCl buffer, pH 8.6, containing 6 M urea, 0.01% sodium azide, 1 mM phenylmethanesulfonilfluoride. Flow rate was 10 mL/h, automatic recorder rate was 6 mm/h. Elution profile was recorded from an ultraviolet (UV) detector 2151 LKB (Sweden). Ion-exchange chromatography was performed with a DEAE-Toyopearl 650 M TSK anion exchanger (Japan) balanced by 0.01 M Tris/HCl buffer, pH 8.6, containing 6 M urea. To produce apoA-I, elution was conducted by the initial buffer with NaCl linear gradient (0 - 0.5 M).

Electrophoresis was conducted in polyacrylamide gel with sodium dodecylsulphate, in Tris/glycine buffer, pH 8.3, at room temperature. The amperage was 10 mA; 3% concentrating and 12.5% separating gels, containing 0.1% sodium dodecylsulfate were used to separate the apolipoprotein mixture. Protein solutions were inserted into the cells of gelby 10-20 μL. Protein bands were visualized with 0.1% Kumassi G-250 in the mixture of methanol and 10% water acetic acid (1:1). Molecular marker-proteins were used as tags: phosphorylase, 94 kDa; albumin, 67 kDa; ovalbumin, 43 kDa; carbonic anhydrase, 30 kDa; trypsin inhibitor, 20.1 kDa; and lactalbumin, 14.4 kDa (Pharmacia, Sweden). According to the electrophoresis data, purified apoA-I contained no other protein admixtures.

Viscosity Measurement

0.02 M K^+,Na^+-phosphate buffer and freshly prepared blood serum lipoproteins were used. Protein content in lipoprotein solution was measured by the method described in [12]. Viscosity of lipoproteins and apoA-I was measured by an automatic device allowing to determine the time of capillary efflux accurate to 0.01 s [13]. Height of a detector was fitted according to the nature of a fluid under study, adjusting the shear stress. In our case, the detector height was 20 mm. At the capillary diameter of 0.56 mm the shear rate was 2.5 s^{-1}. For reliable recording of anomalous region in lipoprotein solutions, their relative viscosity was within 1.2-2 sP. Relative (dynamic) viscosity was calculated by the formula [14]:

$$\eta = \eta_0 \cdot \frac{\rho_x t_x}{\rho_0 t_0},$$

where η_0 is the water viscosity corrected for temperature, ρ_0 is the density, t_0 is the time of water efflux, t_x is the efflux time, and ρ_x is the density of the lipoprotein solution under study.

Viscous flow enthalpy was calculated from the formula [15]:

$$\lg \eta = \lg A' + \frac{\Delta H \cdot R}{2.3 \cdot T},$$

where A ´ is $(hN_A / V)e^{-\Delta S^* / R}$, N_A is the Avogadro constant, h is the Planck constant, V is the molar volume, ΔS^* is the temperature-independent activation entropy, ΔH is the activation heat content (enthalpy), R is the universal gas constant, and T is the absolute temperature. In HDL solutions the protein content was 4 mg/ml, and in LDL and VLDL 1.0 and 0.5 mg/mL, respectively. Concentration of apoA-I was 0.5 mg/mL. The sample volume for viscosity measurements was 0.3 ml. Concentration of H^+ ions was measured with an OP-211/1 Radelkis pH-meter (Hungary) with an accuracy of ±0.05 pH units. Samples were controlled thermostatically at T = 33-41°C at 1°C intervals. Incubation time for each temperature was 5 min.

Conductivity Measurement

Facility used for conductivity measurements has been described previously [14]. Conductometry regime: (1) the cuvette voltage, 2 V; (2) the alternating current frequency, 2000 Hz; (3) the cuvette length, 1.8 cm, and volume of fluid poured into the cuvette, 0.7 ml; and (4) the cuvette contents were thermostated with a U-4 Ultrathermostat (Germany) in the range from 33 to 41°C in a degree interval, incubation time being 5 min. The concentration of hydrogen ions was determined by a OP-211/1 Radelkis pH-meter (Hungary) to an accuracy of ±0.05. The activation energy was calculated according to [16]:

$$\sigma = \sigma_0 e^{-E / k_B T},$$

where σ_0 is the constant depending on the type of the substance, k_B is the Boltzmann constant, E is the activation energy, and T is the absolute temperature. The results were processed statistically; the confidence intervals on η(T) and σ(T) curves were within the limits of the label sizes, $P \geq 0.95$.

Infrared Spectroscopy

Samples for IR spectra recording were prepared as films on a fluorite support. Size of the optical aperture 8 mm in diameter was bounded by a fluoroplastic plate. Lipoprotein samples were deposited uniformly on the support over the whole aperture section in a 50-µL volume. Both of these samples (control lipoprotein and test lipoprotein + cortisol) were simultaneously put into the vacuum case with the initial temperature 4 ± 1°C. Vacuum was brought to 0.1-0.2 atm (104 n/m^2) with a vacuum pump and then dried at this pressure for 80 min. In a deeper vacuum, lipoproteins are partially entrapped by water vapor, and sublimation occurs. After drying, the homogeneous film was obtained over the whole aperture section. Spectra were recorded with an IR Specord-M 80 spectrometer (Germany). Spectrometer with the samples was blown with dry air for 30 min, then the scanning regime was switched on, and spectra of the samples were recorded with the dry air feed not turned off. Repeated recording of differential spectra of the samples (control-control) showed the zero absorption. In our opinion, drying of lipoproteins, i.e., the lipid-containing structures, slightly changes the protein conformation, since only the free water is separated, and the molecules of bonded H_2O do remain and determine the conformation of proteins and lipids. In our samples, the protein structure is stabilized largely through the interaction with lipids [17]. According to IR spectra we obtained, addition of dehydrating agents (methanol, ethanol) to the samples even in low amounts (10^{-6} – 10^{-7}M) significantly affected the disordering of the protein secondary structure. This fact was taken into account in our study, and in the test we added cortisol in methanol (C_0 = 0.01 M) preliminarily diluted with K^+,Na^+-phosphate buffer by 3 and 7 orders of magnitude (4.15×10^{-6}; 1.6×10^{-9}), and in the control added an equivalent methanol amount into the mixture with the same buffer. The Merck reagents were used in the study.

3. Results

Analysis of the Interaction between Lipoproteins and Steroid Hormones

The interaction of steroid hormones with individual classis of lipoproteins and apoproteins was studied by ultracentrifugation, gel-filtration, equilibrium dialysis, and fluorescence spectroscopy. The equilibrium association constant (K_a) for cortisol, corticosterone, testosterone and progesterone was determined by [18]. The distribution of corticosteroids among individual Lp fractions was estimated upon addition of the [^{3}H]-labeled hormones to rats blood serum. Serum was incubated with cortisol, corticosterone, pregnenolone and desoxycorticosterone (DOC) for 30 min at 37°C. This was followed by preparative isolation of Lp by "Optima 90K" Ultracentrifuge (Beckman-Coulter, USA). The obtained VLDL, LDL, HDL and infranatant were assessed for radioactivity content using a β-counter Mark-III (USA). It was shown that the blood serum lipoproteins bind steroid hormones and can serve as their active carriers (Table 1).

Table 1. Distribution of labeled corticosteroids between blood serum Lp during in vivo and in vitro testing (% of the total radioactivity)

Steroids	VLDL	LDL	HDL	Infranatant
In vitro				
Pregnenolone	18 ± 1.1	12 ± 0.5	8 ± 0.2	61 ± 2
DOC	14 ± 0.9	8 ± 0.3	5 ± 0.2	73 ± 7
Cortisol	18 ± 1.1	14 ± 2.5	9 ± 0.2	58 ± 5
Corticosterone	16 ± 0.4	11 ± 2.5	7 ± 0.2	66 ± 4
In vivo				
Corticosterone	7 ± 0.6	4 ± 0.6	4 ± 0.1	85 ± 6
Cortisol	12 ± 0.2	3 ± 0.6	3 ± 0.1	82 ± 5

Note: In each group, 4 experiments were performed.

As shown by intraperitoneal injection of tritium-labeled glucocorticoids in the rats, 15% of corticosterone and 18% of cortisol became bound to blood serum Lp within 30 min. In vitro experiments gave even a higher percentage of steroids bound to Lp fractions. Main part of the label was present in the protein fraction of infranatant.

In rats corticosterone is the main glucocorticoid hormone, their blood cortisol content being less than 1 μg/dL. In this connection, it was interesting to examine the cortisol distribution among rat blood Lp. Distribution of the tritium-labeled hormone after its intraperitoneal injection in rats was similar to that of corticosterone. The only distinction was a higher content of label in the VLDL fraction.

It is commonly accepted that a major part of glucocorticoids (up to 90%) in blood serum are bound to proteins, mainly to corticosteroid-binding globulin (transcortin); the other part of hormones remains free and is regarded as 'biologically active'. Transcortin is a glycoprotein with molecular weight of 52 kDa. The protein is synthesized on hepatocyte ribosomes, which is followed by its glycosylation.

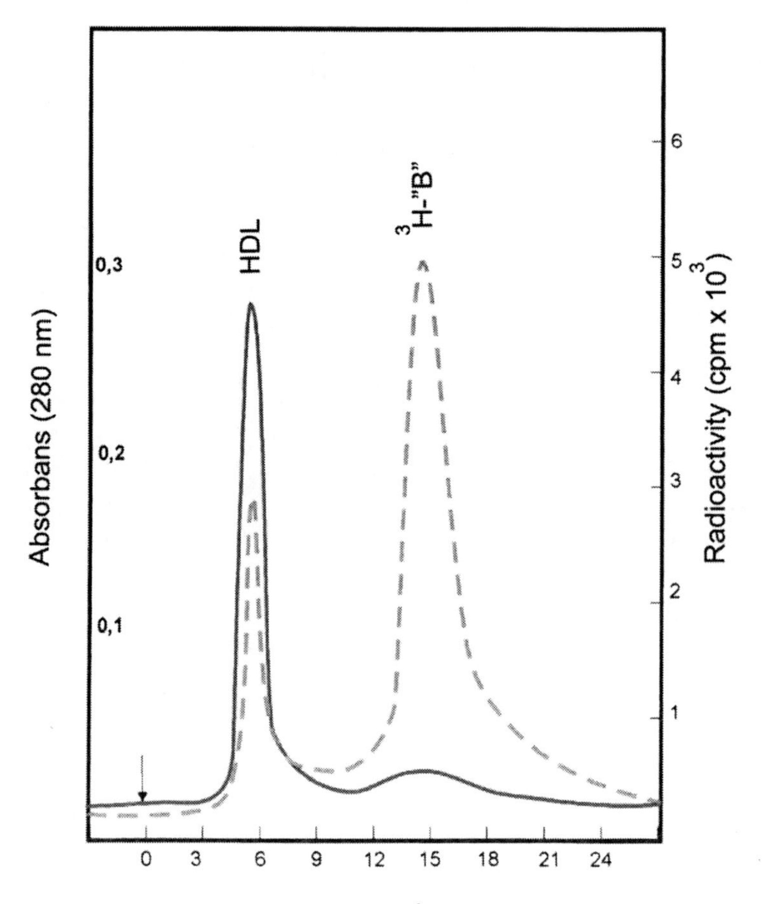

Figure 1. Chromatographic elution profile of HDL and ^{3}H-corticosterone. Protein adsorption - solid line. Distribution of radioactivity - dotted line. Column with Sephadex G-50 (1 x 20 cm). Eluent: 5 mM Tris/acetate, pH 7.4, 0.15 M NaCl.

Since a large part of apolipoproteins is represented by glycoproteins, it was natural to suggest that steroids are bound to Lp via glucosaminoglycan-containing apolipoproteins. On the other hand, a possible binding of steroids by the lipid phase of Lp could not be ruled out. Besides, we took into account that infranatant contains a substantial part of non-structural apolipoproteins. This is the so-called 'free pool' of apoLp, which can also bind steroid hormones.

The possibility of steroid transport by Lp was verified chromatographically. Figure 1 shows the elution profile of HDL fraction and radioactivity. The equal yields of labeled corticosterone and HDL indicate a possible formation of the hormone-Lp particle complex.

To compare different classes of Lp with respect to their binding ability toward steroid hormones, we used also the equilibrium dialysis, which is regarded as the most robust method for studying the processes of complex formation in a receptor-ligand system under thermodynamic equilibrium.

For equilibrium dialysis 0.5 ml of Lp (0.1-0.3 mg protein/ml) was placed in each dialysis tubing (8/32, Serva, Germany). Dialysis was performed for 48-72 h at 4°C against a 5 mM Tris/acetate buffer (pH 7.4), 0.15 M NaCl, containing tritium-labeled cortisol or corticosterone and the growing concentrations of a non-labeled ligand. Non-specific sorption of dialysis membrane measured in the control experiment was 11%.

The corticosterone saturation curve for the rat HDL fraction list the complex-forming process as a function of concentration of excessive non-labeled ligand introduced in the external buffer solution. This is a saturating curve that points to an increase in the number of HDL-corticosterone complexes during the saturation. Based on the calculation of binding parameters in Scatchard coordinates. Table 2 lists the association constant (K_a) for different Lp classes obtained by equilibrium dialysis.

Table 2. Equilibrium association constant (K_a) for lipoproteins interaction with corticosterone and cortisol in Scatchard coordinates

Lp fractions	Corticosterone	Cortisol
VLDL	$(8 \pm 0.1) \times 10^6$ M^{-1}	$(5 \pm 0.1) \times 10^6$ M^{-1}
LDL	$(1 \pm 0.04) \times 10^6$ M^{-1}	$(4 \pm 0.1) \times 10^6$ M^{-1}
HDL	$(9 \pm 0.2) \times 10^6$ M^{-1}	$(8 \pm 0.1) \times 10^6$ M^{-1}

Note. The values are averaged over 5 measurements. The corticosterone experiments were performed with rat serum Lp; the cortisol experiments, with human serum Lp.

High specificity and sensitivity of fluorescence methods underlie their wide application for analysis of the protein-ligand interaction. It is known that of three fluorescent amino acids (tryptophan, tyrosine, phenylalanine) only the tryptophan provides ample information on the structure of macromolecule. Tryptophan is regarded as a natural 'reporter label'; so parameters of its fluorescence give an insight into its microenvironment and transformations caused by certain conditions, in particular, by the formation of transcortin-steroid complexes. Besides, the tyrosine fluorescence (with excitation wavelength of 280 nm) is suppressed by polar groups (CO, NH_2) and adjacent peptide bonds; phenylalanine gives a very low quantum yield.

Examination of fluorescence spectra obtained from different Lp classes showed that they differ both in the form and in characteristic wavelengths of the maxima. HDL had a maximum at 333 nm, the fluorescence maximum of VLDL being in the region of 338 nm. Interestingly, the fluorescence of tryptophan in an aqueous solution reached its maximum at 354 nm. The characteristic short-wave shift of Lp tryptophanyl fluorescence can be attributed to specific features of the microenvironment: polar (aqueous) for tryptophan solution, and nonpolar (lipid) for the protein component of Lp.

The interaction of corticosterone with HDL was accompanied by tryptophan fluorescence quenching (Fig. 2). A decrease in fluorescence intensity was most pronounced for HDL (55-65%) as compared to VLDL (30-35%) and LDL (15-20%). Shapes of the spectra and their half-width remained virtually unchanged. In all cases, a 'blue' 1-3 nm shift was observed, which is attributed to local conformational rearrangements of the protein component of Lp upon its interaction with the hormone.

Titration was performed by adding 1 μl aliquots of the hormone to 2 ml of Lp solution in a thermostated cuvette. The concentration of ethanol at the end of experiment was not higher than 0.5%. Fluorescence quenching caused by ethanol did not exceed 8-10% of the total fluorescence decay. The quenching had a saturable nature and attained its maximum only at a certain molar ratio of steroid and protein.

Our study of the time dependence of fluorescence quenching at the introduction of saturating amounts of corticosterone and cortisol gave nearly the same curves for the two hormones. Total saturation of the binding regions of Lp particles with the hormones was observed within 20-30 minutes after the onset of experiment. The fluorescence decay and the recorded short-wave shift of the fluorescence maximum upon formation of steroid-Lp complex may be caused by quenching of tryptophan luminophores and by a relative increase in the contribution of total protein fluorescence due to tyrosine residues.

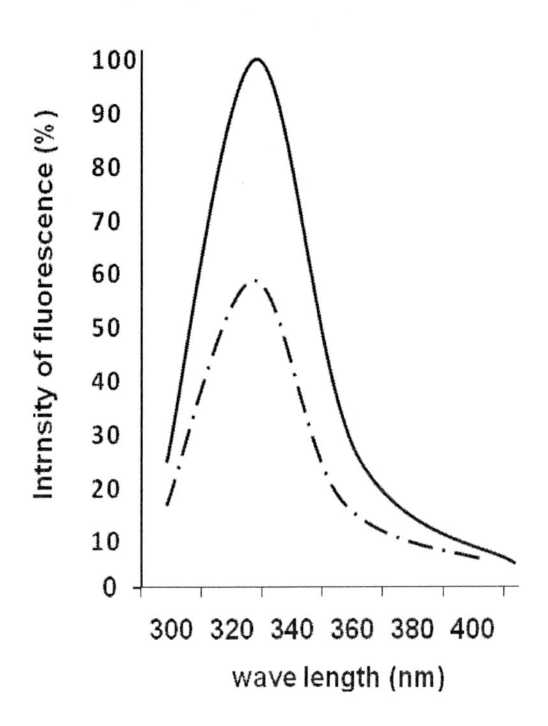

Figure 2. Tryptophan fluorescence quenching of HDL upon its interaction with the corticosterone: — HDL; —·— HDL + hormone.

In the experiments with varying pH, a maximum value of tryptophan fluorescence of apoA-I (0.05 mg/ml) was observed at physiological pH. The addition of corticosterone to apoA-I did not change the spectrum shape, although a decay in tryptophan fluorescence made up nearly 40% of its initial value.

The fluorescence polarization method is highly convenient for the study of Lp binding to steroid hormones. Changes in fluorescence polarization are caused by altering mobility of the entire molecule or its part that contains fluorophores. Natural fluorophores (tyrosine, tryptophan) are present virtually in all proteins. The indole rings of tryptophan residues are remarkably sensitive and complex fluorophores. For some transport proteins, it is known that binding to a ligand increases the fluorescence polarization due to a decrease in fluorophore mobility. Thus, this effect is a sensitive indicator of the complex-forming processes.

In our studies, the extent of fluorescence polarization in HDL fractions was found to increase upon interaction with corticosterone or cortisol (Fig. 3). A segment of fluorescence polarization steadily rising within the first 10 minutes appeared on the recorded curve.

Analysis of the Lp interaction with steroid hormones demonstrated that the most pronounced binding occurs with HDL, where the main structural and functional protein is represented by apoA-I. In this connection, the possibility of corticosterone binding by isolated apoA-I was investigated.

The obtained fluorescence quenching curves were used to calculate the values of equilibrium association constant (K_a) by the method of Attallah and Lata (Table 3). Molecular weights of apoA-I, HDL, LDL and VLDL were taken equal to $2.8x10^4$, $3x10^5$, $1x10^6$ and $5x10^6$ Da, respectively. Equilibrium association constant for the hormones with isolated apoA-I happened to be lower than those for native HDL. This is caused most likely by removal of lipids, which are necessary for stabilization of the protein structure and maintaining a certain conformational interrelations that impart surfactant features to the protein.

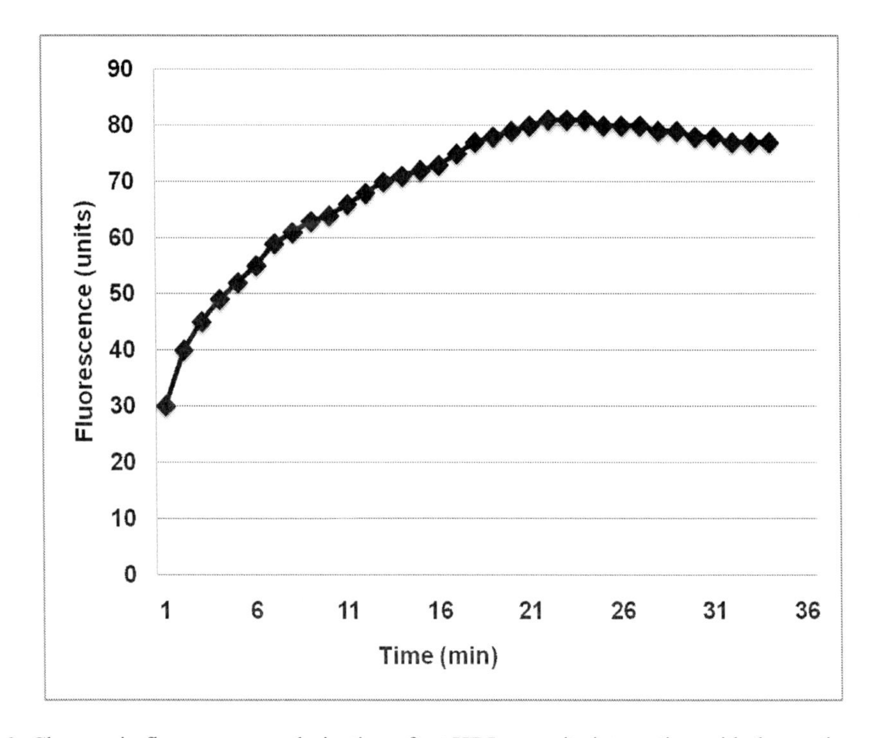

Figure 3. Changes in fluorescence polarization of rat HDL upon its interaction with the corticosterone.

Therefore, ultracentrifugation, chromatography, tryptophan fluorescence quenching and equilibrium dialysis allowed us to demonstrate that the blood serum lipoproteins bind steroid hormones and can serve as their active carriers in the organism. Affinity for steroids was most pronounced in HDL particles. Protein components of Lp were shown to be involved in the formation of Lp-steroid complexes, steroids being adsorbed most likely on the surface of Lp particles. ApoA-I is among the main apolipoproteins that can bind steroid hormones. More fine structural rearrangements of apoLp upon their interaction with steroid hormones were studied by means of IR spectroscopy.

Table 3. Equilibrium association constant (*K*a) for lipoprotein fractions upon interaction with corticosteroids as calculated from fluorescence quenching

Steroids	VLDL	LDL	HDL	ApoA-I
Corticosterone	0.6×10^6 M^{-1}	0.67×10^6 M^{-1}	3.6×10^6 M^{-1}	0.8×10^6 M^{-1}
Cortisol	1.02×10^6 M^{-1}	0.14×10^6 M^{-1}	4.0×10^6 M^{-1}	0.3×10^6 M^{-1}
Testosterone	1.2×10^6 M^{-1}	0.27×10^6 M^{-1}	4.8×10^6 M^{-1}	0.7×10^6 M^{-1}
Progesterone	0.6×10^6 M^{-1}	0.64×10^6 M^{-1}	4.4×10^6 M^{-1}	0.5×10^6 M^{-1}

Note. The values are averaged over 5 measurements.

Effect of Temperature on Lipoprotein Structure

Temperature dependence of viscosity in the rat HDL solutions (0.9% NaCl solution, pH 6.1) in the range of 17-42°C demonstrated the anomalous η change between 22 and 23°C. The subsequent heating of the specimen allowed to reveal one further anomalous region at T = 35-36°C. This region width was ΔT = 2°C (Fig. 4).

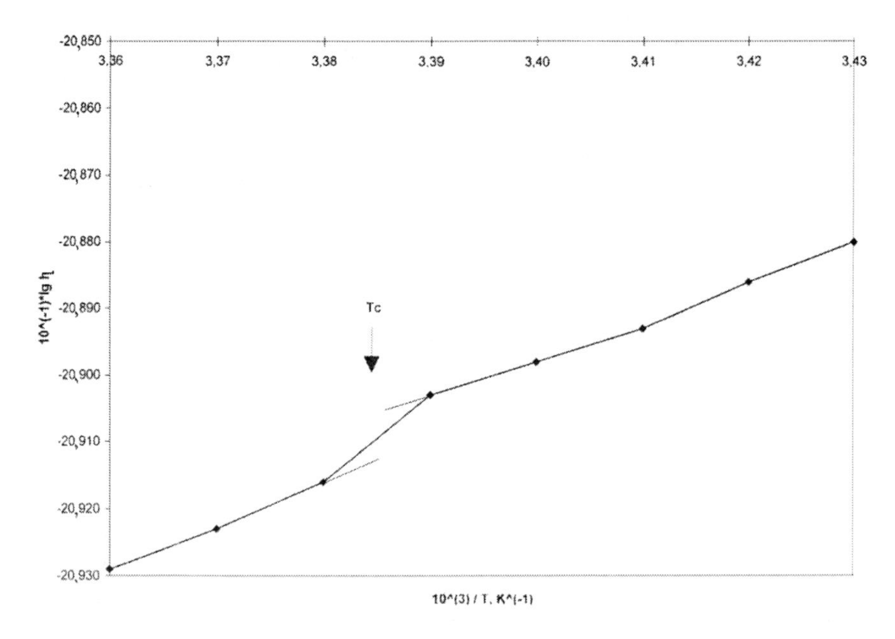

Figure 4. Viscosity of colloidal solutions (kg/m s) in rat lipoproteins measured at a varying temperature: HDL (0.9% NaCl, pH 6.1).

The temperature dependence of HDL viscosity at pH 7.35 in 2 mM K^+,Na^+-phosphate buffer showed the anomalous region to shift higher by 1-2°C. The anomalous region width also comprised 2°C (Fig.5). The value of the viscosity changed in the range of 34-40°C was 5.2%, the viscosity temperature coefficient was 2.93 × 10^{-5} kg/m s deg between 34 and 37°C and 1.07 × 10^{-5} kg/m s deg between 37 and 40°C (Table 4). Study of the temperature dependence of viscosity in NaCl solution and K^+,Na^+-phosphate buffer indicated the absence of anomalies in the range of 17-42°C. Thus, the viscosity temperature coefficients on either side of T_c differed from one another by a factor of 2.75 and depended on variable state of HDL.

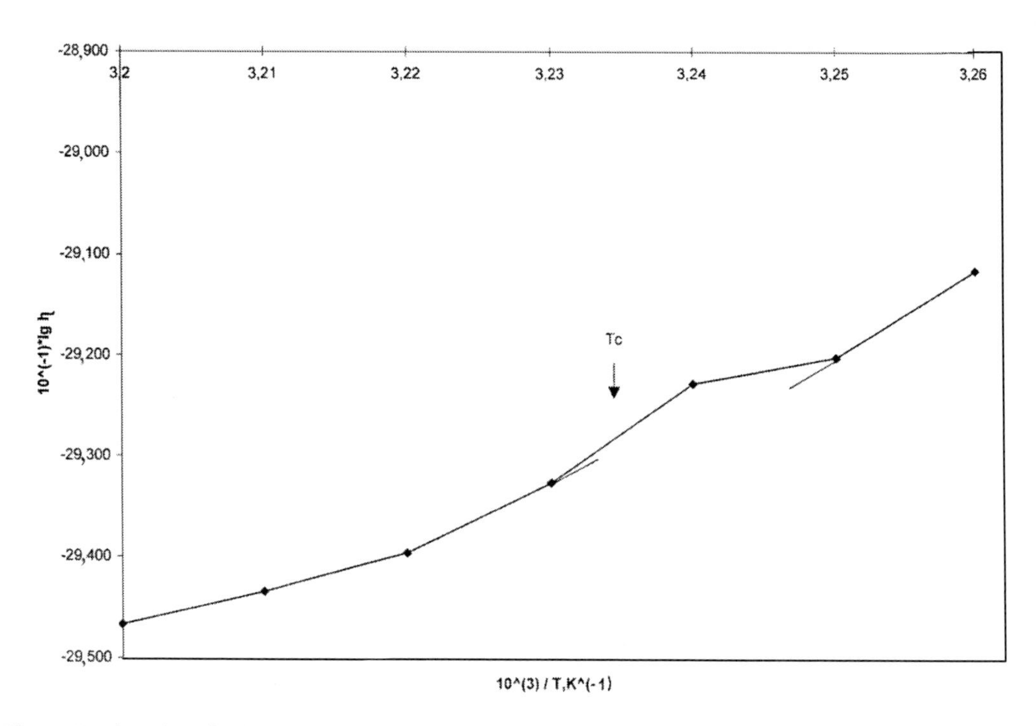

Figure 5. Viscosity of colloidal solutions (kg/m s) in rat lipoproteins measured at a varying temperature: HDL (K^+,Na^+ phosphate buffer, pH 7.35).

The temperature dependence of viscosity in rat HDL apoA-I in the same buffer at pH 7.35 indicated the anomalous area to fall on 36-37°C (Fig. 6). The width of the anomalous area was 2°C.

As in the case of HDL, the shape of $\eta(T)$ function on either side of T_c was different.

The temperature dependence of LDL viscosity is presented in Figure 7. In the case of LDL, the anomalous area is seen to occur at T = 37.5 ± 0.5°C. The intensity of viscosity "splash" is markedly less than that in HDL. The temperature viscosity coefficient was 2.6 × 10^{-5} ± 3 × 10^{-7} kg/m s deg before T_c and 1.0 × 10^{-5} ± 3 × 10^{-7} kg/m s deg after it. These values ratio was equal to 2.6.

The temperature dependence of viscosity in rat and human VLDL showed the anomalous regions to locate, respectively, at 37 and 35-37°C (Fig. 8). The manifestation of anomalous region is less clear than that in HDL and LDL. The viscosity temperature coefficient is 1.71 × 10^{-5} ± 3 × 10^{-7} kg/m s deg before the "inflection" point, and 1.21 × 10^{-5} ± 3 × 10^{-7} kg/m s deg after it (Table 4). The shape of $\eta(T)$ function for LDL and VLDL on either side of anomalous area is different, $\Delta T = 2°C$ (Figs. 7-8).

Kinetic studies of rat HDL viscosity depending on incubation time revealed that at 34°C, η changed only within the next few minutes (Fig. 9). Upon incubation of the sample at 36°C η reliably changed after 25 min thermostating. Finally, incubating the sample at 38°C was not attended by η changing. Clearly the viscosity at 34°C was higher than that at the more elevated temperatures.

However, the viscosity increment between 34 and 36°C was less than between 36 and 38°C, confirming once again the nonmonotonous character of this value changing with the temperature.

Table 4. Temperature coefficient of viscosity in rat lipoproteins attemperatures below and above the transition point T_c (kg/m s deg)

Range of temperature 33-41°C	HDL	LDL	VLDL
$\leq$ Tc	$2.9 \times 10^{-5} \pm 3 \times 10^{-7}$	$2.6 \times 10^{-5} \pm 3 \times 10^{-7}$	$1.7 \times 10^{-5} \pm 3 \times 10^{-7}$
$\geq$ Tc	$1.1 \times 10^{-5} \pm 3 \times 10^{-7}$	$1.0 \times 10^{-5} \pm 3 \times 10^{-7}$	$1.2. \times 10^{-5} \pm 3 \times 10^{-7}$

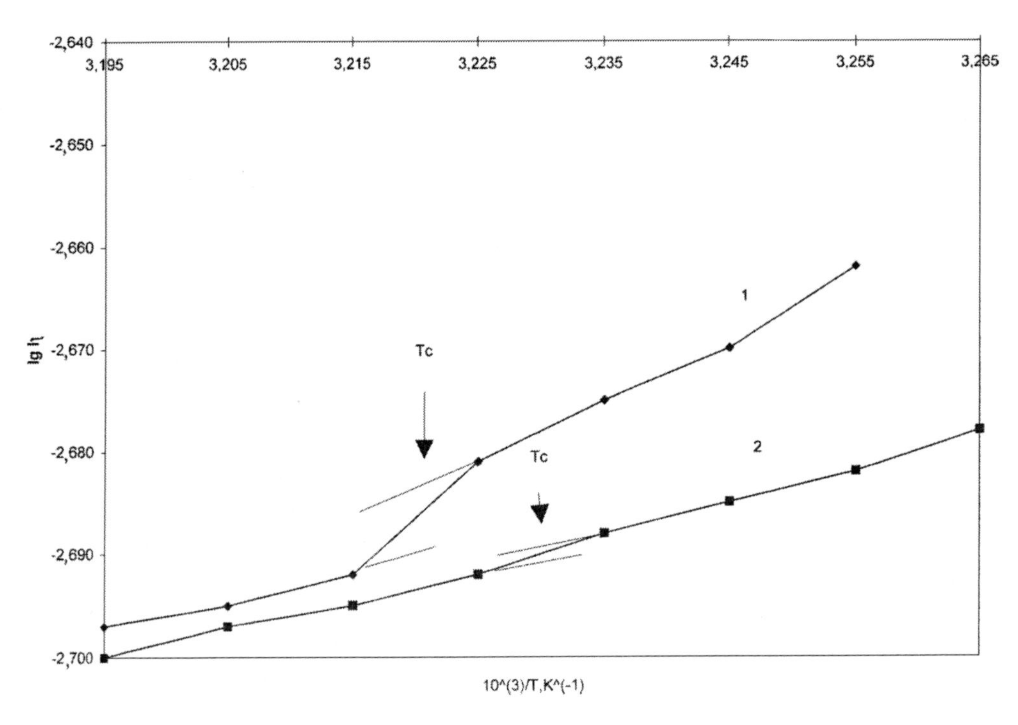

Figure 6. Viscosity of colloidal solutions (kg/m s) in rat lipoproteins measured at a varying temperature: HDL (1) and apoA-I (2).

The kinetic curve variations observed indicated that lipoprotein particles stability changed depending on the temperature and time. Based on the data obtained, it may be concluded that at 34°C lipoproteins are more stable to temperature, probably because their structure is more ordered. At 36°C the HDL structure is stable up to a certain time (the 25-th minute), after which its partial disordering begins, and the viscosity decreases in this case. When 38°C are reached, we obtain a new state of lipoprotein particles structure, less ordered and more hydrated, coming faster to equilibrium with the environment. Kinetic studies of HDL viscosity allow to differentiate their structural stability to temperature.

Calculation of the viscous flow enthalpy for the lipoprotein on either side of T_c and the difference in these values relative to this point (ΔH_{trans}., the conventional heat of transition) demonstrated that with a rather high ionic force of the solution the activation energy of viscous flow was small both before T_c and after it. Of even lower value is ΔH_{trans}. To a larger extent this is revealed in HDL (Table 5).

Table 5. Viscous flow activation heat in rat lipoproteins at temperatures below and above the transition point (T_c) and value of transition enthalpy (ΔH_{trans}.) (kcal/mol)

Range of temperature 33-41°C	HDL (rat)	LDL (rat)	VLDL (rat)	VLDL (human)
$\leq T_c$	1.94 ± 0.03	4.02 ± 0.03	2.87 ± 0.035	0.09 ± 0.01
$\geq T_c$	1.31 ± 0.03	1.24 ± 0.03	2.14 ± 0.035	0.12 ± 0.01
ΔH_{trans}	0.63 ± 0.03	2.78 ± 0.03	0.74 ± 0.035	−0.03 ± 0.01

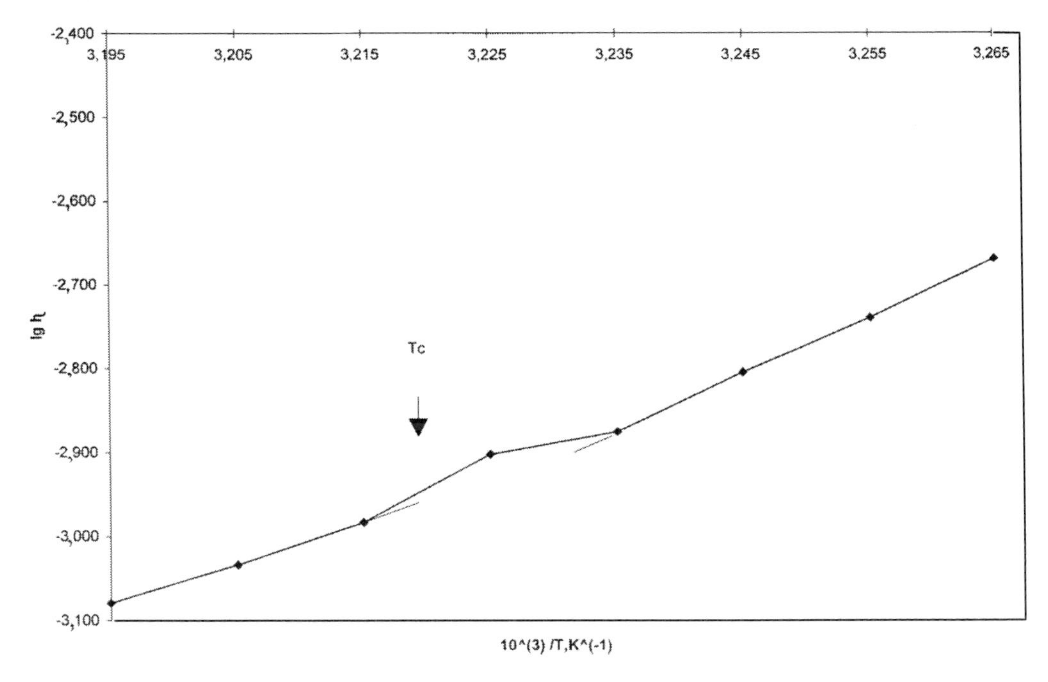

Figure 7. Viscosity of colloidal solutions (kg/m s) in rat lipoproteins measured at a varying temperature: LDL.

Further we studied the temperature dependence of conductivity (σ) in various rat lipoprotein fractions in 0.2 mM K^+,Na^+-phosphate buffer, pH 7.24, in the range from 33 to 41°C. In HDL, the anomalous region was found at 37°C (T_c). The conductivity/temperature coefficients ($\Delta\sigma/\Delta T$) on either side of T_c differed from each other by a factor of 1.5 (Table 6). The integral value of conductivity change in the range from 33 to 40°C was equal to 0.015 (Ω

m)$^{-1}$ (Fig. 10). The anomalous region was detected between 37 and 38°C on the temperature dependence of conductivity for LDL. The manifestation of this region was less than for HDL. The temperature coefficients σ on either side of T_c did not differ essentially. In this case, the integral value $\Delta\sigma$ was 0.019 (Ω m)$^{-1}$. The shape of $\sigma(T)$ function on either side of T_c varied for HDL, but differed little for LDL. The anomalous area width is 2°C (Figs. 10(1), 10(2)). In VLDL, the anomalous region of conductivity changing occurred at 35°C. The integral value $\Delta\sigma$ in the range of 33-40°C was 0.0064 (Ω m)$^{-1}$ (Fig. 10(3)).

Analysis of $\sigma(T)$ dependence for human HDL in 4 mM K^+,Na^+-phosphate buffer at pH 7.24 revealed the anomalous area to occur at T = 36 - 37°C. The temperature coefficient σ on either side of T_c remained practically the same (Table 6). This result differed from the previous one (Table 5). The integral σ changing in the range 34-41°C was 2.85 (Ω m)$^{-1}$ (Fig. 11(1)).

Effect of Cortisol on the Structural Transitions in Lipoproteins

We have noted already that HDL and apoA-I interact with cortisol, which allows them to take in regulation of gene expression [8]. It was of interest to elucidate the effect of this interaction on the protein structure. It was found that the manifestation of anomalous region significantly increased in HDL after their preincubation with cortisol (incubation time was 5 min at 34°C). The integral value $\Delta\sigma$ markedly decreased and was 1.24 (Ω m)$^{-1}$, that is, 2.3 times less as compared to the control. The shape of function on either side of T_c was also different (Fig.11(2)). The temperature coefficient σ below T_c was practically 2 times as high as $\Delta\sigma/\Delta T$ above Tc (Table 6).

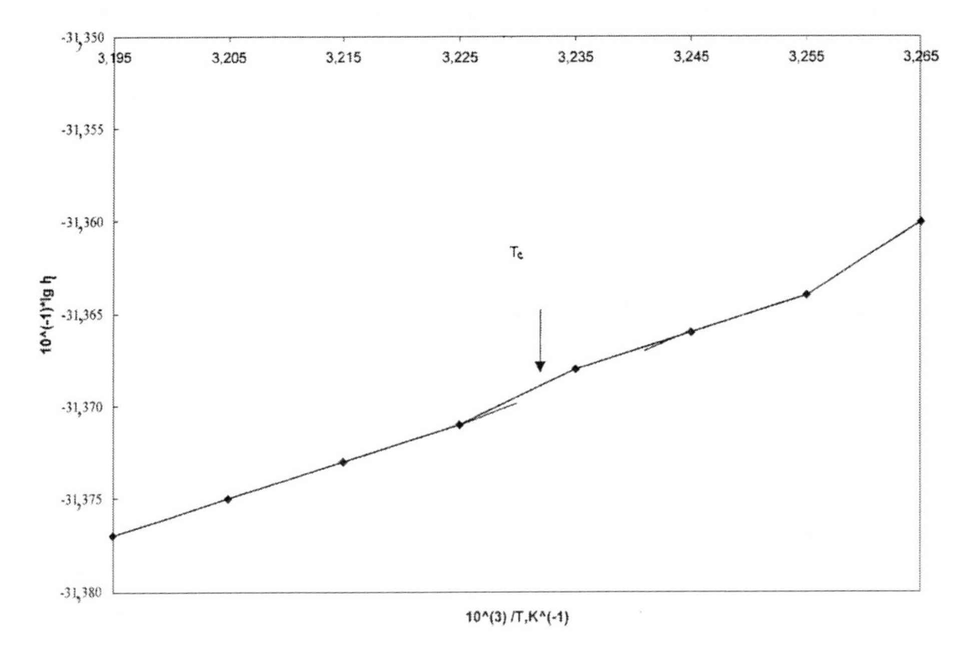

Figure 8. Viscosity of colloidal solutions (kg/m s) in rat lipoproteins measured at a varying temperature: VLDL (20 mM K^+,Na^+-phosphate buffer, pH 7.35).

Table 6. Temperature coefficient of conductivity in rat lipoproteins at temperatures below and above the transition point (T_c), (Ω m deg)$^{-1}$ (K^+,Na^+-phosphate buffer, C = 0.2 mM, pH = 7.24)

	HDL	LDL	VLDL
≤ Tc	$2.11 \pm 0.03 \times 10^{-4}$	$1.51 \pm 0.03 \times 10^{-4}$	$7.73 \pm 0.08 \times 10^{-4}$
≥ Tc	$3.13 \pm 0.03 \times 10^{-4}$	$1.66 \pm 0.03 \times 10^{-4}$	$7.11 \pm 0.08 \times 10^{-4}$

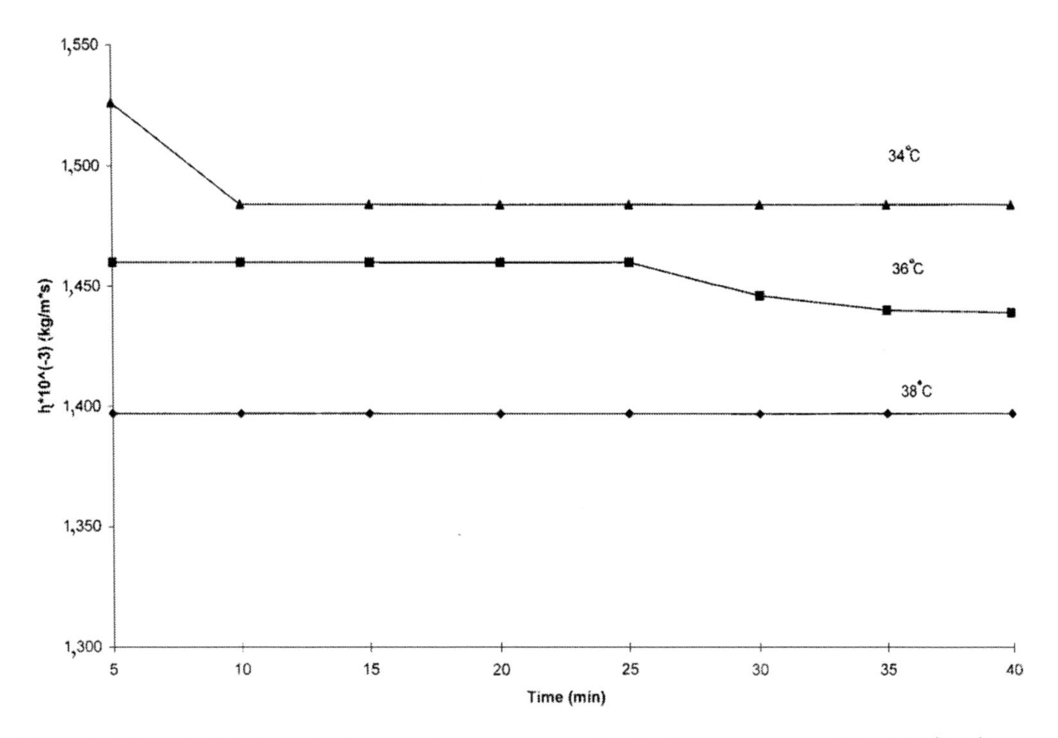

Figure 9. Viscosity in HDL solutions vs. time at a varying incubation temperature (20 mM K^+,Na^+-phosphate buffer, pH 7.35).

Temperature dependencies of conductivity in the rat and human apoA-I solutions exhibited the anomalous σ change in the range 36-38°C. Temperature coefficients σ for human apoA-I on either side of T_c differed 1.67 times. The value of $\Delta\sigma/\Delta T$ in apoA-I differed considerably from $\Delta\sigma/\Delta T$ in native HDL (Figs. 13). The integral σ change was 0.75 (Ω m)$^{-1}$. In the human apoA-I preincubated with cortisol (C = 7.14 × 10^{-7} M), the anomalous region was shifted to 34-35°C, i.e., was 2°C lower. Thermal coefficients σ changed relative to the control. In both cases they differed from one another relative to T_c (Table 7). The shape of $\sigma(T)$ function in these cases was practically linear relative to T_c, but they differed in their slopes.

Calculation of the lipoprotein activation energy by the results of $\eta(T)$ and $\sigma(T)$ dependencies revealed that the activation energy was low on either side of T_c (the buffer concentration was 0.2 and 4 mM, respectively). The difference in these values was even less. This was more pronounced in human HDL and VLDL (Tables 4, 7). The activation energy and transition enthalpy in human HDL and apoA-I increased markedly after their preincubation with cortisol (Table 8). More fine structural transformations of apoproteins upon their interaction with steroid hormones were studied by means of IR spectroscopy.

Table 7. Temperature coefficient of conductivity in human HDL and apoA-I at temperatures below and above the transition point (T_c) before and after the action of cortisol ($C = 1.4 \times 10^{-5}$ M), $(\Omega$ m deg$)^{-1}$ (K^+,Na^+-phosphate buffer, $C = 4$ mM, pH = 7.24)

	HDL	HDL+cortisol	ApoA-I	ApoA-I+cortisol
≤ Tc	$8.87 \pm 0.08 \times 10^{-4}$	$3.24 \pm 0.03 \times 10^{-4}$	$3.20 \pm 0.03 \times 10^{-4}$	$4.80 \pm 0.05 \times 10^{-4}$
≥ Tc	$8.37 \pm 0.08 \times 10^{-4}$	$1.59 \pm 0.03 \times 10^{-4}$	$5.52 \pm 0.08 \times 10^{-4}$	$4.03 \pm 0.05 \times 10^{-4}$

Changes of IR-Spectra

According to IR spectra, interaction between human HDL and cortisol results in a considerable increase in the intensity of absorption bands at 1654 cm^{-1} (amide-I), 1546, 1520 cm^{-1} (amide-II), 3292 cm^{-1} (amide A), and produces a decrease in their half width, which indicates the growing moiety of α-helixes (the tangle $\leftrightarrow$ α-helix transition). The absorption bands at 1696 and 1630 cm^{-1} slightly increased in intensity, which reflected the growth of β-structure moiety (the tangle $\leftrightarrow$ β -structure transition) (Fig. 14).

Table 8. Activation energy of conductivity in human HDL and apoA-I at temperatures below and above the transition point (T_c) and transition enthalpy (kcal/mol): (a) control, (b) sample + cortisol (K^+,Na^+-phosphate buffer, $C = 4$ mM, pH = 7.24)

		HDL	ΔH transition	ApoA-I	ΔH transition
a	≤ Tc	1.70 ± 0.05		5.14 ± 0.15	
	≥ Tc	1.50 ± 0.05	0.20 ± 0.05	5.99 ± 0.15	−0.85 ± 0.15
b	≤ Tc	5.32 ± 0.15		6.93 ± 0.15	
	≥ Tc	2.61 ± 0.15	2.71 ± 0.15	4.15 ± 0.15	2.78 ± 0.15

An increase in the integral intensity of amide I band comprised about ¼ ÷ ⅓ of the control sample. Upon resolving the amide I band of differential spectrum (curve 3) into components, the β-structure moiety exhibited a 10-12% increment, whereas α-helixes accounted for 18-20%. Interestingly, a new absorption band appeared at 3242 cm^{-1}, probably the β-structure amide A [17]. In HDL, upon their interaction with cortisol, the change in the secondary structure was more pronounced than in apoA-I. A considerable increase was noted in the absorption band intensity at 1470 cm^{-1}, corresponding to the deformation oscillations of CH bonds; the splitting of this band also became more pronounced (1468 and 1456 cm^{-1}), which indicated the strengthening of interaction between hydrocarbon chains due to their increasing ordering [19].

The absorption band intensity of the ester lipid group CO bond rose markedly, which also indicated the increased ordering of hydrocarbon chains [14].

In addition, an about 5 cm^{-1} frequency shift of the absorption band of the mentioned bond to the long wave region was observed, which was reflected on the differential curve by the appearance of asymmetry and a weak maximum at 1733 cm^{-1}. This can be explained by an increase in the number of hydrogen bonds that are likely to form between unesterified cholesterol and human apoA-I in native HDL.

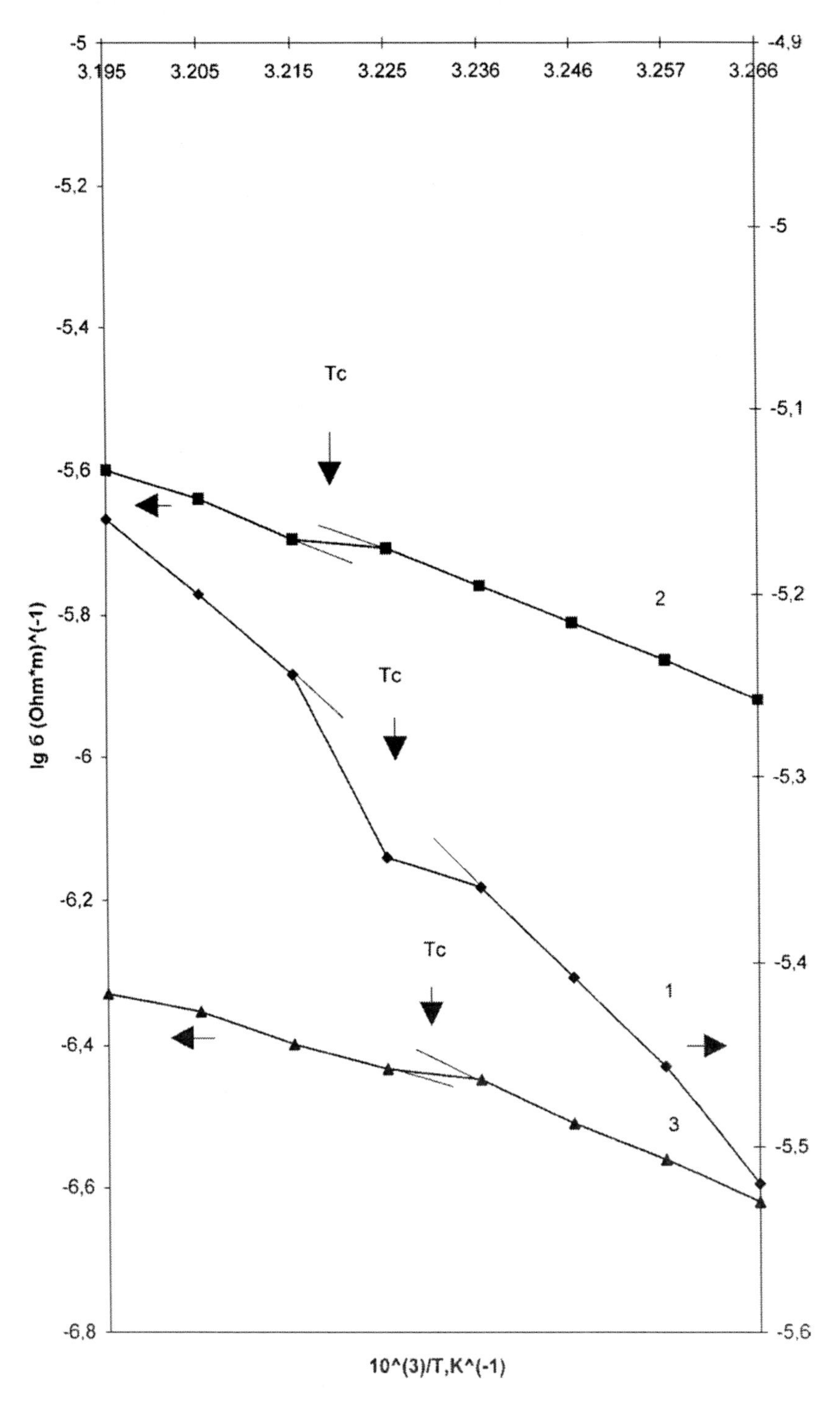

Figure 10. Conductivity at a varying temperature in rat lipoproteins (C_{buffer} = 0.2 mM, pH = 7.24): (1) HDL ($C_{protein}$ = 1 mg/ml), (2) LDL ($C_{protein}$ = 0.6 mg/ml), (3) VLDL ($C_{protein}$ = 0.5 mg/ml).

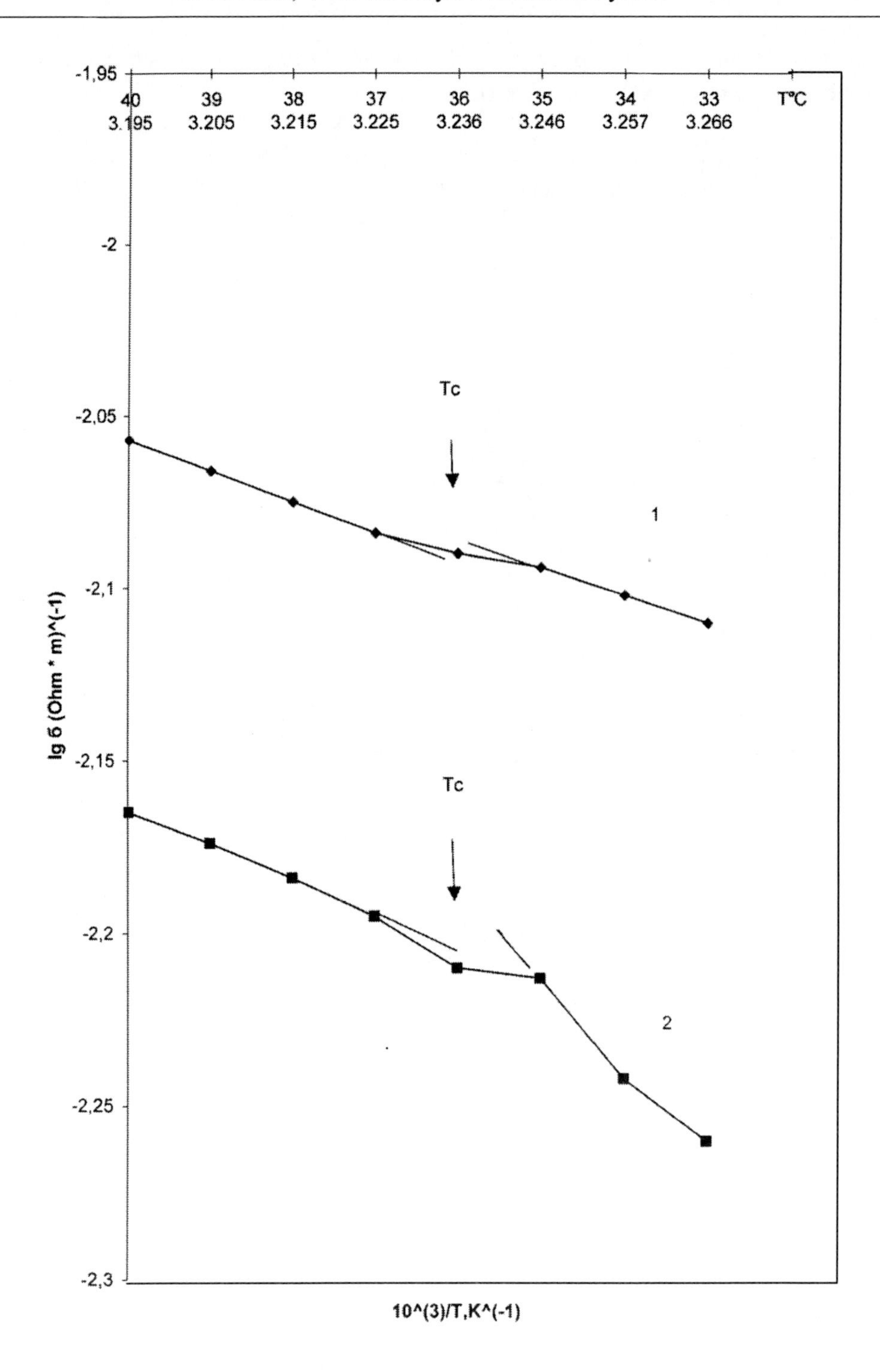

Figure 11. Conductivity at a varying temperature in human lipoproteins preincubated with cortisol: ($C_{cortisol} = 1.4 \times 10^{-5}$M, $C_{protein} = 0.12$ mg/ml, $C_{buffer} = 4$ mM, pH = 7.24): (1) HDL (control), (2) HDL + cortisol.

The intensity of absorption bands at 1244 (P=O bond), 1088 and 1172 cm^{-1} (P—O—C and C— O—C bonds, respectively) increased, as well as the valent oscillations of CH bonds at 2928 and 2852 cm^{-1}. As seen from the spectra, the HDL structure undergoes significant transformations upon interaction with cortisol, with these transformations occurring in both protein and lipid moieties. Redistribution of the secondary protein structure elements under the action of temperature after HDL preincubation with cortisol seems to be related to apoA-I. ApoA-I is known to interact with cortisol with high affinity [11].

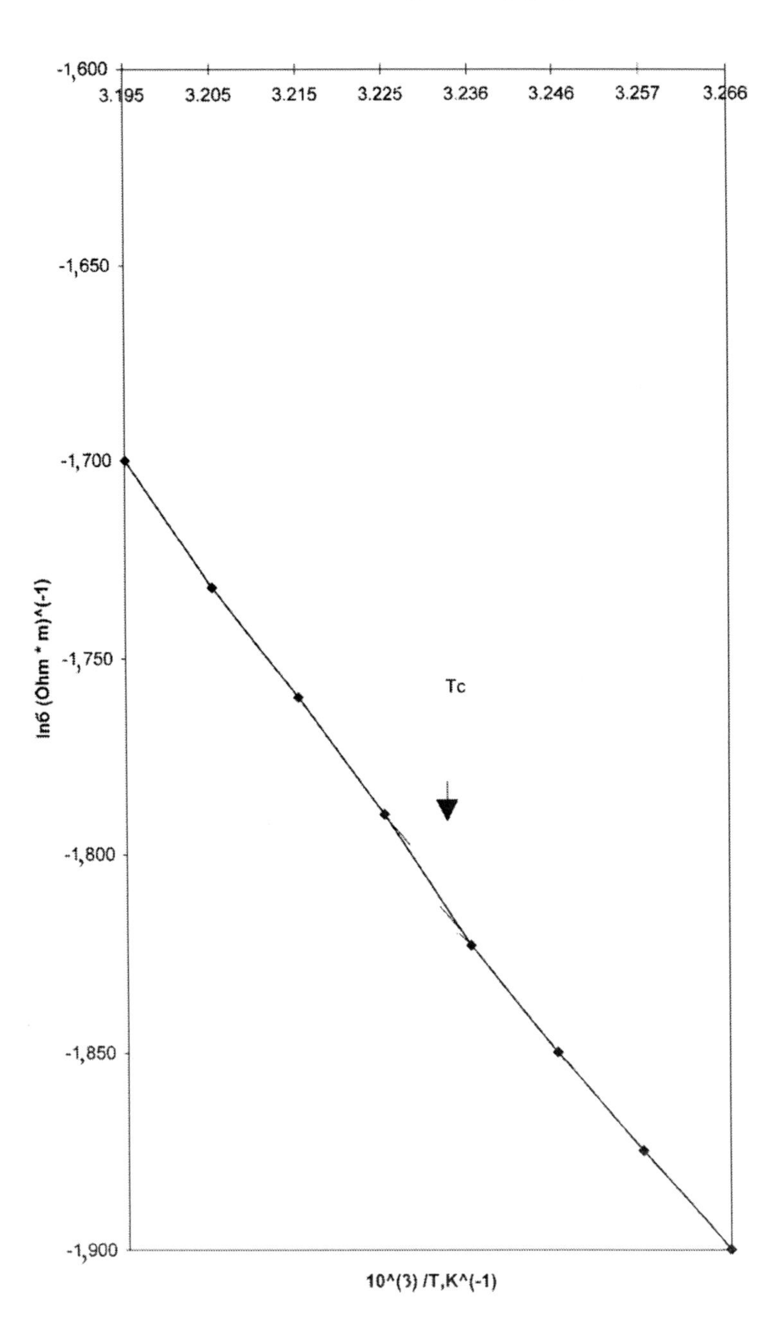

Figure 12. Conductivity at a varying temperature in rat apoA-I ($C_{protein}$ = 0.04 mg/mL, C_{buffer} = 4 mM, pH = 7.24).

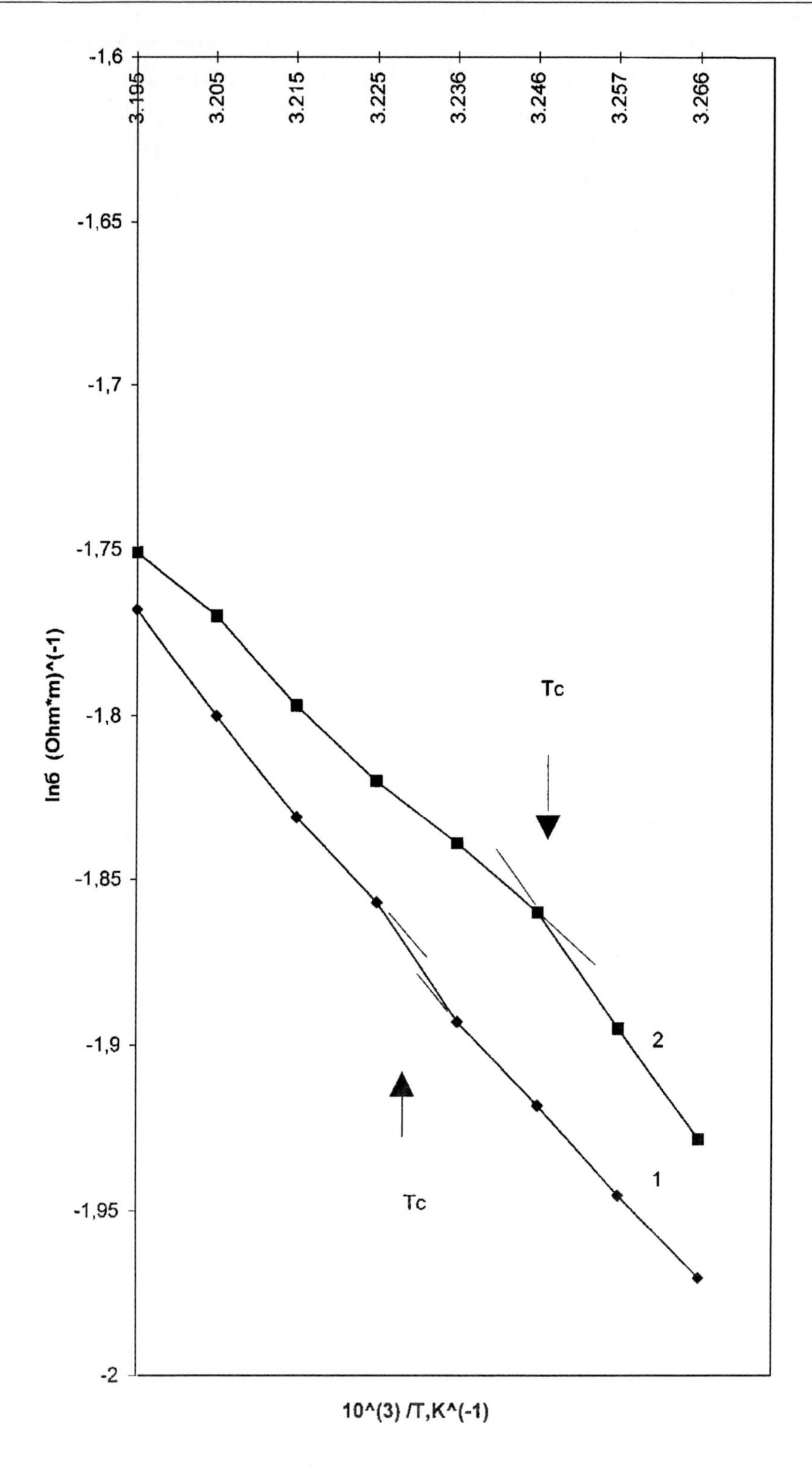

Figure 13. Conductivity at a varying temperature in human apoA-I preincubated with cortisol: ($C_{cortisol}$ = 7 × 10^{-7} M, $C_{protein}$ = 0.04 mg/mL, C_{buffer} = 4 mM, pH = 7.24): (1) apoA-I (control), (2) apoA-I + cortisol

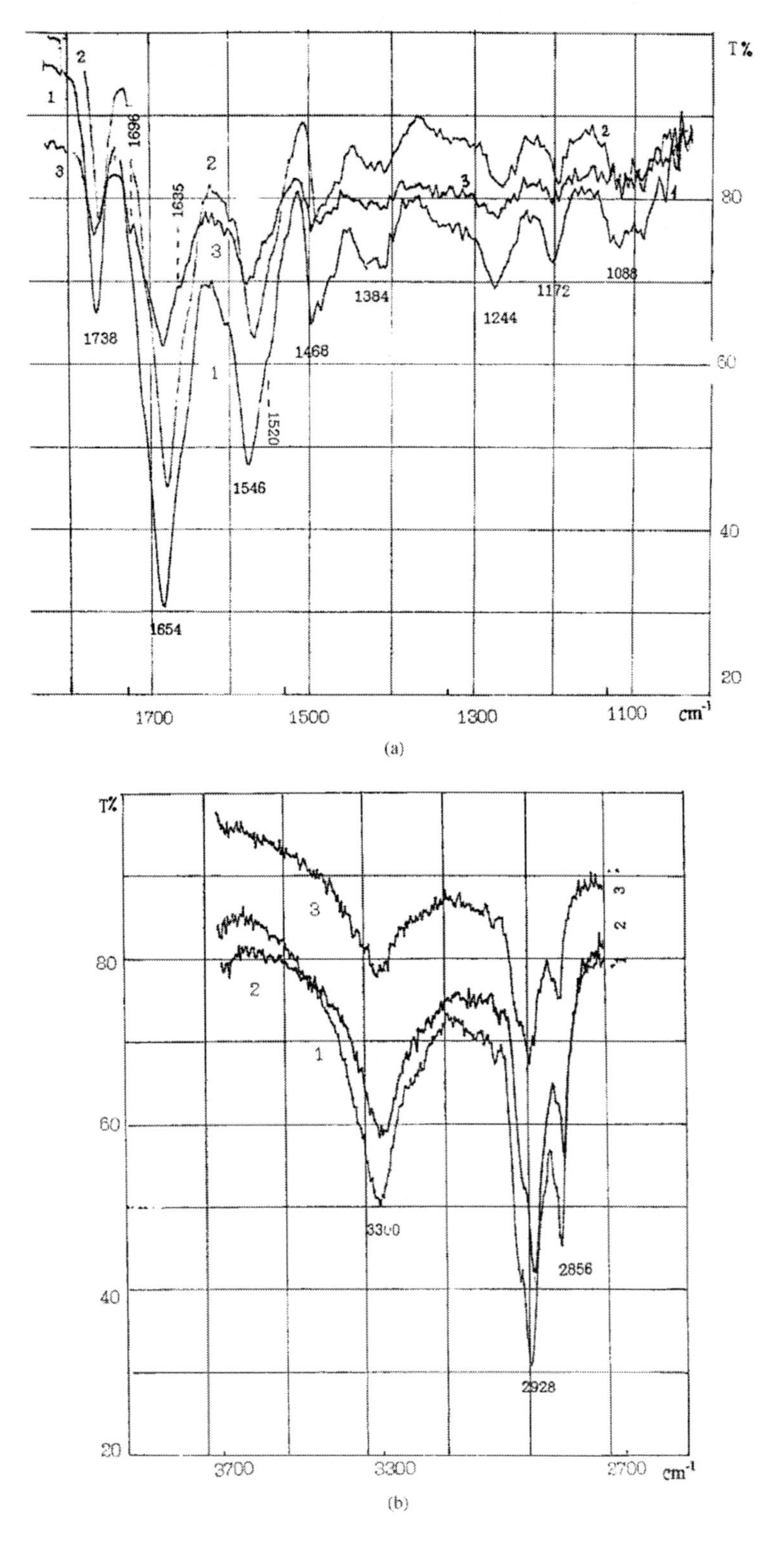

Figure 14. IR absorption spectra of human HDL preincubated with cortisol ($C_{cortisol} = 4.15 \times 10^{-5}$ M, $C_{buffer} = 4$ mM, pH = 7.24): (a) $v = 1000–1800$ cm^{-1} , (b) $v = 2700–3700$ cm^{-1}; (1) HDL + cortisol, (2) HDL (control), (3) differential spectrum.

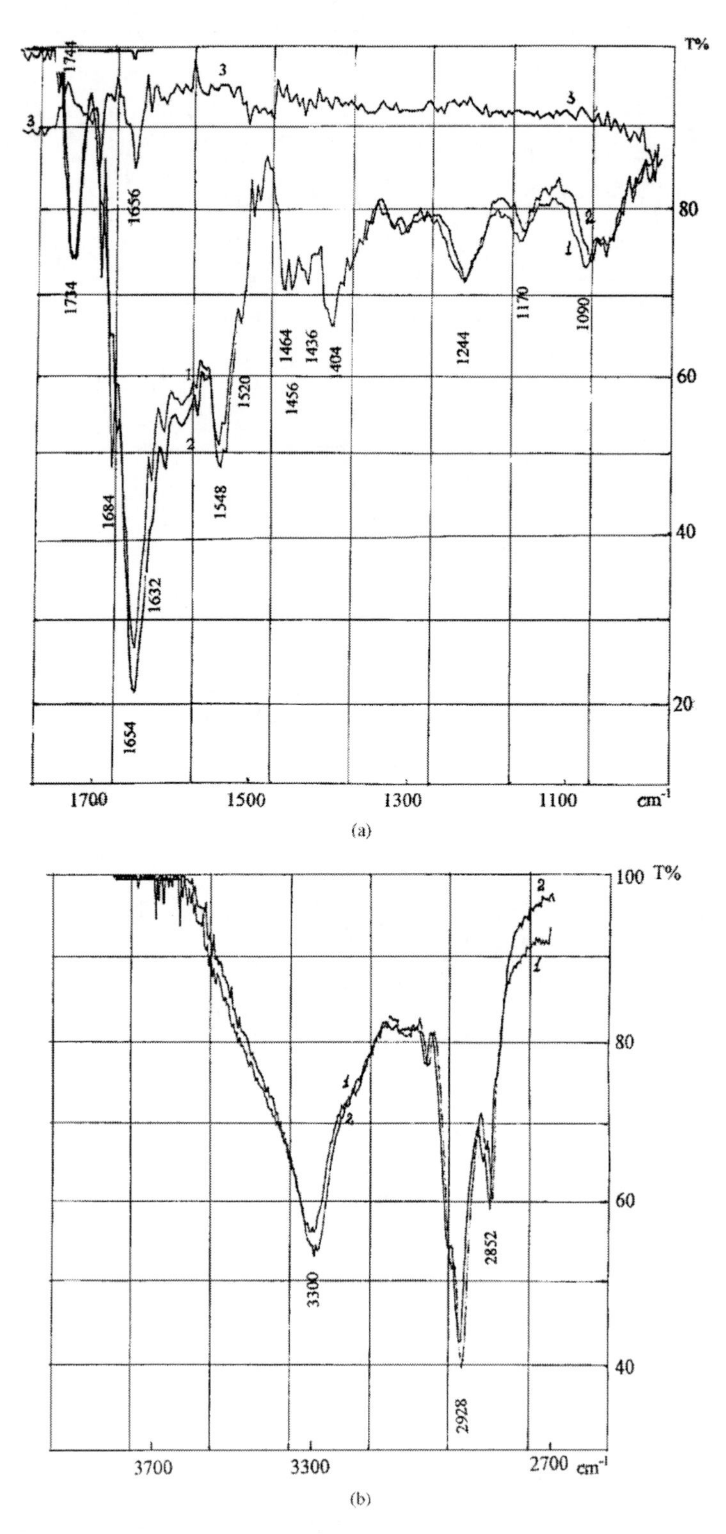

Figure 15. IR absorption spectra of rat HDL preincubated with cortisol ($C_{cortisol} = 4.15 \times 10^{-5}$ M, C_{buffer} = 4 mM, pH = 7.24): (a) v = 1000–1800 cm^{-1} , (b) v = 2700–3700 cm^{-1}; (1) - HDL + cortisol, (2) - HDL (control), (3) - differential spectrum.

Analysis of IR spectra of rat HDL after their preincubation with cortisol (1.6 × 10-9 M) revealed some peculiarities. For example, intensity of the absorption bands at 1654, 1548, and 3296 cm^{-1} decreased, and resolution of the absorption bands at 1696, 1684, 1635, 1624, and 1520 cm^{-1} increased. This gave grounds to suggest a 5-7% decrease in the moiety of α-helix and disordered structure (tangle), since the absorption band at 1654 cm^{-1} is a derivative of the two structures: α-helix and disordered structure [20]. One may assume also some increase in the β-structure moiety. In our opinion, the α-helix → β-structure and tangle → β-structure transitions occur under the action of cortisol in HDL apoproteins, primarily in apoA-I. In addition, a considerable intensity increase was noted in the absorption band at 1744 cm^{-1} corresponding to free oscillations of CO-ester bond in lipids, which indicated the increasing lipid ordering in deeper layers of lipoprotein particle, where contact with proteins is weaker (Fig. 15).

Comparison of IR spectra in human and rat after their preincubation with cortisol allowed to reveal the essential distinctions. In human HDL, upon interaction with cortisol the α-helix and β-structure moiety increases against the background of decreasing moiety of disordered structure, which indicates the occurrence of the tangle → α-helix and tangle → β-structure transitions. At the same time, in rat lipoproteins it was noted the process of despiralization and decreasing of the tangle against the background of increasing β-structure moiety, i.e., the α-helix → β-structure and tangle → β-structure transitions occurred. In addition, the lipid ordering in human proceeded through the formation of new hydrogen bonds, probably of phospholipids with proteins and cholesterol or of cholesterol molecules with apoproteins [21]. At the same time, in rats the lipid ordering was not attended by changing the number of hydrogen bonds, which is likely to be caused by the lipid ordering in deeper layers of the lipoprotein particle, where contact with proteins is weaker or absent.

The features of human and rat HDL interaction with cortisol seem to be related to the differentinitial degree of lipoprotein ordering. Analysis of IR spectra of the cortisol free human and rat HDL showed that the ordering of the HDL is higher in both protein and lipid components, which manifests itself in (i) a decrease in half-width of absorption bands in lipid (1744 cm^{-1}) and protein (1654 cm^{-1}) spectral regions and (ii) a resolution increase in many of them (1696, 1684, 1520 cm^{-1}). Decreasing half-width of absorption band with the maxima at 2928 and 2852 cm^{-1}, corresponding to the valent oscillations of CH bonds, along with a higher resolution of the bands at 1464, 1456, and 1436 cm^{-1} (deformation oscillations of CH_2 groups) additionally indicate a higher ordering of HDL structure in rats than that in human (Fig. 15) [20].

Discussion

Analysis of the results of studying the thermal dependencies of viscosity in colloidal solutions of serum lipoproteins showed that at T = 36-38°C the anomalous change η is recorded, which is arbitrarily called T_c (Figs. 4-9). What is the nature of this anomalous region?

Calculation of the Reynolds number, characterizing the rate of the fluid flow through a viscosimeter capillary, showed that its value was 200, which is an order of magnitude lower than the value at which laminar flow transforms to turbulent one (2300). Hence, we are dealing here with the laminar flow, in which the viscosity of fluid in a capillary is determined by the frictional forces between the fluid layers given a rate gradient.

We consider lipoprotein as a particular case of mesomorphic structures, and so in our reasoning we proceed from a knowledge of the properties and behavior of liquid crystals.

Viscosity of mesomorphic structures (mesophases) is known to reflect the inner ordering of molecular orientation, and its dependence on the temperature is bound to correlate with the order parameter Lubensky demonstrated that Franck's elastic constants [22] vary with the square of orientation order parameter S, which agrees well with experiment. Imura and Okano [23] interpret the viscosity of nematic phase by phenomenological equation relating the Leslie viscosity coefficient [24] to the order parameter S. Theoretical calculations lead to the conclusion that the viscosity coefficient is nearly proportional to the order parameter S. In smectic A-phase, the order parameter is expressed as $S = SN + \delta S$, where S is the order parameter in the nematic phase, and δS is the value allowing for additional orientation ordering upon the smectic phase formation [25]. On this basis one may assume that the anomalous change of viscosity in the physiological temperature range 36-38°C in lipoprotein fractions (Figs. 4-9) is caused by the anomalous change in the order parameter [22]:

$$S = -\frac{1}{2} + \frac{3}{2}\left\langle \cos^2 \vartheta \right\rangle,$$

where ϑ is an angle the long axis of the molecule makes with director n, and the angular brackets denote the statistical average over all molecules. In the nematic phase, parameter S takes an average value $0 < S < 1$ depending on the temperature and concentration of the bar molecules [26].

The anomalous viscosity change on $\eta(T)$ curves in lipoprotein and apoA-I (Figs. 4-9) is determined by the S-shaped segment typical for the structural phase transition. The inflection point of the S-shaped segment corresponds to the point of structural phase transition T_c [16, 26]. The T_c value in lipoproteins is 37.5 ± 0.5°C.

According to the Landau theory of structural phase transitions, the order parameter fluctuations as well as symmetry changes and heat capacity jump are observed near the points of phase transitions [27, 28]. In our opinion, this regularity has also been shown in this case. Structural transitions recorded in HDL, LDL, and VLDL occur at the same temperature, ~37°C, even though these lipoproteins differ in composition, particle size, and structure. It is believed that all lipoproteins have a core consisting of cholesteryl esters and triglycerides [29]. It is known that in the rabbit VLDL core the phase transition of smectic → cholesteric type is recorded in the range of 40-42°C. The lipoprotein upper layer, 2.15 nm thick, consists of apolipoproteins, phospholipids, and cholesterol [29]. In cholesteryl esters extracted from human LDL, the phase transition of smectic → isotropic fluid type is recorded in the range of 10-45°C by polarization microscopy, calorimetry, and X-ray diffraction analysis. Upon addition of about 3% free cholesterol to cholesteryl esters, the smectic → cholesteric transition is initiated. The authors assign the transition near 35°C to the smectic → isotropic fluid transition [2]. We have recorded the transition in rat LDL in the range of $T_c = 37.5$°C with transition enthalpy of 2.78 kcal/mol and transition width $\Delta T = 2$°C. On this basis we assume that in the lipid core of rat LDL, as in human LDL core, the smectic → isotropic fluid transition does occur. Taking into account that transition width is only 2°C, we are dealing most likely with the smectic → nematic transition; enthalpy of such transition being 0.16-2.30 kcal/mol, which agrees well with our results. In our opinion, in rat and human VLDL the smectic → cholesteric transition occurs; enthalpy of such transition being 0.74 ± 0.03 and −0.03 ± 0.01 kcal/mol, respectively, which agrees well with the literature data (ΔH_{trans} = 0.5 - 1.1 kcal/mol). No phase transitions in the range 10-40°C were recorded in HDL core [30].

Numerous works report on the interaction of apoA, C, and E with phospholipids and cholesterol resulting in the change of the secondary protein structure, in particular, the tangle ↔ α-helix transition [21]. Upon interaction of human HDL with cortisol, the tangle → β-structure transition is recorded along with the tangle → α-helix transition [31]. In apoB within human LDL, upon pH shift there occur the tangle → α-helix, tangle → β-structure, and probably α-helix → β-structure transitions. ApoB secondary structure exhibits significant transformations in LDL storage for 1 month.

It is known from the literature that decreasing viscosity of proteins with increasing temperature is related to their despiralization [33]. Thus, the observed viscosity jump at T = 37°C is likely to be due to the jump in change of the secondary structure, i.e., the tangle → α-helix transition. This fact may be assigned to both apoA-I, apoA-II, apoA-IV HDL and apoB, E, and C LDL and VLDL. However, in parallel, the tangle → β -structure transition is also possible [31]. The tangle → α-helix transition is assigned to the phase transition of the first type [34]. The tangle → β-structure transition is connected with the symmetry change, and it is assigned to the phase transition of the second type. For the second type phase transition the transition enthalpy is zero, in the real physical systems the transition enthalpy is often moderate (Tables 6, 9) [16]. The observed viscosity jump is related to the heat capacity jump as well. Heat capacity jump is typical also for the second type phase transition [35].

Thus, phase transition of the first and second types may occur in A, B, C, and E apoproteins under the action of temperature. Small activation energy and transition enthalpy (Tables 4, 7) evidence that in the transition region of lipoprotein the weak interaction forces are manifested: hydrophobic (1-3 kcal/mol), Van der Waals (1 kcal/mol) forces, and partially the hydrogen bonds.

In HDL, the transitions are recorded at 22-23°C and 37-38°C on $\eta(T)$ curves (Figs. 4, 5). In the HDL core [36] and in the cholesteryl esters extracted from HDL, no transition is detected in the physiological temperature range. In view of these circumstances, we assume that phase transition in HDL at 37-38°C is initiated by proteins.

A small value of transition enthalpy in HDL and a narrow T-transition range (ΔT = 2°C) allow to suggest that in the surface layer of phospholipids the transition of smectic A → C type may occur, which is attended with changing slope of hydrocarbon chains relative to the phospholipid surface normal. Enthalpy of the smectic A → C transition in liquid crystals is 0.01-0.66 kcal/mol [16], and our results are within the indicated range (Table 4). In addition, the symmetry change and heat capacity jump occur, which is characteristic for the second-type phase transition [27, 28].

Such a transition may proceed by the following mechanism: apolipoproteins may form clusters on the HDL3 surface. Changing protein conformation upon structural transitions leads to changing intermolecular forces in the protein-lipid system, and hence, to a moderate spatial shift of hydro- carbon chains in phospholipids. This is sufficient to produce their angular displacements relative to the lipoprotein normal. A prominent role in changing the lipoprotein particle structure may be played by cholesterol, which is capable of forming the hydrogen bond with NH_2 protein groups or the ion-dipole interaction with arginine guanidine group of apolipoproteins A and E by hydroxyl group. Another point of interaction with apolipoproteins is the dimethyl group cholesterol at the end of hydrocarbon chain (C_{26}-C_{27}) [21]. The change in protein conformation may lead to the interaction between perhydrophenanthrene core cholesterol with phospholipids located at the near to proteins of the boundary layer. This may result also in angular displacement of phospholipid clusters. Cholesterol has long been known to decrease considerably the transition enthalpy in phospholipids and biological membranes by raising their phospholipid ordering. Actually, it shows the condensing effect upon its introduction into phospholipids. But at a moderately increasing cholesterol content in phospholipids, this phenomenon may be related also with the smectic → cholesteric [2] and A → C-smectic transitions, since their transition enthalpy is lower than that for other transitions, which may occur in biostructures when cholesterol is absent in them or present only in small amounts.

Analysis of the mechanism of cholesterol interaction with apoA-I and apoE proposed in [21] allowed to suggest that in HDL cortisol interacts with apoA-I or apoA-II involving carbonyl group (C=O) to form the hydrogen bond with NH_2 protein groups or the ion-dipole interaction with guanidine group of arginine. The CH_2OH group of cortisol aliphatic chain interacts with methyl groups of leucine (Fig. 16).

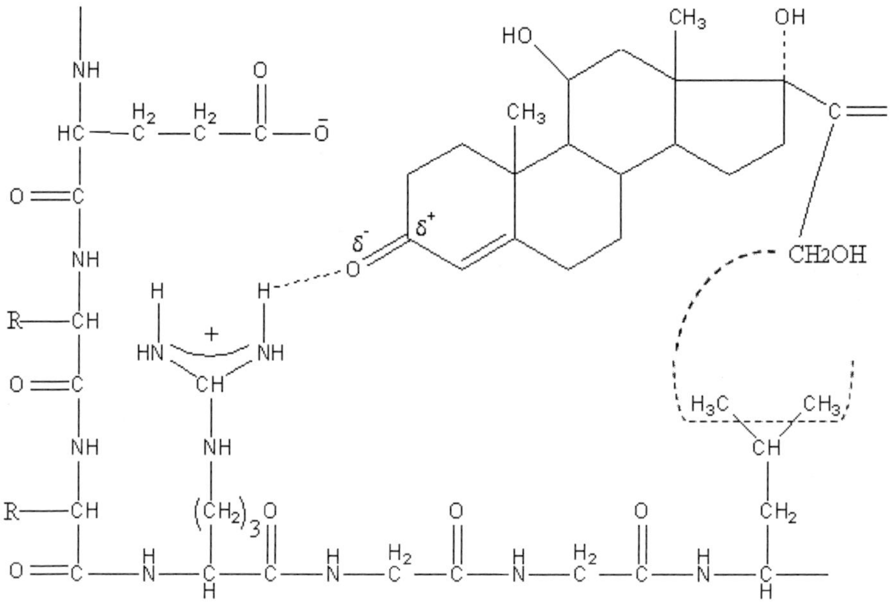

Figure 16. Scheme of cortisol interaction with apoA-I.

It is worth mentioning that activation energy and transition enthalpy in human VLDL is lower than that in rats (Table 5), which is likely to relate with the difference in apoB nature, protein, and cholesterol content. Structural transition in rat LDL and VLDL in the physiological temperature range manifests itself weaker as compared to HDL, which may be connected with their markedly lower protein content [36].

Analysis of the data of Tables 3, 4 showed that the hormone profoundly changed the $\Delta\sigma/\Delta T$ coefficient calculated for HDL. In particular, this coefficient decreased 2.75 times below T_c , and 5.26 times above T_c. Whereas in apoA-I it increased 1.59 times below T_c, and decreased 1.25 times above it. The presented numerical values evidence that the hormone produces more profound changes in native HDL than in isolated protein. Similar distinction in $\Delta\sigma/\Delta T$ coefficient changing can be explained by a lower hormone concentration in the samples with apoA-I, but in this case attention must be given to the character of $\Delta\sigma/\Delta T$ changing relative to Tc. This allowed us to make some conclusions: (1) conformational state of isolated apoA-I differs from its state in native HDL, since phospholipids profoundly increase the α-helix moiety, (2) the cortisol binding constant is higher in HDL than in apoA-I, and (3) electrophoretic mobility in HDL and apoA-I varies in a different manner relative to T_c. The latter conclusion follows from the relation [22]

$$\sigma = Q \cdot v \cdot u,$$

where Q, is the particle charge, v is the number of particles in a volume unit, and u is the electrophoretic mobility. A significant decrease in HDL $\Delta\sigma/\Delta T$ in the samples with cortisol indicates a substantial fall in u, with the effect being more pronounced in the region above T_c. Whereas in apoA-I u increases below T_c and decreases above it.

Calculation of activation energy in the lipoprotein by the results of $\sigma(T)$ dependencies on either side of T_c revealed that ΔH was low on either side of T_c (concentration of the buffer was 4 mM). Even lower was the difference in these values, ΔH_{trans} (conventional transition heat). It was more pronounced in human HDL (Table 7).

Small value of transition heat in human HDL (Table 7) allows to suggest the smectic A $\leftrightarrow$ C transition occurrence in the lipid moiety of the particle surface layer, which is attended with a synchronous changing in the slope of hydrocarbon chains relative to the lipoprotein surface normal for a great number of phospholipid molecules. Enthalpy of the smectic A $\leftrightarrow$ C transition is within the range of 0.01-0.66 kcal/mol [16]. The smectic A $\leftrightarrow$ C transition is assigned to the second type of phase transition [35].

Interestingly, after the hormone interaction with HDL and apoA-I ($C = 1.4 \times 10^{-5}$ M and 7×10^{-7} M, respectively) the activation energy increased 3.1 times in HDL and 1.5 times in apoA-I below T_c. Far less changes were observed in the region above T_c , in particular, in HDL the activation energy increased only 1.63 times, and in apoA-I it decreased 1.67 times. Finally, this resulted in a profound elevation of transition heat. In HDL it increased 13 times, and even more in apoA-I. Character of the transition changed due to a considerable elevation of the transition heat. Such a transition must be assigned to the first type of phase transition [16].

Increasing activation energy and transition enthalpy in the rat lipoprotein at a low ion force [6] is somewhat related to the essential differences caused by a different phospholipid content, variations in the primary structure of apoB, etc. [36]. Energy parameters are greatly contributed by an increase in the particle charge and ξ potential at a low ion force of electrolyte, which in this case is proportional to $1/\sqrt{C}$ [37]. Hence, an increase in activation energy and transition enthalpy at a low ion force of electrolyte is partially connected with the essential differences as well as with appearance of additional intramolecular interaction forces of electrostatic nature.

Conductivity is closely related to viscosity by the equation of Smolukhovsky and Valden [16, 22]. Actually, the obtained temperature dependencies of viscosity in lipoprotein and apoA-I unambiguously demonstrate this interrelation. The anomalous changing of viscosity with temperature is recorded in the range of 36-38°C. Conductivity is also closely related with dielectric permittivity. According to, permittivity fluctuations are related to the order parameter fluctuations. Hence, the conductivity fluctuations are caused by the order parameter fluctuations. According to the Landau theory, the order parameter fluctuations are observed near the points of phase transitions [27, 28]. However, symmetry is also at the basis of the second type of phase transition theory. At the smectic A $\leftrightarrow$ C phase transition the symmetry change does occur [35]. In addition, we suggest that the symmetry change appears also in apolipoproteins due to changing of their secondary structure. In our opinion, the anomalous conductivity change in lipoproteins is related not only to the phase transition in lipids but also to the changing secondary structure of their proteins.

Conclusions

Study on temperature dependencies of viscosity (η) and conductivity (σ) of blood lipoproteins (HDL, LDL, and VLDL) and A-I apolipoprotein (human and rats) has revealed the presence of the anomalous region at temperature 35-38°C ± 0.5°C (T_c). Transition width is 2°C. Viscous flow enthalpy, activation energy (ΔH), and transition enthalpy (ΔH_{trans}) as well as thermal coefficients $\Delta\eta/\Delta T$ and $\Delta\sigma/\Delta T$ on either side of T_c have been calculated. Transition heat is very low in human HDL, VLDL, and apoA-I, and in LDL it is higher by a factor of 4-5. Some mechanisms of the cortisol interaction with HDL and apoA-I have been studied by IR spectroscopy and conductometry. The hormone has been found to strengthen the tangle $\rightarrow$ α-helixes and tangle $\rightarrow$ β-structures transitions and increase the ordering of lipids. Therewith, ΔH_{trans} rises markedly (13 and more times), and at the same time, the anomalous region is shifted by 1-2°C in apoA-I. The anomalous change of viscosity and conductivity in the physiological temperature range for all lipoprotein fractions and apoA-I seems to be due to the structural phase transition in both proteins and lipids. In view of the heat capacity jump and a low value of ΔH_{trans} in human HDL, one may assume the phospholipids of these particles to exhibit the orientation transition of smectic A $\leftrightarrow$ C type, which is assigned to the second type of phase transition. The structural transition into apoA-I includes the elements of both the first type of phase transition due to the occurrence of tangle $\rightarrow$ α-helix transition, and the second type, since it is related to the changing symmetry due to the tangle $\leftrightarrow$ β-structure transition and has a low transition enthalpy. In human and rat VLDL, the smectic $\rightarrow$ cholesteric phase transition is likely to occur. Enthalpy of viscous flow and structural transition in VLDL is higher for rats than for humans. The pH shift of the medium to the neutral region (pH 6.1) results in shifting the anomalous region by 1-2°C.

References

[1] Panin, LE; Panin VE. Effect of the "chessboard and mass transfer in interfacial media of organic and inorganic nature. *Physical Mesomechanics*, Dec. 2007. V.10. No. 6, pp. 5-20.

[2] Deckelbaum, RJ; Shipley, GG; Small, DM. Structure and interactions of lipids in human plasma low density lipoproteins. *J. Biol. Chem.*, 1977 Jan 25. V.252. No. 2, pp. 744-754.

[3] Morrisett, JD; Gaubatz, JW; Tarver, AP; Allen, JK; Pownall, HJ; Laggner, P; Hamilton, JA. Thermotropic properties and molecular dynamics of cholesteryl ester rich very low density lipoproteins: effect of hydrophobic core on polar surface. *Biochemistry*, 1984 Oct 23.V.23. No. 22, pp. 5343-5352.

[4] Kroon, PA, Krieger, M. The mobility of cholesteryl esters in native and reconstituted low density lipoprotein as monitored by nuclear magnetic resonance spectroscopy. *J. Biol. Chem.*, 1981 Jun 10. V.256. No. 11, P.5340-5344.

[5] Scanu, A; Pollard, H; Hirz, R; Kothary, K. On the conformational instability of human serum low density lipoprotein: effect of temperature. *Proc. Natl. Acad. Sci. USA, 1969* Sep. V.62. No. 1, pp. 171-178.

[6] Panin, LE; Kunitsyn, VG. Mechanism and thermodynamics of multilevel structural transitions in liquid crystals under external actions. *Physical Mesomechanics*, Nov – Dec. 2008. V.11. No. 6, pp. 61-68.

[7] Panin, LE. *Determinate systems in physics, chemistry and biology*. Novosibirsk: Siberian university publishing House; 2008.

[8] Panin, LE; Maksimov, VF; Usynin, IF; Korostyshevskaya, IM. Activation of nucleolar DNA expression in hepatocytes by glucocorticoids and high density lipoproteins. J. Steroid Biochem. *Mol. Biol.,* 2002 May. V.81. No. 1, pp. 69-76.

[9] Panin, LE; Kunitsyn, VG; Tuzikov, FV. Changes in the secondary structure of highly polymeric DNA and CC(GCC)n - type oligonucleotides under the action of steroid hormones and their complexes with apolipoprotein A-I. *J. Phys. Chem. B*, 2006 July 13. V.110. Issue 27, pp. 13560 -13571.

[10] Panin, LE; Mokrushnikov, PV; Kunitsyn, VG; Zaitsev, BN. Interaction mechanism of cortisol and catecholamines with structural components of erythrocyte membranes. *J. Phys. Chem. B,* 2010 July 29. V.114, Issue 29, pp. 9462-9473.

[11] Polyakov, LM; Sumenkova, DV; Knyazev, RA; Panin, LE. The analysis of interaction of lipoproteins and steroid hormones. *Biochemistry (Mosc) Suppl. Series B: Biomedical Chemistry,* 2010. V.4. No. 4, pp. 362-365.

[12] Lowry, OH; Posebrough, NJ; Farr, AL; Randall, RJ. Protein measurement with the Folin phenol reagent. *J. Biol. Chem.*, 1951 Nov. V.193. No. 1, pp.265-275.

[13] Kunitsyn, VG; Monastyrev, IA; Panin, LE. (Inventors). Russian patent No. 2080583 *Automatic device for determining the viscosity of blood, suspensions of cells and membranes*. Publication data 27.05.1997. Intern. Class G01N11/06.

[14] Panin, LE; Kunitsyn, VG; Nekrasova, MF. Changes in the structure of erythrocyte membranes and the activity of Na^+, K^+-ATPase in participants of the Soviet-Canadian trans-ski trip. *Space biology and aerospace medicine* (Russian), 1991. V.25. No. 5, pp.12-15.

[15] Gul, VE; Kuleznev, VN. *Structure and Mechanical properties of polymers*. Moscow: Vysshaya Shkola; 1979 (in Russian).
[16] Kapustin, AP. *Experimental Studies of Liquid Crystals*. Moscow: Nauka; 1978 (in Russian)
[17] Kunitsyn, VG. *Structural phase transitions in erythrocyte membranes, lipoproteins, biologically active macromolecules in normal and in certain diseases*. Ph.D.Thesis, Novosibirsk; 2002 (in Russian).
[18] Attallah, NA; Lata, GF. Steroid-protein interactions studied by fluorescence quenching. *Biochim. Biophys Acta*, 1968 Oct 21. V.168. No. 2, pp.321-333.
[19] Gribov, LA. *Theory of infrared spectra of polymers*. Moscow: Nauka; 1977 (in Russian).
[20] Chirgadze, YN. *Infrared spectra and structure of polypeptides and proteins*. Moscow: Nauka; 1965 (in Russian).
[21] Klimov, AN; Kozhevnikova, KA; Kliueva, NN; Belova, EV. *Binding of cholesterol by apolipoproteins A-I and E. Biokhimiia*, 1992 Apr., V.57. No. 4, pp. 555-565. (in Russian).
[22] Lubensky, TC. Hydrodynamics of Cholesteric Liquid Crystals. Phys. Rev. A, July 1972. Vol.6. Issue 1, pp. 452-470.
[23] Imura, H; Okano, K. Temperature Dependence of the Viscosity Coefficients of Liquid Crystals. Jpn. J. Appl. Phys., Oct. 1972. V.11. No.10, pp.1440-1445.
[24] Leslie, FM. Some constitutive equations for liquid crystals. Arch. Rat. Mech. Anal., Jan. 1968. V. 28. No. 4, pp. 265–283.
[25] Voyutsky, SS. Course of colloidal chemistry. Moscow: Khimiya; 1976 (in Russian).
[26] *de Gennes,* PG; Prost, J. *Physics of liquid crystals.* Oxford: Clarendon Press; 1993.
[27] Landau, LD; Lifshitz, EM. Statistical Physics. Part. 1. Moscow: Nauka; 1976 (in Russian).
[28] Patashinsky, A.Z; Pokrovskii, VL. Fluctuation Theory of Phase Transitions. Moscow: Nauka; 1982 (in Russian).
[29] Klimov, AN; Nikulcheva, NG. Metabolism of lipids and lipoproteins and its disorders. St. Petersburg: Piter; 1999 (in Russian).
[30] Tall, AR; Deckelbaum, RJ; Small, DM; Shipley, GG. Thermal behavior of human plasma high density lipoprotein. Biochim. Biophys. Acta, 1977 Apr. 26. V.487. No. 1, pp. 145-153.
[31] Kunitsyn, VG; Panin, LE; Poliakov, LM. Anomalous change in the specific conductivity in lipoproteins at around physiologic temperatures. *Biofizika*, Sep-Oct. 1999. V.44. No. 5, pp. 861-869 (in Russian)
[32] Singh, S; Lee, DM. Conformational studies of lipoprotein B and apolipoprotein B: effects of disulfide reducing agents, sulfhydryl blocking agent, denaturing agents, pH and storage. *Biochim. Biophys. Acta*, 1986 May 21. V.876. No. 3, pp. 460-468.
[33] Ooi, T.; Itsuka, A.; Onari S. et al. *Biopolymers*. Imanisi Y. (Ed.). Moscow: Mir; 1988 (in Russian).
[34] Konev, SV. *Structural lability of biological membranes and regulatory processes.* Minsk: Nauka i Tekhnika; 1987 (in Russian).
[35] Anisimov, MA. *Critical phenomena in liquids and liquid crystal*s. Moscow: Nauka; 1987 (in Russian).

[36] Kholodova, YD; Chayalo, PP. *Blood lipoproteins*. Kiev: Naukova Dumka; 1990 (in Russian).

[37] Fridrihsberg, DA. *Course of Colloidal Chemistry.* Leningrad: Khimiya; 1989 (in Russian).

In: Protein Structure
Editors: Lauren M. Haggerty, pp. 155-203

ISBN: 978-1-61209-656-8

Chapter 9

A Multiparametric Approach in Analysis of Protein Structure and Unfolding-Refolding Reaction

Vladimir N. Uversky
Department of Biochemistry and Molecular Biology, and the Center for Computational Biology and Bioinformatics, Indiana University School of Medicine, Indianapolis, Indiana 46202, USA;
Institute for Biological Instrumentation, Russian Academy of Sciences, 142292 Pushchino, Moscow Region, Russia;
Molecular Kinetics, Inc., Indianapolis, IN 46268, USA

Abstract

This chapter analyzes the advantages of a multiparametric approach to gain exhaustive information related to the structure and folding of globular proteins. The principal bases of the some widespread contemporary physico-chemical methods for studying the structural properties of the globular protein and their changes during the unfolding-refolding reaction are considered. It is discussed what kind of information can be obtained using different physico-chemical methods and what advantages a researcher might have applying several techniques simultaneously.

Keywords: multiparametric approach, protein folding, protein self-organization, structural changes, partially folded conformations, conformational stability.

Introduction

The appearance of a functional protein in the living cell involves two major processes, biosynthesis of the polypeptide chain and its folding into the native three-dimensional (3D-) structure. Although the peculiarities of the protein biosynthesis are relatively well studied, the precise molecular mechanisms underlying the protein folding process are not known as yet. The structure of the native protein is formed and stabilized by physical forces of a different nature: hydrogen bonds, hydrophobic, electrostatic, van der Waals interactions, etc. The successive (depending on the external conditions or time) incorporation of these interactions determines the emergence of elements of secondary structure, the appearance of compactness and the formation of a unique 3D-structure. It is believed that contrarily to the biosynthesis, which is known to heavily rely on complex machineries (including translation and transcription apparatuses, ribosomes, etc.), protein folding might generally occur spontaneously and does not require the presence of ribosomes or any other additional factors. In other words, all information required for a given protein to fold into the functional machine is encoded in its amino acid sequence. This conclusion is based on several key observations briefly outlined below. It has been shown that many proteins can refold *in vitro* from the completely unfolded state [1]. This follows from the classic experiments performed by Anfinsen and co-workers in the middle sixties [2], where it has been shown that urea-unfolded ribonuclease A with the reduced disulfide bonds can spontaneously restores the native spatial structure (with native SS-bonds and biological function) after the denaturant removal. Based on these observations it has been concluded that the protein amino acid sequence contains information sufficient for correct folding of the protein molecule [2]. On the other hand, it is known that homologous proteins from different sources, possessing a similar spatial structure, can have different amino acid sequences. This means that the information on the protein 3D-structure is most likely encoded in some common physico-chemical properties of the polypeptide chain rather than being dependent on the structural peculiarities of separate amino acid residues. To account for the potential role of the ribosome and the difference in the timing of the polypeptide biosynthesis, David C. Phillips has suggested in 1967 an attractive hypothesis that the biosynthesis of the nascent polypeptide chain from the N-terminus might play a key role in the folding of the protein molecule [3]. This hypothesis was based on the analysis of then known 3D-structure of the hen egg white lysozyme using an interesting approach of considering the conformation of progressively larger sections of the protein molecule, starting at the terminal amino end [3]. It has been emphasized that among the first 40 residues of lysozyme there are many α-helix-promoting side chains, with a number of hydrophobic residues distributed in such a way that the resulting α -helices possess partially hydrophobic surfaces. In the finished molecule this part of the sequence forms a compact globular unit with hydrophobic side chains providing the contacts between two α -helices. The following 12 residues constitute the comparatively hydrophilic group and form the antiparallel pleated α -sheet, implying that in an aqueous environment there is no need for these residues to aggregate with the rest of the folded chain until the hydrophobic residues 55 and 56 require shielding, when the chain folds back on itself, forming the pleated sheet and burying these residues in the existing hydrophobic pocket. Residues 56-86 were shown to be folded rather irregularly in contact with the pleated sheet so that residues 1-86 constitute a structure with two wings. The gap between these wings is gapped by a rather irregular α-helix with a partially hydrophobic surface formed by residues 88-100. Finally the last 20 residues are folded around the globular unit built up by the first 40, so that it is clear at least that in this molecule the amino end of the chain must fold before the carboxyl end [3]. However, later it was shown that this hypothesis contradicts the correct folding of a protein with a closed chemical structure, whose N- and C-termini were chemically linked [1]. Based on these observations it has been concluded that although such a nascent state of the protein chain is definitely interesting by itself [4] it does not necessary play any essential role in the process of protein folding (at least *in vitro*).

Main Definitions

The *self-organization* of protein *in vitro* is understood here as the folding of a polypeptide chain into the native molecule, which is capable of carrying out specific biological function. A natural criterion of the *native* state of the protein molecule is its ability to perform a specific biological function. From the structural viewpoint, the native molecule might be considered as an aperiodic crystal in which the location of virtually all the atoms is fixed in space. It is also natural to assume that the *denatured state* is such a form of the protein molecule in which it has lost its biological function. In this case the disruption of the unique 3D-structure of the protein molecule occurs. If under certain experimental conditions the protein molecule completely loses all (or almost all) the elements of tertiary and secondary structure, then it is said that the *complete unfolding* has occurred. It is obvious that denaturation of the protein molecule is far from being always accompanied by its complete unfolding. Very often, different *partially folded intermediate* states (intermediates) appear. Currently, two such states have been found and characterized in detail under the equilibrium conditions: the *molten globule* [5, 6] and the molten globule precursor (*pre-molten globule*) [7-12].

The Essence of the Multiparametric Approach

The vast majority of initial studies on protein de- and renaturation were performed using single-domain globular proteins, such as ribonuclease A and lysozyme [13, 14]. For these simple systems it was shown that in response to changes of the environmental conditions, a cooperative transition (monitored by characteristic changes in studied structural parameters) took place. This transition is well described by the sigmoidal curve, suggesting that the process of a globular protein denaturation might proceed according to the all-or-none principle. The validity of using of such an approximation to describe these processes was analyzed in detail in 1966 in the theoretical work of Lumry, Biltonen and Brandts [15], who has established that it is rather difficult to distinguish a two-state transition from the transitions of other types based only on the analysis of the denaturation curve shape. However, a more-or-less unambiguous answer to the posed question can be obtained analyzing the results of several physico-chemical methods sensitive to the different levels of structural organization of the protein molecule [15]. In essence, this work can be considered as the foundation of a new experimental (multiparametric) approach in studies of de- and renaturation of globular proteins.

The idea of simultaneous application of several physico-chemical methods to describe the structural properties of the protein molecule and to study the changes of these properties in response to changes in the environmental conditions was very fruitful. By the end of the sixties, using this approach it has been established that the denatured states of a given protein are structurally non-identical and strongly depend on the denaturing conditions [14]. It was also demonstrated that unfolding of some globular proteins might be characterized by non-symbatic (non-simultaneous) changes of different structural parameters. Thus, the first indications were obtained that the unfolding of globular proteins does not always represent an all-or-none transition, but might be a complex multi-stage process, being characterized by the formation of intermediate states [14]. Later, this supposition was confirmed and substantiated experimentally (see, e. g., [6, 16]).

Thus, the multiparametric approach, being an experimental method in which several physico-chemical tools sensitive to the different structural levels of a protein molecule are simultaneously used, might provide researchers with the most complete and unambiguous information on the structural changes which a protein underwent in both the kinetic and equilibrium processes of de- and renaturation. Some methods constituting the basis of this experimental approach are considered below in more details.

Thermodynamic Analysis of Protein De- and Renaturation

First, the analysis of the denaturation curves has to be briefly discussed. Proteins are known to possess a remarkable ability to adopt a strictly defined conformation. Various stable forms (ordered, disordered and partially ordered) can exist under certain conditions, and a polypeptide chain can be forced to undergo reversible transitions between these conformations induced by the corresponding changes in the environmental conditions. Such conformational transitions are, in essence, a response of a system to changes in such external conditions as temperature, pH, pressure, or composition of the solvent. They often occur in a cooperative manner; i.e., within the narrow interval of the environmental changes. The question then arises of how stable is the rigid structure of globular proteins from the thermodynamic viewpoint.

It has been pointed out that the problem of protein stability is not as simple as it seems at the first sight [17]. In fact, what does the stability of a very exactly defined structure mean? It is known that the native protein does not have an absolutely rigid structure, possessing fluctuations of different scales at different structural levels [18]. Must all deviations from the crystallographic structure be considered as different structures or different conformations? If the answer is yes, then from evident considerations the stability of a protein at any temperature above 0 K is always equal to zero. Alternatively it may be assumed that not all the changes of the protein structure are important? Then, which changes are important? Since there is a border separating the native (functional) state from the denatured one (i.e., a form with the lost functionality), it can be asserted that denaturation is connected with some important changes in the structure of the protein molecule [14]. How to divide the essential structural changes from the non-essential ones?

Structural changes accompanying conformational transitions usually occur in a narrow interval of changes in the environmental conditions. A typical denaturation curve is shown in Figure 1. Using the simple model of two states (the native and denatured) to describe the process of denaturation, the effective equilibrium constant of the given process can be determined from the data of Figure 1:

$$K^{eff} = \frac{Y_N - Y_X}{Y_X - Y_D} \tag{1}$$

where Y_N and Y_D are values of any studied parameter characterizing the purely native and the purely denatured state, respectively. Y_X is the value of this parameter at the defined conditions. Studies of the dependence of this effective equilibrium constant on such external conditions as temperature, pressure and ion activity can determine the effective parameters characterizing the denaturation process. If these parameters are determined using the equation of equilibrium thermodynamics, they can be physically characterized as changes in Gibbs energy (ΔG^{eff}), enthalpy (ΔH^{eff}), entropy (ΔS^{eff}), volume (ΔV^{eff}), and a number of bound ligands ($\Delta \nu^{eff}$). These parameters are defined as follows:

$$\Delta G^{eff} = -RT \ln(K^{eff}) \tag{2}$$

$$\Delta H^{eff} = RT^2 \frac{\partial \ln(K^{eff})}{\partial T} \tag{3}$$

$$\Delta S^{eff} = \frac{\Delta H^{eff} - \Delta G^{eff}}{T} \tag{4}$$

$$\Delta V^{eff} = -RT \frac{\partial \ln(K^{eff})}{\partial p} \tag{5}$$

$$\Delta \nu^{eff} = \left[\frac{\partial \ln(K^{eff})}{\partial \ln a} \right]_{a=a_t} \tag{6}$$

where a is the of the denaturant and a_t is its activity at the mid-transition point. This possibility of quantitative description of the process of denaturation in physical terms was first suggested by Anson and Mirsky in 1934 [19, 20] and is widely used at the present.

It should be emphasized that the indicated approach gives a quantitative estimation of the measure of stability of the protein molecule by the difference of Gibbs energy between the native and the denatured states calculated from eq. (2).

Another interesting consequence of such an analysis is the possibility to establish the nature of conformational transitions in small globular proteins induced by strong denaturants [21, 22]. It is known that the steepness of the phase transition occurring in the small system increases proportionally to the increase in the size of the system [23]. Following from the fact that for transitions N → U, N → MG, or MG → U (where the symbols N, U, and MG refer to the native or unfolded or to the molten globule state) the value of $\Delta \nu^{eff}$ was proportional to the molecular mass of the protein, it was suggested that these transitions possess the all-or-none character and thus can be considered as the intramolecular analogues of the first order phase transition [21, 22].

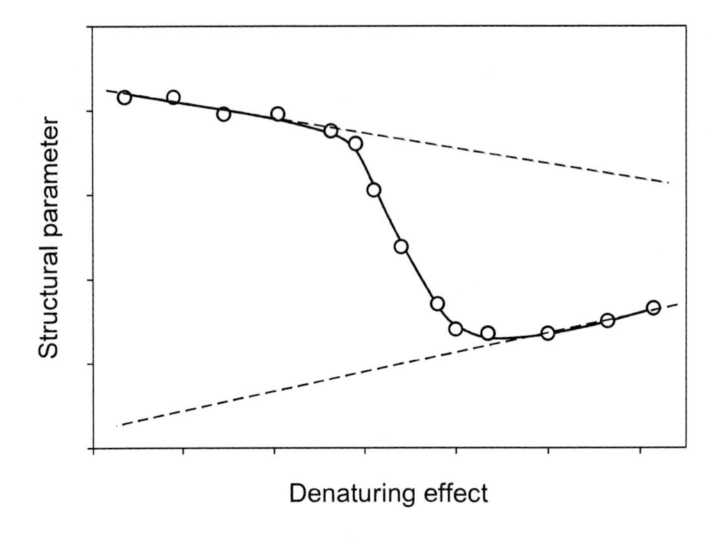

Figure 1. A typical sigmoidal curve describing structural changes in a globular protein induced by changes in the environment. The change of the strength of the denaturing effect is denoted on the abscissa axis and the changes in the studied parameter, sensitive to protein molecule denaturation are denoted on the ordinate. The range of linear dependence (preceding the denaturation transition and following it) reflects changes of this parameter in the native and denatured states, respectively.

Scanning Microcalorimetry

Differential scanning microcalorimetry opens a unique possibility of direct measurement (and not calculating) of the denaturation enthalpy [17]. Analysis of the calorimetric curve can determine the following parameters [17]: 1) the temperature of denaturation T_D; 2) the heat of the denaturation process Q_D; 3) the heat capacity change associated with denaturation $\Delta_D C_p(T_D) = C_p^D(T_D) - C_p^N(T_D)$; 4) the temperature dependence of denaturation enthalpy.

Importantly, both the calorimetric (ΔH^{cal}) and the effective or the van't Hoff enthalpy (ΔH^{eff}) can be measured calorimetrically. The first of these parameters represents the heat expended on the denaturation of one protein molecule, whereas ΔH^{eff} is the heat required to disrupt the cooperative unit. The indicated feature of microcalorimetry obviously yields direct information on the nature of the studied denaturation process. In fact, the whole protein molecule is the cooperative unit of the denaturation process described by the two-state model.

Consequently, the effective enthalpy of this process should be equal to the calorimetric and the relationship ΔH^{cal} / ΔH^{eff} should be close to unity [17]. Interestingly, such an effect actually takes place during the heat denaturation of small globular proteins [17]. Another picture is observed at studies of heat denaturation of large globular proteins constituted of two or more independent domains. In these cases analysis of the calorimetric curve indicates that the value of the relationship ΔH^{cal} / ΔH^{eff} is much greater than unity [24].

Furthermore, the formalism of equilibrium thermodynamics was successfully applied to predict the existence of so-called cold denaturation; i.e., the transition of a protein from the native state to the denatured state induced not by heating, but by cooling the solution. This phenomenon was experimentally revealed for such proteins as metmyoglobin, apomyoglobin, staphylococcal nuclease and phosphoglycerate kinase [25, 26].

To conclude this section it should be noted that scanning microcalorimetry is an important component of the multiparametric approach. Besides the unique capability to estimate the thermodynamic parameters from he single experiment, the presence or absence of the peak of cooperative heat absorption on the calorimetric curve, obtained for the given protein at studied experimental conditions, is a simple and convenient criterion indicating the presence or absence of a rigid cooperatively melting tertiary structure [6, 17]. Such an approach is especially convenient in cases where information on the presence of a specific tertiary structure in the protein molecule cannot be obtained by spectral methods (e. g., when the protein molecule does not contain aromatic amino acid residues or has a fluctuating structure in their nearest spatial surrounding).

Spectral Methods

Absorption Spectroscopy

The chromophore can change significantly its absorption spectrum in response to the changes in the environmental conditions (pH of the medium, polarity of the solvent or neighboring groups and relative orientation of neighboring chromophores). These factors are utilized in application of the absorption spectroscopy to characterize proteins [27].

It has been shown that the tyrosine residue ionization is a good probe for conformational changes in proteins, as the pK value of these residues is strongly depended on their micro-surroundings [28]. Importantly, the tyrosine ionization is known to be accompanied by the dramatic changes of the absorption spectra in the ultraviolet region. This effect was used for the analysis of the rapid conformational changes induced in the pepsin by the pH-jump from pH~13 to pH~10 [29]. Here, as in the case of other proteins (human serum albumin [29], amilase from *Bacillus subtilus* [30], α-lactalbumin [31] and many others), the conformational changes were monitored by the characteristic changes in the absorption spectra observed at the tyrosine deprotonation.

Absorption of heme and bound co-factors. The *heme proteins* are easily identifiable because of their red to red-brown color. The intense color of heme proteins is due to the heme group, more specifically the transitions within the π–electron system of the porphyrin [32, 33]. These transitions are influenced by the peripheral substitution pattern of the individual porphyrins, by the oxidation state of the iron and by the axial ligation of the iron, as well as by the polarity of the heme environment (or heme pocket) within the folded protein chain. Each heme protein consists of a polypeptide (protein) and the heme – a complex of iron (usually in the +2 or +3 oxidation state) and the organic ligand porphyrin. For hemoglobin and myoglobin, the oxygen transport and storage proteins, the porphyrin is protoporphyrin IX (1,3,5,8-methyl-2,4-vinyl-6,7-propionic acid porphine). In hemoglobin and myoglobin, the imidazole side chain of a histidine residue is bonded axially to the heme iron, anchoring it within the protein. In cytochrome *c*, both a histidine (N donor) and a methionine (S donor) bond axially to the heme iron. In addition, cytochrome c heme is bonded to the protein at the peripheral 2 and 4 positions via a thioether linkage to cysteine amino acid residues [32, 33].

Heme protein spectra show two characteristic bands: one is in the 390 to 450 nm wavelength range and is called the "Soret" or "Q" band; the other (usually a group of several) lies in the 450 to 700 nm range and is called the "visible" or "A" and "B" band(s). The Q band is typically 5 to 10 times more intense than the visible bands. Both are "allowed" transitions of the $\pi \rightarrow \pi^*$ designation with molar extinction coefficients in the 10^5-10^6 range. The visible (A and B) bands are found to be the more sensitive to such changes as the oxidation state and/or ligation of the iron [32, 33]. For example, the oxygenation level of hemoglobin can be readily determined by recording the absorption spectrum in the range 450 to 600 nm. The relative ratio of oxidized to reduced cytochrome *c* can also be determined from the absorption spectrum in this region. It has been pointed out that the electronic spectrum of cytochrome *c* provides very useful information regarding the heme environment and its coordination structure [34-38]. Although the absorption spectrum of cytochrome *c* is quite characteristic of oxidation state, there is little variation in this spectrum as a function of pH within a range which is considered physiologically relevant, say pH 6 to 9. Much can be learned about the structure and stability of this heme protein, however, by studying the pH induced variation of this spectrum cover a much wider pH range. Since the heme group for cytochrome *c*, unlike that of hemoglobin, is covalently attached to the protein it is not easily displaced, but merely changes environment as the protein chain folding and/or axial ligation are altered due to pH changes. Within a pH range that such changes are reversible, we can learn something about these changes by recording and analyzing the absorption spectra as a function of pH [32-38].

Similarly, bound co-factors often possess characteristic absorption spectra, following which might provide important information about function and structure of the given protein. For example, it has been shown that the absorption spectrum of the purified 4-amino-4-deoxychorismate lyase (ADCL) from *E.coli* (en enzyme catalyzing the elimination of pyruvate from 4-aminodeoxychorismate with concomitant aromatization of the cyclohexadiene ring) exhibited major peaks at 278, 414, and 334 nm with absorbance ratios of 1:0.18:0.05 [39]. The absorption maximum at 414 nm was attributed to the ketoenamine tautomer of the bound pyridoxal 5'-phosphate (PLP), whereas the maximum at 334 nm was attributable to the enolimine tautomer of the bound PLP or to a substituted aldamine, carbinolamine, or pyridoxamine 5'-phosphate (PMP) [39]. It has been shown that the addition of D-alanine to holo-ADCL resulted in a time-dependent decrease in the absorption peak at 414 nm, with a concomitant increase in the absorbance at 328 nm. This was interpreted as an indication of the fact that the PLP form of the enzyme is converted to the PMP form by a half-transamination reaction with D-alanine [39].

Absorption of chromophore in GFP-like proteins. Members of the green fluorescent protein (GFP) family are characterized by a very bright fluorescence in the visible region. GFP-like proteins are unique in that their fluorophores are not the separately synthesized prosthetic groups but are composed of the modified amino acid residues within the polypeptide chain. For example, in GFP from the jellyfish *Aequorea victoria*, the chromophore is a p-hydroxybenzylidene-imidazolidinone derivative formed by an autocatalytic, posttranslational cyclization and oxidation of the polypeptide backbone, involving Ser-65, Tyr-66, and Gly-67 residues, with the cyclized backbone of these residues forming the imidazolidone ring [40-42]. The chromophore of the wild type GFP (wt-GFP) possesses peculiar spectroscopic properties: two absorption peaks at 395 and 475 nm, usually attributed respectively to the neutral and anionic forms of the chromophore [43], and one emission peak at 508 nm, with a high fluorescence quantum yield of 0.79 [44, 45]. The substitution of one or more amino acids within the chromophore or in its immediate proximity has allowed the development of GFP mutants with specific absorption and emission properties [46-48].

Another convenient tool for studying conformational and structural changes in proteins is a *differential spectroscopy* [28, 49, 50]. This variety of absorption spectroscopy was suggested in 1962 [49] and is now widely used to study conformational changes in proteins. In particular, differential spectroscopy was used to study equilibrium strong denaturant-induced unfolding of bovine carbonic anhydrase B [51], β-lactamase from *Staphylococcus aureus* [52] and numerous other proteins, as well as the conformational changes undergone by carbonic anhydrase B [53], β-lactamase [54-57] and other proteins upon change of the solvent pH. Differential spectroscopy is also widely used in the kinetic studies on protein de- and renaturation [58, 59].

Infrared (IR) Spectroscopy

IR spectra appear as a result of intramolecular movement (bond stretching, change of angle values between bonds and other complicated types of motion) of various functional groups (e.g., methyl, carbonyl, amide, etc) [60]. The merits of spectral analysis in the IR region result from that the mode of vibration of each group is very sensitive to changes of the chemical structure, conformation and environment. In the case of proteins and polypeptides, there are three most interesting infrared bands connected with the vibrational transitions in the peptide backbone and reflecting the normal oscillation of simple atom groups. These characteristic bands correspond to (*i*) the stretching of the N-H bonds, (*ii*) the stretching of the C=O bonds (amide I band) and (*iii*) the deformation of N-H bonds (amide II band). The frequency ν_{max}~3300 cm^{-1} is characteristic for the band describing the stretching of N-H bonds, whereas the amide I and amide II bands have the characteristic frequencies in the ranges ν_{max} ~1630-1660 and 1520-1550 cm^{-1}, respectively [61]. The utilization of IR for the analysis of protein secondary structure is based on the fact that the formation of a hydrogen bond results in the changes of the band position. Thus, analysis of IR spectra allows determination of the relative content of α-helices, β-structures and random coil in proteins by the deconvolution of the amide I and II bands.

The main advantage of using IR spectroscopy to determine the secondary structure of protein is that this method is based on a simple physical phenomenon, the change of vibrational frequency of atoms upon formation of hydrogen bonds. Thus, there is a unique possibility to calculate the parameters of normal vibrations of the main asymmetrical units in the secondary structure elements and to compare these calculated values with the experimental data. Such calculation of normal vibrational parameters for α-helices and β-sheets was performed by Miyazawa [62], and in the early sixties it was shown that there was a good correlation between the calculated and experimental data for the amide I and amide II bands [61]; i.e., it was shown that such calculations serve as a rational basis for the interpretation of the polypeptide and protein IR spectra. The first attempts to apply IR spectroscopy to evaluate protein secondary structure were based on several assumptions related to the protein structure: 1) protein consists of a limited set of different types of secondary structure elements (α-helix, parallel and antiparallel β-structures, β-turns and the random coil); 2) the IR spectra of a given protein is a sum of the spectra of these structures taken with the weights corresponding to their content in the protein; 3) the spectral characteristics of secondary structure are the same for all proteins and for all the structural elements of one type in the protein. Following with these assumptions, several methods were developed to evaluate the protein secondary structure content by IR spectroscopy [63, 64]. To do this, reference spectra of 100% secondary structure elements were obtained from spectra of polypeptides [63, 64] and proteins. The spectrum of the studied protein was represented as a linear combination of these reference spectra in such a way as to achieve the best coincidence between the experimental and model curves.

It should be mentioned that for quite a long time, despite its attraction, IR spectroscopy did not find wide application in protein studies. This was due, primarily, to bulky equipment and an inadequate recording system. However, the situation was changed due to the development of a principally new approach to measure IR spectra, the so-called Fourier-transform infrared (FTIR) spectroscopy. Modification of the method, as well as the fact that IR spectroscopy exhibits a high sensitivity to the conformational state of macromolecules, resulted in a number of studies where this approach was used for analysis of protein molecule structure and investigation of the process of protein denaturation and renaturation [65-76].

It has been emphasized that although *transmission FTIR* is the most frequently used IR method, there are a number of technical difficulties associated with this technique that can make collection of high-quality spectra difficult [76]. For example, high protein concentrations (~30 mg/ml) are required for measurements in water as H_2O possesses an intense absorption band at 1640 cm^{-1}. Furthermore, to avoid total absorption by water in the amide I region, transmission cells have to have a path length of 10 μm or shorter. The alternative solvent for proteins, D_2O, allows path lengths in excess of 50 μm. However D_2O is known to have different H-bonding properties than H_2O and its use requires that amide protons in the protein be completely exchanged for deuterons (see below).

Attenuated total reflectance-Fourier-transform infrared spectroscopy (ATR-FTIR) is more versatile than transmission FTIR because it is possible to collect spectra of nontransparent samples, to use samples of very low protein concentration (<0.3 mg/ml), and to study proteins in the presence of strongly absorbing solutes (such as denaturants). ATR-FTIR has been shown to be useful for studying the secondary structure and other properties of proteins in a variety of environments [77-85]. Two ATR methods have been developed for studying soluble proteins. The first involves drying the protein to a hydrated thin film on the surface of an internal reflection element, IRE, (TF-ATR). The other strategy for collecting ATR spectra is to allow proteins or peptides to adsorb to the surface of an IRE from solution (Ads-ATR). In the case of TF-ATR, the resulting film has an extremely high protein concentration and thus provides a high signal-to-noise ratio. Thin-film spectra of reasonable quality can be collected with as little as 10 μg of protein with this method [76]. Although it had been shown that the formation of a thin film does not disrupt the native structure of most proteins [79, 82, 86], there is always a potential for the secondary structure of an unstable protein to be altered by the evaporation of solvent, even though it remains hydrated in the film [87]. Thus, secondary structure determinations made with TF-ATR require independent verification from solution FTIR, CD, NMR, or X-ray analyses [76].

In the case of Ads-ATR it has been shown that the protein–IRE interaction is strong enough to keep the protein bound to the IRE surface, even in the presence of solvents other than water [88]. Ads-ATR has been used to study conformational changes induced in the polypeptides by environmental changes or ligand binding, as well as to measure the rates of protein binding to surfaces [76-85, 89]. It has been noted that the immobilizing interaction between peptides and IRE materials can also be expected to have an effect on the structure of these molecules, but this has not previously been well characterized [76].

With the introduction of methods for enhancing spectral resolution, it is possible to obtain high quality spectra of proteins in different environments and under different conditions [90-92]. Furthermore, as it has been already mentioned above FTIR spectroscopy can be used in the study of protein secondary structure in the dissolved, aggregated, and solid state [93, 94]. For the vast majority of proteins analyzed so far, the intermolecular interactions leading to aggregation involve β-sheet-like interactions [95]. The main advantage of FTIR in comparison with circular dichroism is that this method is much more sensitive to the β-structure. The combination of these two factors, the ability to analyze aggregated material and high sensitivity to β-structure, makes FTIR a powerful tool for the study of conformational changes occurring during the protein aggregation and the formation of insoluble fibrils [85, 96-106]. In FTIR studies of protein aggregation it has been established that a common feature was the appearance of a low-frequency band around 1620 cm^{-1}, which was often accompanied by a weak band around 1685 cm^{-1}. These bands were attributed to the formation of an intermolecular β-sheet structure in the protein aggregates [107-109]. Another characteristic of spectral alterations in protein aggregates is the gradual disappearance of ordered secondary structures, which is accompanied by the appearance of a broad and weakly featured band at 1645 cm^{-1}, which was assigned to unordered structure [92, 110-112].

Finally, it is necessary to emphasize that IR spectroscopy gives another unique possibility to keep track of the conformational changes in proteins, via the H→D exchange. It has been established that upon *deuteration* (i.e., at the substitution of hydrogen atoms by deuterium), many bands change their frequency [113-116]. It is also considered a benefit that the amide I signal frequencies of deuterated random coil and α-helical structures are well separated. Since it is well known to which functional group a given band belongs (the carbonyl, oxy- or amino group), one can identify the group where the H→D exchange occurred simply by observing which bands changed their position in the FTIR spectrum at deuteration. The rate of deuteration, as a rule, depends on the accessibility of a given functional group to the solvent, which, in turn, can change upon conformation changes. Thus, by tracking the change of the rate of deuterium-exchange at different experimental conditions the unique information on the conformational transformations in the given protein can be obtained. This technique has been applied for conformational studies on numerous proteins, including bovine α-lactalbumine [5, 117], bovine carbonic anhydrase B [118], and immunoglobulin light chain LEN [119].

Raman Spectroscopy

When light is scattered from a molecule most photons are elastically scattered; i.e., the scattered photons have the same energy (frequency) and, therefore, wavelength, as the incident photons. However, a small fraction of light (approximately 1 in 10^7 photons) is scattered at optical frequencies different from, and usually lower than, the frequency of the incident photons. The process leading to this inelastic scatter is known as the Raman effect [120]. Although Raman scattering can occur with a change in vibrational, rotational or electronic energy of a molecule, the vast majority of studied were devoted to the vibrational Raman effect. It has been emphasized that the Raman scattering spectrum provides essentially the same type of information as the IR absorption spectrum, namely, the energies of molecular normal modes of vibration [121]. However, it has been shown that a fundamental difference in selection rules leads to an important advantage of Raman over IR for the study of large molecules and assemblies: for a localized molecular vibration to generate a Raman band, there must be an associated change in polarizability; i.e., a distortion of electron density in the vicinity of the vibrating nuclei. Thus, localized vibrations of multiply bonded or electron-rich groups (C=O, C=N, C=C, S−S, S−C, S−H, etc) generally produce more intense Raman bands than do vibrations of singly bonded or electron-poor groups. Accordingly, the Raman spectrum of a protein is largely dominated by bands associated with the peptide main chain, aromatic side chains, and sulfur-containing side chains [121]. In contrast, IR absorption intensities depend on an oscillating dipole moment with vibration, and therefore vibrations of all types of nonsymmetrically bonded atoms can contribute substantially to the IR spectrum. This leads to the fact that IR absorption spectra of proteins are more complicated by overlapping bands from many types of residues than corresponding Raman spectra [121].

It is known that the Raman spectrum of a protein consists of numerous discrete bands representing molecular normal modes of vibration and serves as a sensitive and selective fingerprint of three-dimensional structure, intermolecular interactions, and dynamics [121]. Similar to ATR-FTIR, Raman spectroscopy has been proven to be an efficient technique for characterizing highly scattering and opaque samples as well as samples in different physical states (solutions, suspensions, precipitates, gels, films, fibers, single crystals, amorphous solids, etc) [121]. This suggests that the Raman spectroscopy is a unique tool that can be exclusively used to analyze protein aggregation and fibril formation in solution [122, 123]. At deep ultraviolet excitation around and below 200 nm, Raman scattering is resonantly enhanced from the amide chromophore, a building block of a polypeptide backbone, providing quantitative information about the protein secondary structures [124]. Recently, a new deep UV Raman spectrometer has been described, which requires only a 100-μL sample with protein concentration of >0.1 mg/mL [123]. These and other recent improvements in instrumentation, coupled with innovative approaches in experimental design, dramatically increase the power and scope of the Raman spectroscopy as a unique method for investigations of proteins and their large supramolecular assemblies.

Circular Dichroism (CD)

There are two types of optically active chromophores in proteins, the side groups of aromatic amino acid residues and the peptide bonds [125]. CD spectra in the near ultraviolet region (250-350 nm), also known as the aromatic region, reflects the symmetry of the environment of aromatic amino acid residues and, consequently, characterize the protein tertiary structure. However, CD spectra in the far ultraviolet (UV) region (peptide part of the spectrum, 180-250 nm) reflects the symmetry of the peptide bond environment and, thus, indicate the content of secondary structure in the protein molecule. An analysis of protein secondary structure content can be done with a sufficient level of precision using the CD spectra in the peptide region and several reference spectra for synthetic polypeptides and proteins with known secondary structure [27, 126-137].

Currently, CD represents one of the key methods for the unambiguous detection of the molten globule state formation. In fact, this intermediate is known to be characterized by the lack of rigid tertiary structure and the preservation of the native-like secondary structure [6]. These characteristics are visualized by the CD spectroscopy as flat, feature-less near-UV CD spectra and pronounced, native-like far-UV CD spectra. The comparison of the far-UV CD spectra for nine globular proteins in their native and pH-induced molten globular states revealed that the secondary structure of proteins in the molten globule state is indeed close to that in the native state [138]. In fact, using the singular value decomposition it has been shown that the spectra of molten globules could be described as a superposition of at least three independent components (most likely α-, β- and irregular structures). A self-consistent procedure of CD spectra analysis revealed the existence of a clear correlation between the shape of the molten globule spectra and the content of secondary structure elements in the corresponding native proteins, as determined from X-ray data. A mathematical expression of this correlation in terms of the Pierson coefficient amounts to the value of 0.9 for both the α-helix and the β-structure [138].

It has been established however that in some cases there is a strong distortion of the far-UV CD spectra as a result of the contribution of aromatic side groups in this region of the spectrum [9, 134, 139-145]. The circular dichroism spectra of human and bovine carbonic anhydrases B [9, 142-145], retinol-binding protein [146] and *Y. pestis* capsule protein Caf1 [147, 148] are striking illustrations of this phenomenon. Particularly, it has been shown that the far-UV CD spectra measured for the molten globule state of these proteins were more intensive than those of the native proteins, whereas their near-UV CD spectra were absent. These observations evidence that disruption of the rigid tertiary protein structure does not only reduce the CD spectrum in the near-UV region but is also accompanied by a cardinal decrease of the contribution of aromatic groups in the peptide region of the spectrum. It was also shown that aromatic groups can have both a positive and a negative contribution to the CD spectra in the far-UV region [139].

Based on these and similar observations, an approximation method for decomposing CD spectra in the peptide region into aromatic and peptide group contributions was suggested by Bolotina [134, 140]. The application of this method to the spectra of such proteins as α-lactalbumine and carbonic anhydrase B showed that the difference of the CD spectra in the far-UV region of these proteins, in the native state and in the molten globule state, is mainly the result of contribution of the aromatic group side chain to the native protein spectrum

[140]. A significant step toward the understanding this phenomenon was the study by Manning and Woody [141] where the circular dichroism spectra of pancreatic trypsin inhibitor in the peptide region were calculated taking into account the existence of the transitions in the peptide groups. The calculated peptide group CD spectra completely differed from the experimental ones. However, the authors managed to obtain a spectrum near to the experimental one in the peptide region after taking into account the contribution of phenylalanine and tyrosine residues, forming a cluster in the native protein [141].

^{1}H-, ^{13}C-, and ^{15}N-NMR Spectroscopy

The special advantage of NMR is that the theory of this method is so well-developed that the positions of individual atoms in a small molecule can, in principle, be determined from analysis of the spectrum [149-155]. The parameters of the protein NMR spectrum (position and number of bands, area, width and intensity of a separate line, probability of its splitting) have a high sensitivity to changes in both the amino acid sequence and the conformation of the protein molecule. The spectra obtained for large proteins, as a rule, are too complicated to attribute separate signals to concrete amino acid residues. However, well dispersed regions, where defined types of protein groups or interactions are exhibited, can be distinguished even in NMR spectra for such proteins. In fact, the ^{1}H-NMR spectrum of a native protein is strongly different from the sum spectrum of its composite amino acid residues [149-153, 156]. This circumstance, even despite the absence of a more detailed attribution, yields averaged information on the behavior of defined types of protein groups. Thus, signals from the aromatic group protons are exhibited in the region from 8 to 6 million parts (ppm), proton signals of the CH-groups of polar amino acid residues (lysines and arginines) are mainly in the region of 3.5 to 2.5 ppm, the CH-groups of methionines in the region of 2 ppm, and that of alanines in the region of 1.4 ppm [149-153, 156]. Protons of the methyl groups of hydrophobic amino acid residues (valines, leucines, isoleucines and threonines) are exhibited in the range from 1.2 to 0.7 ppm. It should also be noted that a well-expresses high field region (between -0.7 and +0.8 ppm) is, as a rule, presented in the ^{1}H-NMR spectrum of native protein, containing a significant number of specific resonances reflecting the interaction of protons of the CH-groups of the hydrophobic amino acids residues with the aromatic group circular currents [149-153, 156]. Such kinds of interactions are intrinsic only for the proteins with rigid 3D-structures, whereas upon the denaturation and unfolding this region of the spectrum undergoes essential changes, where virtually all the resonance lines disappear [156]. Furthermore, for some amino acid protons there are characteristic chemical shifts induced by the position of amino acids within elements of the secondary structure, the α-helices and the β-sheets. Thus, NMR is a powerful technique capable of yielding information on the structure of biopolymers in different conformational states, on the interaction between molecules and on the intramolecular mobility of proteins in different conformations [117, 143, 157-164].

Great hopes are currently placed on *two-dimensional NMR (2D-NMR)* and, more generally, *multidimensional NMR*, which, in principle, can be used for the determination of the precise 3D-structure of proteins via the assignment of all the peaks. Spectra have been analyzed, the peaks have been assigned and thus, 3-D structures in solution were determined for such small proteins as ribonuclease A [165], cytochrome *c* [166], hen egg lysozyme [167], human lysozyme [168], thioredoxin [169], insulin-like mini-growth hormone [170], leech derived tryptase inhibitor [171], turkey lysozyme [172] and many-many other proteins. It is interesting that such an approach can also determine the spatial structure for highly fluctuating polypeptide chains in the partially folded proteins. An example of such kind of analysis is the 130-amino acid residues-long fragment of a nuclease protein from *Staphylococcus aureus*, which is virtually in a completely unfolded state at neutral pH values of the medium, and whose residual structure was determined using the NMR spectroscopy [173].

The next step of application of NMR-spectroscopy for determination of the protein unique tertiary structure is a heteronuclear NMR. In this case modified proteins are studied where the carbon (or nitrogen, or both) atoms of defined groups are substituted by ^{13}C (or ^{15}N). Already first studies revealed that the application of this method highly facilitates the assignment of the separate resonance lines [174-177]. At the present, heteronuclear magnetic resonance is one of the most popular methods for determining the tertiary structure of proteins in solution (including relatively large proteins such as carbonic anhydrase B [178]).

An extremely convenient and promising approach for studying the protein folding and dynamics has been elaborated at the end of the eighties by the laboratories of Baldwin [179] and Englander [180]. This method is based on a combination of deuterium-exchange with 2D-NMR spectroscopy. It is known that 2D ^1H-NMR spectroscopy not only records signals from amide protons of the main chain, but can also distinguish between the proton and deuteron signals. The essence of the suggested approach is that the time courses of the deuterium-exchange reactions for the amide protons of the main chain are studied by 2D-^1H-NMR. It has been shown that the rates of exchange for protons participating in the formation of hydrogen bonds of secondary structure (α-helices and β-structure) both in the kinetic intermediates and in the completely folded conformation are significantly lower ($\sim 10^9$) than these in the coil-like state where such bonds are absent. Thus, if the amide proton is initially deuterated in the coil-like polypeptide chain dissolved in D_2O, then the degree of resistance of this deuteron to exchange with solvent protons at a defined moment of refolding can be determined by the degree to which this place is protonated after addition of H_2O [179, 180].

Electron Paramagnetic Resonance (EPR) Spectroscopy

Biomolecules, as a rule, do not contain unpaired electrons and, thus, are diamagnetics. However, EPR-spectroscopy can be applied to characterize biological systems using chemically added stable free radicals, known as the spin labels [181-183] . Nitroxil radicals are usually used as organic free radical labels since they can be obtained in the pure form, are stable under storage and are resistant to most reagents. Furthermore, spin labels containing various reaction groups at certain distances from the radical can be synthesized, thus specifically joining the labels to the biomolecules. The spin label is sensitive to the physical properties of its biomolecular microenvironment and as a consequence its EPR-spectrum can yield detailed information of such environment and indirectly characterize the biomolecular structure and dynamics. The shape of the EPR-spectrum depends on the mobility of the spin label [184].

The EPR-spectrum of the nitroxyl radical, present in a low concentration in a low viscosity solvent, consists of three decomposed sharp lines of about equal intensity. The characteristics of the spectrum are the distance between the spectrum lines (A, constant of hyperfine splitting), position of the central line of the spectrum (g-factor) and the spectrum line widths [184]. As a rule, EPR-spectra are sensitive to changes in the polarity of the label environment, to movements of the label and its orientation relative to the applied magnetic field [184]. Furthermore, the EPR-spectrum of the nitroxyl radical can be modified by the presence of other paramagnetic compounds and the spin label itself can increase the relaxation rate, being close to the nucleus. With the increase in the solvent polarity A decreases, whereas g-factor increases [185].

Often EPR-spectra can yield information on the mobility of the nitroxyl label as the increase of solvent viscosity results in widening of the EPR-lines with a loss of spectral symmetry [184]. Consequently, the absence of the label movement or its retardation can be detected via the specific features of the EPR-spectrum [185, 186]. For example, changes of the ribonuclease EPR-spectrum were used to determine the site of spin label binding to this protein [186]. Studies of spin-labeled hen egg lysozyme [187] using NMR spectroscopy and distance geometry calculations demonstrated that the EPR-label is in the hydrophobic groove of the protein [188].

Since the attachment of spin labels, as a rule, does not result in the deactivation of proteins [189], there is a possibility, in principle, to use EPR-spectroscopy to analyze the processes of de- and renaturation in globular proteins. Carlsson and colleagues in 1975 made the first attempt to use the changes in the characteristics of EPR-spectrum to study protein self-organization [190]. They demonstrated that the mobility of the label at refolding of carbonic anhydrase B decreases in less than 0.1 sec after the beginning of the process of protein renaturation [190]. A subsequent development of this work was done by Semisotnov and co-authors [191] who used the spin labels for studying the rapid compactization processes at the globular protein renaturation. It was shown that the compact molten globule-like intermediate state is formed extremely fast upon re-folding of carbonic anhydrase B ($t_{1/2}$ = 0.03±0.01 sec) [191].

Fluorescence Spectroscopy

Intrinsic Protein Fluorescence

With the exception for the discussed above GFP-like polypeptides, proteins contain only three intrinsic fluorescent chromophores, residues of tryptophan, tyrosine and phenylalanine (enumerated in order of the decrease is quantum yield). The most commonly used in practice is the fluorescence of tryptophan, since the quantum yield of phenylalanine fluorescence is extremely low and tyrosine fluorescence is strongly quenched in the majority of cases due to its ionization or location near the amide or carboxyl groups, or tryptophan [192].

The main advantage of intrinsic protein fluorescence in protein conformation studies is that the parameters of the tryptophan fluorescence (intensity and position of the maximal fluorescence) depend dramatically on the environment (solvent polarity, pH, presence of the quencher) [192]. For example, the fluorescence spectrum of the solvated tryptophan residue has a maximum in the vicinity of 350 nm, whereas the embedding this chromophore into the non-polar environment of a globular protein results in a characteristic shift of its fluorescence maximum to the shorter wavelengths (the so-called blue, or Stokes, shift) [192]. This makes intrinsic fluorescence an indispensable tool for the investigation of protein conformational changes.

Important information can be obtained analyzing the peculiarities of the *dynamic quenching* of intrinsic fluorescence with small molecules. Fluorescence quenching data are typically analyzed using the general form of the Stern-Volmer equation, taking into account not only dynamic, but also static quenching [193, 194]:

$$\frac{I_0}{I} = (1 + K_{SV}[Q])e^{V[Q]} \tag{7}$$

where I_0 and I are the fluorescence intensities in the absence and presence of quencher, K_{SV} is the dynamic quenching constant (Stern-Volmer constant), while V is a static quenching constant, and $[Q]$ is the total quencher concentration.

The information obtained from experiments on dynamic quenching is close to that from studies on deuterium exchange, since it reflects the accessibility of defined protein groups to molecules of quencher contained in the solvent. However, in distinction to the deuterium exchange, this method evaluates the dynamic process in much wider range of the amplitudes and times by using the quencher molecules of different size and charge. One of the smallest and most efficient quenchers of intrinsic protein fluorescence is the neutral oxygen molecule. The classical works of Lackowicz and Weber [195, 196] illustrated well the applicability of this method to study the intramolecular mobility of the protein molecule. The authors have shown that the rate of diffusion of oxygen within the protein molecule is only 2-4 times slower than in the aqueous medium. It has been also established that the oxygen can affect even those tryptophan residues, which according to X-ray structural analysis should not be accessible to O_2, suggesting that some large-scale structural fluctuations (se-called conformational breathing) take place even for the well-folded proteins. Studies of dynamic quenching of proteins with acrylamide were initiated by Eftink and Giron [197-199]. Acrylamide, being neutral in charge, significantly exceeds oxygen in size. This difference in size results in the significant decrease in the fluorescence quenching rates: the quenching rate of acrylamide in globular proteins was shown to be two orders below the corresponding value for oxygen [195, 196, 198]. A detailed analysis of the possibility of permeation of quenchers of different size, polarity and charge (oxygen, nitrite, methylvinylketone, nitrate, acrylate, acrylamide, acetone, and methylethylketone) into the different globular proteins, such as mellitin, parvalbumin, ribonuclease T, human serum albumine, aldolase and equine alcohol dehydrogenase was also performed [200]. The simultaneous application of two different quenchers whose molecules strongly differ in charge and size (acrylamide and succinimide) gave exhaustive information on the mobility of the internal tryptophan residue of the yeast phosphoglycerate kinase molecule [201].

In studies of acrylamide quenching of tryptophan-containing polypeptide chains with different structural organizations, Eftink and Giron showed that quenching can decrease by two orders of magnitude at the transition from free tryptophan and unstructured polypeptides to globular proteins, and that the activation energy of dynamic quenching can increase approximately threefold [197, 199]. The peculiarities of the conformational transition induced in the fragment of the tryptophan synthase from *E. coli* by the co-factor were analyzed using changes in the efficiency of the intrinsic fluorescence quenching by acrylamide [202]. Furthermore, it has been established that the degree of shielding of tryptophan residues by the intramolecular environment in the molten globule state is close to that determined for native proteins [6].

Additional information on the mobility of macromolecules in solution can be gained using *fluorescence polarization*. If excited light is polarized fluorescence will be only partially polarized or completely depolarized. The degree of fluorescence depolarization depends on the following factors characterizing the structural state of the protein molecule: 1) mobility of the emitting group which to strong degree depends on the density of its environment and, 2) the energy transfer between the chromophores [203-205]. It has been established that the relaxation times of the tryptophan residues determined from the polarized luminescence data are a reliable test for the compactness of the polypetide chain. For example, several authors have noted that the retention of intact disulfide bonds at denaturation in proteins containing disulfide bonds often resulted in a non-essential decrease of intrinsic fluorescence polarization [203-205], whereas disruption of the SS-bridges led to a dramatic decrease in the luminescence polarization to values approximately equal for all studied proteins [205, 206]. The relaxation times of tryptophan residues determined by fluorescence polarization for α-lactalbumin [5, 117] and bovine carbonic anhydrase B [143] evidenced the high (and comparable) degree of protein compactness both in the native and in the molten globule state.

Along with intrinsic protein fluorescence, the *fluorescence of introduced chromophore groups* is widely used in conformational studies. According to the type of interaction with the protein molecule the introduced chromophores are divided into covalently attached labels and non-covalently interacting probes. *Fluorescence labels* are indispensable means for studying *energy transfer* between two chromophore groups. The essence of this phenomenon is that at interaction of two oscillators at a small distance the electromagnetic field of the exited oscillator (donor) can induce oscillation with the same frequency in the non-excited oscillator (acceptor) [207-209]. It should be noted that the excitation energy transfer between the donor and the acceptor originates only after the fulfillment of several conditions: 1) the absorption (excitation) spectrum of the acceptor overlaps with the emission spectrum of the donor; 2) a sufficient spatial proximity of the donor and the acceptor is necessary, they must be at distance not exceeding a few dozen Angströms; 3) a sufficiently high quantum yield of donor luminescence is also necessary. All these conditions are satisfied for intrinsic protein chromophores (aromatic amino acids, prostetic groups attached to protein) as well as for non-protein chromophore groups cross-linked to the protein polypeptide chain by covalent bonds [210]. In order to analyze the efficiency of the energy transfer from protein chromophores to external fluorescent probes, protein labeling can be done both non-specifically (along the protein surface) and site-directed (at defined side groups, by choosing the sites of the donor and acceptor beforehand if the 3D-structure of protein is known). The great importance of the resonance energy transfer experiments is that the efficiency of the energy transfer from one luminescenting region of a biopolymer to another depends on their mutual location. According to Förster theory [207], the probability of energy migration (number of transitions per second) from the excited donor molecule (D) to the non-excited acceptor molecule (A) located from the first one at a distance R_{DA}, is proportional to the inverse sixth power of this distance. In practice, the direct energy transfer efficiency, E, is calculated as the relative loss of donor fluorescence due to interaction with the acceptor [192, 207, 209]:

$$E = 1 - \frac{\phi_{D,A}}{\phi_D} \tag{8}$$

Here, ϕ_D and $\phi_{D,A}$ are the fluorescence quantum yields of the donor in the absence and presence of the acceptor, respectively. Further, the energy transfer efficiency is directly related to the donor-acceptor distance R_{DA} and the characteristic donor-acceptor distance R_o:

$$E = \frac{1}{1 + \left(\frac{R_{DA}}{R_0}\right)^6} \tag{9}$$

The latter quantity is given by the well-known equation [209]:

$$R_0^6 = \frac{9000 \ln 10}{128\pi^6 N_A} \frac{\langle k^2 \rangle \phi_D}{n^4} \int_0^\infty F_D(\lambda)\varepsilon_A(\lambda)\lambda^4 d\lambda \tag{10}$$

where n is the refractive index of the medium, N_A is Avogadro's number, and λ is the wavelength. $F_D(\lambda)$ is the fluorescence spectrum of the donor with the total area normalized to unity, and $\varepsilon_A(\lambda)$ is the molar extinction coefficient of the acceptor. These data were obtained directly from independent experiments on each conformational state of interest. The $<k^2>$ represents the effect of the relative orientations of the donor and acceptor transition dipoles on the energy transfer efficiency. For a particular donor-acceptor orientation this parameter is given as:

$$k = (\cos\alpha - 3\cos\beta\cos\gamma) \tag{11}$$

where α is the angle between the transition moments of the donor and the acceptor, and β and γ are the angles between the donor and acceptor transition moments and the donor-acceptor vector, respectively. In an unfolded protein, a large number of random donor-acceptor orientations are possible, resulting in $<k^2> = 0.67$ as a statistical average [209, 211].

It is apparent that any structural changes in a protein molecule resulting in the changes of this distance must strongly affect the value of this parameter. It was shown that urea-induced unfolding of several proteins results in a significant (up to 30-50%) decrease in the efficiency of energy transfer from protein aromatic amino acids to the fluorescent label, dansyl chloride [212]. This approach was also successfully applied in the kinetic studies of bovine carbonic anhydrase B refolding [191]: it has been shown that the compactization of the protein molecule proceeds extremely fast (with a characteristic time of $t_{1/2} = 0.03\pm0.01$ sec) [191].

Fluorescent probes. The compact state of globular proteins is in a significant degree stabilized by hydrophobic interactions. Hydrophobic fluorescent probes (which are usually the compounds exhibiting intensive fluorescence upon interaction with protein, but with low fluorescence intensity in water) can be used to detect the hydrophobic regions in protein molecules exposed on the solvent. One of the most well studied hydrophobic fluorescent probes is 8-anilino-1-naphthalene sulfonate (ANS), which is most often used for analysis of protein structural properties [58, 213-215]. An interest to this probe reached its highest point when it was shown that there is a predominant interaction of ANS with the equilibrium and kinetic intermediates accumulating during the folding of globular proteins in comparison with their native and completely unfolded states [57, 143, 191, 216, 217]. The interaction of ANS with the protein molecule in the molten globule state is easy to detect due to the significant increase in the probe fluorescence intensity and a characteristic blue-shift of its maximal fluorescence [216, 217]. These properties were successfully applied to visualize the formation of partially folded intermediates during the process of protein re-folding [216, 217]. The interaction of ANS with proteins is also accompanied by the characteristic changes in the fluorescence lifetime of the probe [218]. The fluorescence decay of free ANS was well-described by the monoexponential law, whereas formation of complexes of this probe with proteins resulted in a more complicated dependence [218]. Analysis of the ANS fluorescence lifetimes in a number of proteins revealed that at least two types of ANS-protein interactions might exist. At interaction of the first type (characterized by fluorescence lifetime of about 1-5 ns) the probe molecules are bound to the surface hydrophobic clusters of the protein molecule and are in a relatively good contact with the solvent. At interaction of the second type (characterized by fluorescence lifetime of about 10-17 ns) the probe molecules are embedded into the protein molecule and are poorly accessible to the solvent [218].

The use of an analogous to ANS hydrophobic probe, TNS (2-toluidinylnaphthalene-6-sulfonate), revealed that similar to ANS, TNS interaction with proteins is accompanied by a blue shift of the fluorescence spectrum and an increase of the quantum yield [219]. TNS virtually does not interact with the coil-like and α-helical conformation of polylysine [220], but has a strong affinity to the β-structural conformation. These observations were interpreted as a high hydrophobicity of β-structure in comparison with α-helices. The similarity of the spectra of TNS, bound to β-structure, with the spectra of TNS in ethanol confirm the presence of non-polar regions in the β-structure of polylysine. Finally, it has been shown that the affinity of TNS to the polylysine β-structure strongly depends on the molecular weight of the polypeptide and on the excess of lysine residues over the probe. The requirement of an appropriate environment at the sites of TNS binding (hydrophobic regions) was demonstrated in studies of TNS binding with peptides. Studies of polylysine β-structure formation showed that probe binding occurs rapidly with a time of less than 0.1 sec [220].

Hydrodynamic Methods

Viscometry

The dependence of characteristic viscosity of a spherical molecule $[\eta]$ on its Stokes radius, R_S, was established by Einstein (e.g., see [14]):

$$[\eta] = \frac{4}{3}\pi \frac{2.5N_A}{M} R_S^3 \tag{12}$$

where N_A is the Avogadro's number and M is the molecular mass. For polymer molecules the relationship between $[\eta]$ and M is expressed by the formula:

$$[\eta] = KM^a \tag{13}$$

where K and a are the constants depending on solvent. For completely unfolded proteins in 6 M GdmCl these constants are well known from numerous publications, $K = 0.716$ and $a = 0.66$ [155]. Thus, the value of characteristic viscosity of unfolded proteins depends on the molecular mass. However, native globular proteins are characterized by the relatively constant value of this parameter ($[\eta] \sim 3$ g/cm^3) independent of the protein molecular size [14, 221]. In other words, the $[\eta]$ value for the compact form of a given protein is essentially lower than that for the random coil of a similar molecular mass [14, 221]. The widespread application of viscometry for protein de- and renaturation studies is based on this observation. For example, it has been shown that the molten globule states are almost as compact as the native conformations of the corresponding proteins [5, 51, 117, 143].

Analytical Ultracentrifugation (AUC)

Analytical ultracentrifugation is a means for determining molecular weight and the hydrodynamic and thermodynamic properties of a protein or macromolecule. Sample purity, molecular weight determination in the native state, analysis of associating systems, determination of sedimentation and diffusion coefficients and ligand binding can all be determined using this versatile instrument. With analytical centrifugation, the two basic types of experiments are sedimentation velocity and sedimentation equilibrium. Sedimentation is a general term used to denote movement of particles in the field of centrifugal force.

Sedimentation equilibrium is a method for measuring protein molecular masses in solution and for studying protein-protein interactions. In sedimentation equilibrium the sample is spun in an analytical ultracentrifuge at a speed high enough to force the protein toward the outside of the rotor, but not high enough to cause the sample to form a pellet. As the centrifugal force produces a gradient in protein concentration across the centrifuge cell, diffusion acts to oppose this concentration gradient. Eventually an exact balance is reached between sedimentation and diffusion, and the concentration distribution reaches equilibrium. This equilibrium concentration distribution across the cell is then measured while the sample is spinning, using absorbance detection. The key advantage of the sedimentation equilibrium is that the concentration distribution at equilibrium depends only on molecular mass, and is entirely independent of the shape of the molecule. These data give information on the molecular weight, density and composition of macromolecules [27]. Furthermore, it gives important information on the presence of association and/or changes of the dimensions of a given protein upon changes of its environment [51, 117].

Sedimentation velocity. The solute particles pellet at the bottom of the cell, producing a depletion of solute near the meniscus and the formation of a boundary between the depleted region and the uniform concentration of the sedimenting solute. This determines the rate of movement of a solute under a centrifugal field. The rate of movement is equal to the sedimentation coefficient(s), which depends directly on the mass of the particle and inversely on the frictional coefficient. This gives a measure of effective size of the particle. Sedimentation velocity allows measuring the following parameters: sedimentation coefficient(s); diffusion coefficient (if the sedimenting components are well separated); effective mass of solute components; shape information; molecular mass. In the sedimentation velocity experiments, the diffusion coefficient is measured by the change of concentration of the dissolved substance. There are many methods for determining both the concentration of the dissolved substance and the concentration gradient at any point of the sample. The most direct method is the measurement of the distribution of optical density or light absorption at various moments of time. Among other methods widely used to measure the differences in the refractive indices are the Rayleigh interferometer and Schlieren optics.

A *Rayleigh interferometer* exploits the interferences of two beams that are divided into upper and lower part such that the differences between the upper and lower part of interference pattern, instead of just one single intensity value on the interference pattern, are analyzed so that the external influences can be canceled out. The Rayleigh interferometer can measure the concentration of molecules with high accuracy due to its sensitivity to the refractive index changes. The Rayleigh interferometer is insensitive to vibration of its environment because the phase shift is calculated by comparing the upper part and lower part of the same fringe. Any vibration will affect both the upper and lower part of fringes equally. Another advantage is that the degree of the resolution on the changes of the refractive index can be modified by changing the sample path length and the magnification of the fringe patterns. Finally, the two beams can be widely separated as long as the magnification of the cylindrical lens is large enough [222].

Schlieren optics is the special optics technique which is extremely sensitive to deviations of any kind that cause the light to travel a different path. In other words, this is an optical system that records inhomogeneities within a medium by detecting the energy refracted by that portion of the medium in which the inhomogeneity occurs. The image appears in the form of a shadowgram, where the Schlieren optical system gives the refractive index gradient (related to the concentration gradient) as a function of radial position in the ultracentrifuge cell. The area under a 'Schlieren peak' provides a measure of the sedimenting concentration.

Elastic (Dynamic) Light Scattering (DLS)

The method of elastic (dynamic) light scattering (DLS) has been developed in recent years for rapid and precise studies of macromolecular diffusion. In the dynamic light scattering experiments, the scattering light is measured at the wavelengths only a little differing from those in the incident beam of coherent emission. When a beam of light passes through a colloidal dispersion, the particles or droplets scatter some of the light in all directions. If the light is coherent and monochromatic, as from a laser for example, it is possible to observe time-dependent fluctuations in the scattered intensity using a suitable detector such as a photomultiplier capable of operating in photon counting mode. These fluctuations arise from the fact that the particles are small enough to undergo random thermal (Brownian) motion and the distance between them is therefore constantly varying. Constructive and destructive interference of light scattered by neighboring particles within the illuminated zone gives rise to the intensity fluctuation at the detector plane which, as it arises from particle motion, contains information about this motion. Analysis of the time dependence of the intensity fluctuation can therefore yield the diffusion coefficient of the particles from which, via the Stokes Einstein equation, knowing the viscosity of the medium, the hydrodynamic radius or diameter of the particles can be calculated. Thus, DLS directly determines the diffusion coefficient of macromolecules in solution.

The method of dynamic light scattering was successfully applied to determine dimensions of the human α-lactalbumin in different conformational states, the native, the molten globule and the coil-like [223]. It has been demonstrated that the efficient linear dimensions of this protein in the acidic and temperature-denatured forms exceed the respective parameter characteristic for native protein by ~10%. However, the difference between the native and the completely unfolded protein with intact disulphide bonds was ~40% [223].

Small Angle X-Ray Scattering

Information on the size of scattering molecule can be obtained from the small angle part of the X-ray scattering curve. In this case the radius of gyration (R_g) can be calculated from the Guinier approximation [224]:

$$\ln I(Q) = \ln(I(0)) - \frac{R_g^2 Q^2}{3}) \quad (14)$$

where Q is the scattering vector determined as $Q = \frac{4\pi}{\lambda}\sin\theta$ (2Θ is the scattering angle and Λ is the X-ray wavelength); $I(0)$, the forward scattering amplitude; i.e., the scattering amplitude from the sample, in the direction opposite to that of the incident beam, is proportional to the square of the molecular weight of the molecule [224]. The degree of association of a protein can be determined from the $I(0)$ values measured by SAXS, as $I(0)$ for a pure N-mer sample is N-times that for a sample with the same number of monomers because each N-mer will scatter N^2 times as strongly, but the number of particles will be $1/N$ as many as in the pure monomer sample. At the present this approach is widely used in protein structural studies [11, 224-229].

Analysis of the small angle X-ray scattering data yields information not only on the size of scattering particles and their association state, but also on a number of other important structural characteristics [224, 225]. Particularly, it allows the evaluation of the degree of protein molecule globularization; i.e., the determination of the presence of tightly packed core in a protein molecule [229]. To this end, the representation of the X-ray data in the form of the Kratky plot is used [224]; i.e., the dependence of $I(Q)\times Q^2$ on the scattering vector Q. The shape of the Kratky plot is sensitive to the conformational state of the scattering protein molecule [11, 226-229]. It has been shown that the scattering curve in Kratky coordinates has a characteristic maximum in the case when the protein molecule is in the native state or in the molten globule state (i.e., it has a globular structure). The value of the parameter $I(Q)\times Q^2$ in this maximum will be $3I(0)/e^{-1}R_g^2$. If the protein molecule is in the completely unfolded state (has no globular structure), such a maximum will be absent on the respective scattering curve

[11, 226-229]. This approach proved to be very useful in describing the structural properties of the protein molecule in the pre-molten globule state. It was shown that the unique property of the polypeptide chain in this conformation is the absence of globular structure in the presence of residual secondary structure and noticeable compaction [10-12].

Chromatography

By taking into account the fact that chromatographic methods are used to separate multicomponent mixtures whose ingredients differ from each other in different physical properties: 1) the hydrodynamic dimensions (gel filtration); 2) the degrees of hydrophobicity (reverse phase chromatography); 3) the electric charges (ion-exchange chromatography); 4) the affinities to other molecules (affinity chromatography) and others, one can understand why these methods have attracted the great attention of researchers. Furthermore, chromatography is a unique tool, which, in principle, allows physical separation of conformers. This separation of conformers, being associated with the appearance of individual peaks at the output of the chromatographic system, usually takes place with the fulfillment of three evident conditions [230, 231]: 1) analyzed system is under conditions corresponding to the conformational transition region; 2) different conformational states strongly differ from each other in their physical properties and/or in their affinity to the column matrix; 3) the exchange rate between the conformers is essentially lower than the characteristic time constant of the given chromatographic process. In the opposite case the movement of only one peak containing rapidly exchanging conformers will be observed.

In my opinion such a feature of the chromatographic method as the capability to obtain a real physical separation of the conformational states presents special interests, since in this way we obtain a unique possibility to observe the bimodal distribution by some physical parameter in the region of the conformational transition. It is obvious that the presence of such a bimodal distribution is a direct indication that the studied process proceeds according to the all-or-none principle [144, 145, 230]. Let us dwell in more detail on one of the chromatographic methods, namely gel filtration. The principal possibility of applying this method, which is sensitive to the molecular size changes, for the protein unfolding studies was substantiated as early as the seventies [232-234]. However, the available at that time column matrixes based on dextrans, polyacrylamide or agarose soft gels showed high resistance to flow and had a low chemical and mechanical stability. All this taken together made it impossible to use this method for studies of protein denaturation in solution with a high concentration of urea or GdmCl. Attempts to attain this goal by using a hard carrier or glass beads of different pore size, though they increased carrier mechanical and chemical stability, resulted in very low resolution [235]. An ideal possibility of using gel filtration for studies of all kinds of denaturation processes in proteins appeared only with the development of new types of column matrixes (in particular, of gel series TSK SW for high pressure liquid chromatography [236]), possessing high chemical and mechanical stability and giving excellent calibration curves for proteins both in the presence and in the absence of denaturant [237].

What kind of information can be gained from the analysis of gel filtration elution profiles? First, the dimensions of the macromolecule and the scale of their changes can be readily evaluated [230, 232-234, 238]. To this end, a gel-filtration column should be calibrated with a set of proteins with known hydrodynamic dimensions. This will result in a calibration curve coupling the elution volume of the given protein (or its retention coefficient on the column, K_r) with its hydrodynamic radius. The retention coefficient is determined from the relationship:

$$K_r = \frac{V - V_0}{V_t - V_0} \quad (15)$$

where V is the elution volume of the protein of the given molecular mass, V_0 is the void column volume determined by the elution volume of blue dextran and V_t is the total column volume determined by the elution volume of some low molecular compound. It is important to remember that the precision of the Stokes radius determination with the gel filtration column directly depends on the number of proteins used for its calibration.

Second, localization of the all-or-none transition can be determined from analysis of the elution profile by plotting the weight content of each of the conformational states in the reaction mixture from the value of the denaturing effect [7, 9, 144, 145, 230, 239]. Evidently, such kind of information can be only gained in the case when the exchange rate between the conformers is low in comparison with the rate of the chromatographic process.

It is known that denaturation and unfolding of globular proteins is accompanied not only by the changes in their hydrodynamic sizes but also by altering of some other physico-chemical characteristics of a protein molecule. The use of different varieties of chromatography yields a sufficiently complete description of conformational changes occurring in a protein upon its denaturation and unfolding. One of the first studies in this direction was carried out in 1987 [240]. The urea-induced unfolding of bovine serum albumin, lysozyme and trypsin was studied using gel filtration, reverse phase and ion-exchange chromatography. It has been established that the alterations in proteins tertiary structure are manifest chromatographically by highly reproducible changes in peak height, retention, and appearance of multiple peaks. Denaturation equilibria and kinetics obtained by classical physical methods, such as fluorescence intensity measurements for bovine serum albumin and enzyme activity for trypsin, were correlated to particular changes in chromatographic behavior. Furthermore, it has been established that the denaturation curves produced by these chromatographic methods correlated well with the curves recorded by the traditional methods (change of intrinsic fluorescence spectra, loss of enzymatic activity and disappearance of the CD signal in the near-UV region) [240].

Thus, the chromatographic methods are selective toward specific structural changes and can be used to monitor the independent denaturation steps via multiphasic kinetics. The correlation of chromatographic behavior with traditional measures of conformational changes indicates that non-denaturing high-performance liquid chromatography can be a useful tool to detect and quantitate perturbations of protein structure associated with protein denaturation and unfolding [240].

Electrophoresis

Electrophoresis is most frequently used either to determine the molecular mass or to elucidate the charge difference of proteins. However, the suggested by Creighton [241] modification of this method, *electrophoresis in the urea gradient*, not only opens the possibility to obtain the denaturation curves directly [241-243], but also to determine the kinetic parameters of unfolding-refolding reaction [241-243]. Urea gradient gels are slab gels prepared with a horizontal gradient of urea concentration, usually 0 - 8 M. A single sample of protein is applied across the top of the gel and electrophoresed in a direction perpendicular to the urea gradient. As the protein is electrophoresed, molecules at different positions across the gel are exposed to different urea concentrations. At positions where the urea concentration is high enough to promote unfolding, the mobility of the protein decreases because of the greater hydrodynamic volume of the unfolded form. When the gel is stained, or otherwise visualized, the electrophoretic pattern can be interpreted directly as an unfolding curve. Although urea gradient gels generally provide only qualitative or semi-quantitative information, they offer several advantages for studying protein folding/unfolding transitions, since they are: easy to implement, with only relatively inexpensive equipment; require only small amounts of protein (typically 50 micrograms); applicable to the analysis of complex mixtures; able to detect conformational or covalent heterogeneity in protein samples. This method represents special interest for studies of artificial or mutated proteins with the introduced radioactive label. It was shown that authentic information on the stability of modified protein can be obtained by electrophoresis in the urea gradient at very low concentrations of the goal product (at the level of several nanograms). Here there is no need to preliminarily purify the studied radioactive-labeled protein [244, 245].

As it has been mentioned above, urea gradient electrophoresis might help estimation of the kinetic parameters of unfolding-refolding reaction [241]. In fact, if a continuous, S-shaped band is observed after staining the gel, one can assume that the folding time is incomparably smaller than the running time of electrophoresis. Furthermore, the experiment can be started both with the folded and unfolded protein, sampled on the gel. When the time required for the folding process is smaller than the running time of the electrophoresis, the bands will not differ from each other, as within the transition region the protein molecules will fold and unfold many times during the electrophoretic process. As a result, at a given urea concentration they will move with the same average mobility. The different situation is observed when the folding rate is in the range of the electrophoresis running time. In this case, the electrophoretic bands will be different for initially folded and initially unfolded protein samples: the curve describing unfolding will shift to much higher urea concentrations, whereas the refolding curve will appear at lower urea concentrations than the true transition region (because the kinetic rates are dependent on the urea concentration). From the difference between the bands one can make conclusions on the folding and unfolding kinetic rates. If the folding time is larger than the electrophoresis running time, then the transition region will be smeared out, as there will be a population, that can fold within the running time, and there will be a population, that won't be able to do it so. Therefore the transition region will be smeared out, with two sharper bands at the edges [241].

Based on the success of the urea gradient electrophoresis, some other modifications of the polyacrylamide-gel-electrophoresis method were elaborated. For example, the *thermal-*

gradient electrophoresis has been developed that permitted the analysis of conformational changes induced by heating [246]. To this end, a stable transverse temperature gradient was produced in an aluminium heating jacket clamped around a vertical polyacrylamide slab gel. After temperature equilibration, gels were loaded with a layer of protein solution and electrophoresis begun. At the end of the run the gels were stained and the effect of temperature on mobility observed. The technique proved informative both for the irreversible unfolding of proteins (Drosophila alcohol dehydrogenase and lactic acid dehydrogenase) and for a protein that was reversibly denatured by heat (β-lactamase). In the latter case a clear transition between the native enzyme and a slower-migrating denatured state was observed [246]. The patterns obtained were analogous to those described above for the urea gradient electrophoresis.

Another very useful modification of the polyacrylamide-gel-electrophoresis method is the *pH-gradient electrophoresis*. It has been pointed out that the migration across a stationary pH gradient results in the electrophoretic titration of a protein's dissociable groups. From the resulting curves, some properties of the protein may be derived, including overall amino acid composition and type of mutation between polymorphic variants, as well as range of stability or, for enzymes, of catalytic activity. Analysis with this technique is a straightforward purity criterion; other applications allow the study of interacting systems based on the differences in surface charge [247]. The most versatile application of the pH-gradient electrophoresis is to measure protein charge (pI) and to separate proteins according to their charges (pI); i.e., *the isoelectric focusing (IEF)*. IEF was introduced by Alexander Kolin in mid fifties [248, 249]. Since that time this method underwent dramatic development, starting as a somewhat shaky two-step technique involving generation of a pH gradient by diffusion followed by a rapid electrokinetic protein separation and ending with present-day immobilized pH gradients [250]. The range of modern applications of IEF is very wide. For example, it has been established that an excellent agreement is observed between computed and experimental isoelectric point (pI) values when proteins of known sequence are focused under denaturing conditions on immobilized pH gradient IPG slabs. The discrepancies between the calculated and measured pI values are ascribed to post-transcriptional modifications, e.g., glycosylation and phosphorylation or to some chemical modifications, e.g., carboxymethylation, carbamylation, deamidation and thiol redox reactions [251].

Finally, *two-dimensional polyacrylamide gel electrophoresis (2-DE)* should be briefly discussed. This methodology was developed by Patrick H. O'Farrell in 1975 [252]. Proteins are separated here according to isoelectric point (pI) by isoelectric focusing in the first dimension, and according to molecular weight by sodium dodecyl sulfate electrophoresis in the second dimension. It has been emphasized that since these two parameters are unrelated, it is possible to obtain an almost uniform distribution of protein spots across a two dimensional gel. Thus, 2-DE has a unique capacity for the resolution of complex mixtures of proteins, permitting the simultaneous analysis of hundreds or even thousands of gene products. In fact, already in the first application of this methodology, 1,100 different components from *Escherichia coli* were resolved [252]. Furthermore, this technique possesses an outstanding sensitivity and proteins differing in a single charge can be resolved. Combination of the methods with autoradiography opens possibility to detect proteins at really low quantities, e.g., proteins containing as little as one disintegration per min of either ^{14}C or ^{35}S can be identified [252]. These unique capabilities combined with the reasonable reproducibility (each spot on one separation can to be matched with a spot on a different separation) suggested that this technique provides a method for estimation of the number of proteins made by any biological system [252]; i.e., it really represents a basement of the modern proteomics. Currently, 2-DE combined with protein identification by mass spectrometry represents the workhorse for proteomics [253]. In fact, 2-DE is currently the only technique that can be routinely applied for parallel quantitative expression profiling of large sets of complex protein mixtures such as whole cell lysates. Depending on the gel size and pH gradient used, 2-DE can now resolve more than 5,000 proteins simultaneously (approximately 2000 proteins routinely), and detect and quantify < 1 ng of protein per spot [253].

Biological Activity

The changes in tertiary structure, at least in the region of the active centre, can be visualized via the suppression of the protein biological activity in the course of its de- or renaturation. That is why protein functional assays are widely used in the equilibrium and kinetic unfolding and refolding studies. For example, the equilibrium unfolding studies of carbonic anhydrase B [9, 143-145] and β-lactamase [7] revealed an excellent correlation between the denaturation curves monitored by decreasing the activity of these proteins and those followed by change in their spectral characteristics (near-UV CD spectra). Similarly, it has been shown that the restoration of carbonic anhydrase B activity in the kinetic refolding studies coincides in time with the restoration of its near-UV CD spectra [191]. A similar behavior was also described for creatine kinase [59] and several other enzymes.

Methods Based on Protein-Protein Interactions

It is evident that the conformational changes induced in a protein by amino acid substitutions or changes in the environment can result in essential alteration of a protein surface and, as a consequence, can alter the efficiency of the studied protein interaction with its binding partners, e.g., other proteins. This section of the chapter will consider methods based on studies of different protein-protein interactions.

Limited Proteolysis

It is known that proteolysis occurs predominantly in unfolded or disordered regions of proteins. For example, trypsin ought to cleave polypeptide at nearly every lysine-X and arginine-X bond (with the partial exception of proline at X), assuming that about 5–10% of the peptide bonds in a typical protein must be susceptible to proteolytic attack. However, in native (or near-native) conditions trypsin will cut only a limited number of such bonds (or on occasions none at all) in a native protein fold. This means that the structure and dynamics of the substrate protein are playing crucial roles in limiting the proteolysis [254, 255]. Interestingly, the analysis of X-ray crystallographic structures of proteases with small protein inhibitors, such as BPTI, revealed that the inhibitor reactive site loop bound into the enzyme active site in the manner of a 'perfect' substrate [256, 257]. Using this canonical conformation of the inhibitor reactive site loops as a template, Thornton and co-workers showed that limited proteolytic sites are quite different in structure from the idealized inhibitor loops, and they must therefore undergo a conformational change in order to enter the proteinase active site [258]. Furthermore, for several proteins it has been shown that local unfolding of at least 13 residues is needed for a set of observed cut-sites to properly fit into trypsin's active site [254, 255]. It has been also established that limited proteolytic sites are typically found at flexible loop regions (as indicated by crystallographic temperature factors or B-values) that are also exposed to the solvent [258-260], and are notably absent in regions of regular secondary structure, especially β-sheets [254, 261, 262]. Fontana and co-workers showed that apomyoglobin is digested many orders of magnitude faster than myoglobin and the sites of digestion from several different proteases all mapped to a region of intrinsic disorder [261]. Similarly, it has been shown that while the holo-form of cytochrome c is fully resistant to proteolytic digestion and the apoprotein is digested to small peptides, the non-covalent complex of the apoprotein and heme exhibits an intermediate resistance to proteolysis, in agreement with the fact that the more folded structure of the complex makes the protein substrate more resistant to proteolysis [263].

Thus, protease digestion is much faster in the disordered regions. Furthermore, it has been shown that the denaturation of globular protein is accompanied by a strong increase of the rate of proteolysis and increase of the number of proteolytic fragments [264]. These observations create a basement for the application of proteases for studies of conformational changes in globular proteins. Studies on several mutant forms of dihydrofolate reductase from *E. coli* revealed that there was an excellent agreement between the resistance of mutant proteins to trypsinolysis and their stability toward the temperature effect or urea-induced unfolding [265]. It has also been established that for many proteins the proteolytic resistance and the resulting proteolytic maps change significantly upon interaction with natural ligands. Thus, the resistance of troponin C to trypsinolysis essentially increased as a result of the Ca^{2+} addition to the apoprotein solution [266-269], and a change of the site of proteolitic attack was also observed. In fact, trypsinolysis of holo-protein (18 kD) resulted at first in the formation of a large (~16 kD) troponin C fragment which then was cleaved into two peptides (8.5 and 7.5 kD, respectively). The removal of Ca^{2+} resulted in the dramatic changes of the proteolytic map: the first products of proteolysis were two fragments with molecular masses of 11.0 and 7.5 kD.

At the later step, the large fragment degraded to a peptide with a molecular mass of 9.7 kD. Soon after the reaction was completed by the extensive cleavage of the two fragments into short peptides [266-269]. A noticeable increase of the proteolysis rate upon removal of the ligand was noted also for such proteins as perch parvalbumin [270], intestinal calcium-binding protein [271], phospholipase A [272], adenylate kinase [273] and many others. It was also suggested that limited proteolysis can be utilized in the kinetic unfolding/refolding experiments. As a matter of fact, this approach was used to describe the process of formation of the lactate dehydrogenase monomer upon the exposure of this tetrameric protein to the acidic solutions [274].

Immunochemical Methods

Stability of the antibody-antigen complexes is determined by the same forces, which are responsible for the maintaining of the unique protein 3D-structure: hydrogen bonds, hydrophobic interactions and van der Waals forces. It is also noteworthy that the antibodies obtained against the given protein-antigen are specific to the different levels of the macromolecular organization: protein primary [275, 276], secondary [277] or tertiary structure [275-277]. In the latter case the antigenic determinants are both the neighboring in the chain (loops) [275, 276] and the spatially dispersed structures [277]. Furthermore, it was shown that there are antibodies in the immune serum possessing a high affinity to the internal elements of the antigen structure [277]. All this allows the successful using of antibodies to study the structural changes undergone by the protein-immunogene upon changes of the media conditions. For example, the antibodies obtained against the Ca^{2+}-saturated fragment of prothrombin did not interact with the apo-form of this protein [278]. An analogous effect was observed in the case of osteocalcine [279]. It has also been noted that antibodies are capable of recognizing mutant protein forms in which replacement occurred of only one amino acid [280, 281].

The production of monoclonal antibodies distinguishing the antigen tertiary structure opened a unique possibility to study the protein molecule 3D-structure under the variety of non-native conditions. Thus, the refolding kinetics of the tryptophan syntase subunit was studied using this approach [282, 283]. One of the conformational antibodies (IgG-19) was shown to specifically recognize the protein-immunogen chain long before the formation of its rigid tertiary structure [282]. Subsequent studies have shown that the earliest immuno-reactive intermediate, recognized by IgG-19, is the kinetic molten globule [283].

Interaction with Molecular Chaperons

Significant factual material has been accumulated evidencing that the specific protein factors do exist in the cell, playing an extremely important role in the formation and maintenance of the protein native conformation [284]. The idea was even formulated that together with spontaneous self-organization an assistant folding also occurs in the cell [285, 286]. This concept assumed that although folding and self-organization of the protein molecule is generally a spontaneous process, there are critical steps at which the participation of cell factors may be required. It is thought that the main role of these protein factors, called molecular chaperones, consists in the maintenance of optimal conditions in the course of self-organization of the protein molecule by removing obstacles and incorrect contacts, e.g., preventing non-specific association and aggregation of folding intermediates [284-286]. The class of molecular chaperons includes numerous proteins. The most representative members of this family are the so-called heat-shock proteins, whose biosynthesis in the cell increases multifold under the stress conditions induced by temperature increase. The most widely studied heat-shock protein is known as GroEL, or chaperonin-60. This protein is a large oligomeric complex consisting of two rings with seven identical subunits in each ring. The molecular mass of one subunit is ~60 kD [287]. It has been established that the functional activity of GroEL is regulated by interaction with the protein co-chaperone, GroES, consisting of seven identical subunits each of 10 kD, and also forming a ring structure [288].

A detailed conception of chaperone-assisted protein folding is considered in several recent reviews [289-291]. The main characteristic property of chaperones is their unique capability to recognize and bind partially folded conformations of target. In fact, in the refolding experiments performed for of several proteins *in vitro*, it has been established that GroEL predominantly interacts with the protein molecule in the molten globule-like intermediate [289, 291-299]. Thus, the protein-chaperone GroEL can be considered as a unique special antibody against non-native conformations of the protein molecule. The promises of this approach in studies on protein folding in complex systems become further evident from the fact that the formation of the GroEL-protein complexes can be readily recorded (e.g., by chromatography).

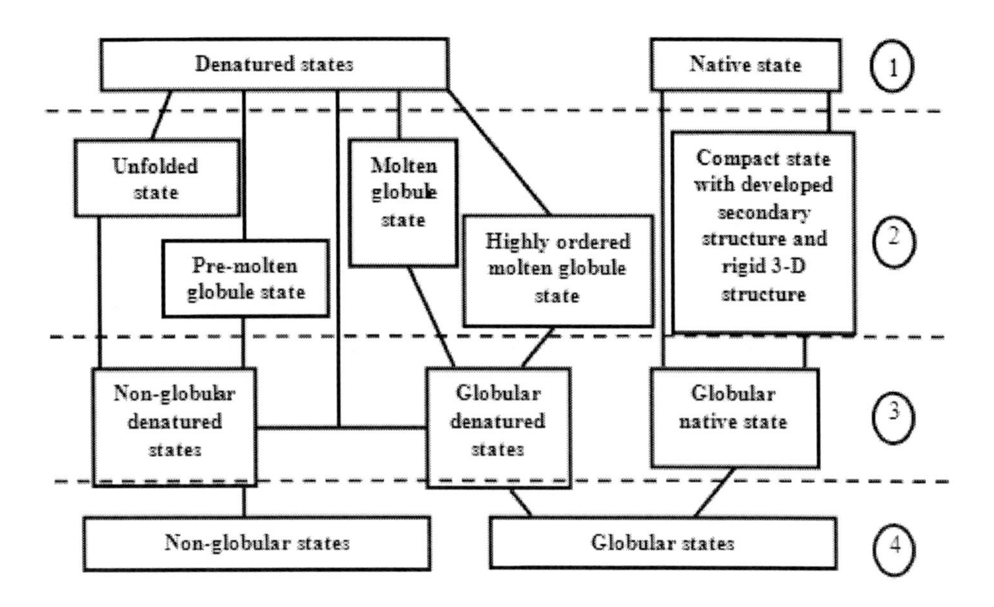

Figure 2. Scheme of conformational states of a typcal globular protein. Different struyctural criteria were used on making the scheme: 1) the presence or absence of biological activity only; 4) the presence or absence of globular structure only; 3) conformations appearing on consideration of only two structural parameters (biological activity and globularity); 2) picture derived on consideration of many structural parameters (activity, secondary structure, compactness, globularity, etc.).

Conclusion

I would like to conclude this chapter by illustrating the thesis that the description of the structural properties of a protein molecule should not be relied on only one (even very specific) structural parameter. This statement is exemplified by Figure 2, which represents a set of known equilibrium conformational states accessible by a protein at different environmental conditions. Figure 2 shows that the complexity of the observed picture and the validity of the result interpretations strongly depend on what structural criteria are taken into account. Importantly, consideration of the studied system from the viewpoint of presence or absence of some specific characteristics, or globular structure (level 2), or any other structural elements) gives a very simplified picture. For example, by observing the absence of biological activity (level 1) one can conclude that the protein is denatured. However, at least four denatured forms are known for many proteins, highly ordered molten globule, classic molten globule, pre-molten globule and coil, and the knowledge of the missing biological activity does not allow their discrimination. Similarly, a set of rigid and molten globular conformations can be discriminated only from a pre-molten globule/coil set by the considering the presence or absence of the globular structure (level 2). The situation is no better even in the case when two specific parameters are considered, activity and globularity (see Figure 2): e.g., if a protein does not have biological function and is not globular, it might accommodate either pre-molten globule or random coil conformation. Only a detailed analysis of the system using the several techniques sensitive to the different structural levels of a protein can yield a more-or-less complete picture and assign a protein to the correct region of the conformational space (see Figure 2). Thus, long life for the multiparametric approach!

References

[1] Creighton, T. E. (1983) *Proteins. Structure and Molecular Properties.*, W. H. Freeman, New York.

[2] Anfinsen, C. B., Haber, E., Sela, M. and White, F. H., Jr. (1961) The kinetics of formation of native ribonuclease during oxidation of the reduced polypeptide chain. *Proc Natl Acad Sci U S A 47*, 1309-14.

[3] Phillips, D. C. (1967) The hen egg white lysozyme molecule. *Proc Natl Acad Sci U S A 57*, 484-495.

[4] Lim, V. I. and Spirin, A. S. (1985) [Stereochemistry of the transpeptidation reaction in the ribosome. The ribosome generates an alpha-helix in the synthesis of the protein polypeptide chain]. *Dokl Akad Nauk SSSR 280*, 235-9.

[5] Dolgikh, D. A., Gilmanshin, R. I., Brazhnikov, E. V., Bychkova, V. E., Semisotnov, G. V., Venyaminov, S. and Ptitsyn, O. B. (1981) Alpha-Lactalbumin: compact state with fluctuating tertiary structure? *FEBS Lett 136*, 311-5.

[6] Ptitsyn, O. B. (1995) Molten globule and protein folding. *Adv Protein Chem 47*, 83-229.

[7] Uversky, V. N. and Ptitsyn, O. B. (1994) "Partly folded" state, a new equilibrium state of protein molecules: four-state guanidinium chloride-induced unfolding of beta-lactamase at low temperature. *Biochemistry 33*, 2782-91.

[8] Ptitsyn, O. B., Bychkova, V. E. and Uversky, V. N. (1995) Kinetic and equilibrium folding intermediates. *Philos Trans R Soc Lond B Biol Sci 348*, 35-41.

[9] Uversky, V. N. and Ptitsyn, O. B. (1996) Further evidence on the equilibrium "pre-molten globule state": four-state guanidinium chloride-induced unfolding of carbonic anhydrase B at low temperature. *J Mol Biol 255*, 215-28.

[10] Uversky, V. N., Karnoup, A. S., Segel, D. J., Seshadri, S., Doniach, S. and Fink, A. L. (1998) Anion-induced folding of Staphylococcal nuclease: characterization of multiple equilibrium partially folded intermediates. *J Mol Biol 278*, 879-94.

[11] Tcherkasskaya, O. and Uversky, V. N. (2001) Denatured collapsed states in protein folding: example of apomyoglobin. *Proteins 44*, 244-54.

[12] Uversky, V. N. (2003) Protein folding revisited. A polypeptide chain at the folding-misfolding-nonfolding cross-roads: which way to go? *Cell Mol Life Sci 60*, 1852-71.
[13] Tanford, C., Pain, R. H. and Otchin, N. S. (1966) Equilibrium and kinetics of the unfolding of lysozyme (muramidase) by guanidine hydrochloride. *J Mol Biol 15*, 489-504.
[14] Tanford, C. (1968) Protein denaturation. *Adv Protein Chem 23*, 121-282.
[15] Lumry, R. and Biltonen, R. (1966) Validity of the "two-state" hypothesis for conformational transitions of proteins. *Biopolymers 4*, 917-44.
[16] Kim, P. S. and Baldwin, R. L. (1982) Specific intermediates in the folding reactions of small proteins and the mechanism of protein folding. *Annu Rev Biochem 51*, 459-89.
[17] Privalov, P. L. (1979) Stability of proteins: small globular proteins. *Adv Protein Chem 33*, 167-241.
[18] Cooper, A. (1976) Thermodynamic fluctuations in protein molecules. *Proc Natl Acad Sci U S A 73*, 2740-1.
[19] Anson, M. L. and Mirsky, A. E. (1932) The effect of denaturation on the viscosity of protein systems. *J. Gen. Physiol. 15*, 341-350.
[20] Anson, M. L. and Mirsky, A. E. (1934) The equilibrium between active native trypsin and inactive denatured trypsin. *J Gen Physiol 17*, 393-298.
[21] Ptitsyn, O. B. and Uversky, V. N. (1994) The molten globule is a third thermodynamical state of protein molecules. *FEBS Lett 341*, 15-8.
[22] Uversky, V. N. and Ptitsyn, O. B. (1996) All-or-none solvent-induced transitions between native, molten globule and unfolded states in globular proteins. *Fold Des 1*, 117-22.
[23] Hill, T. L. (1963/1964) *Thermodynamics of Small Sustems*, Wiley, New York.
[24] Privalov, P. L. (1982) Stability of proteins. Proteins which do not present a single cooperative system. *Adv Protein Chem 35*, 1-104.
[25] Privalov, P. L. (1989) Thermodynamic problems of protein structure. *Annu Rev Biophys Biophys Chem 18*, 47-69.
[26] Griko Yu, V., Venyaminov, S. and Privalov, P. L. (1989) Heat and cold denaturation of phosphoglycerate kinase (interaction of domains). *FEBS Lett 244*, 276-8.
[27] Freifelder, D. (1982) *Physical Biochemistry: Applications to Biochemistry and Molecular Biology*, WH Freeman, New York.
[28] Donovan, J. W. (1973) Ultraviolet difference spectroscopy--new techniques and applications. *Methods Enzymol 27*, 497-525.
[29] Kitagishi, K. and Hiromi, K. (1986) *Agric Biol Chem 50*, 1113-1115.
[30] Hiromi, K., Onishi, M., Kanaya, K. and Matsumoto, T. (1975) The pH jump study of enzyme proteins. I. Liquefying alpha-amylase from Bacillus subtilis. *J Biochem (Tokyo) 77*, 957-63.
[31] Kuwajima, K., Ogawa, Y. and Sugai, S. (1979) Application of the pH-jump method to the titration of tyrosine residues in bovine alpha-lactalbumin. *Biochemistry 18*, 878-82.
[32] Dickerson, R. E. and Timkovich, R. (1975) Cytochromes c. in *The Enzymes* (Boyer, P. O., Ed.) pp 397-547, Academic Press, New York.
[33] Adar, F. (1978) Electronic absorption spectra of hemes and heme proteins. in *The Porphyrins* (Dolphin, D., Ed.) pp 167-209, Academic Press, New York.
[34] Margoliash, E. and Schejter, A. (1966) Cytochrome c. *Adv Protein Chem 21*, 113-286.

[35] Tsong, T. Y. (1975) An acid induced conformational transition of denatured cytochrome c in urea and guanidine hydrochloride solutions. *Biochemistry 14*, 1542-7.

[36] Brems, D. N. and Stellwagen, E. (1983) Manipulation of the observed kinetic phases in the refolding of denatured ferricytochromes c. *J Biol Chem 258*, 3655-60.

[37] Yoshimura, T. (1988) A change in the heme stereochemistry of cytochrome c upon addition of sodium dodecyl sulfate: electron paramagnetic resonance and electronic absorption spectral study. *Arch Biochem Biophys 264*, 450-61.

[38] Babul, J. and Stellwagen, E. (1972) Participation of the protein ligands in the folding of cytochrome c. *Biochemistry 11*, 1195-200.

[39] Jhee, K. H., Yoshimura, T., Miles, E. W., Takeda, S., Miyahara, I., Hirotsu, K., Soda, K., Kawata, Y. and Esaki, N. (2000) Stereochemistry of the transamination reaction catalyzed by aminodeoxychorismate lyase from Escherichia coli: close relationship between fold type and stereochemistry. *J Biochem (Tokyo) 128*, 679-86.

[40] Shimomura, O. (1979) Structure of the chromophore of Aequorea green fluorescent protein. *FEBS Lett 104*, 220-222.

[41] Cody, C. W., Prasher, D. C., Westler, W. M., Prendergast, F. G. and Ward, W. W. (1993) Chemical structure of the hexapeptide chromophore of the Aequorea green-fluorescent protein. *Biochemistry 32*, 1212-8.

[42] Heim, R., Prasher, D. C. and Tsien, R. Y. (1994) Wavelength mutations and posttranslational autoxidation of green fluorescent protein. *Proc Natl Acad Sci U S A 91*, 12501-4.

[43] Warren, A. and Zimmer, M. (2001) Computational analysis of Thr203 isomerization in green fluorescent protein. *J Mol Graph Model 19*, 297-303.

[44] Ward, W. W. and Bokman, S. H. (1982) Reversible denaturation of Aequorea green-fluorescent protein: physical separation and characterization of the renatured protein. *Biochemistry 21*, 4535-40.

[45] Niwa, H., Inouye, S., Hirano, T., Matsuno, T., Kojima, S., Kubota, M., Ohashi, M. and Tsuji, F. I. (1996) Chemical nature of the light emitter of the Aequorea green fluorescent protein. *Proc Natl Acad Sci U S A 93*, 13617-22.

[46] Palm, G. J., Zdanov, A., Gaitanaris, G. A., Stauber, R., Pavlakis, G. N. and Wlodawer, A. (1997) The structural basis for spectral variations in green fluorescent protein. *Nat Struct Biol 4*, 361-5.

[47] Wachter, R. M., Elsliger, M. A., Kallio, K., Hanson, G. T. and Remington, S. J. (1998) Structural basis of spectral shifts in the yellow-emission variants of green fluorescent protein. *Structure 6*, 1267-77.

[48] Ito, Y., Suzuki, M. and Husimi, Y. (1999) A novel mutant of green fluorescent protein with enhanced sensitivity for microanalysis at 488 nm excitation. *Biochem Biophys Res Commun 264*, 556-60.

[49] Herskovits, T. T. and Laskowski, M., Jr. (1962) Location of chromophoric residues in proteins by solvent perturbation. I. Tyrosyls in serum albumins. *J Biol Chem 237*, 2481-92.

[50] Schmidt, F. X. (1989) Spectral methods of characterizing protein conformation and conformational changes. in *Protein Structure. A Practical Approach* (Creighton, T. E., Ed.) pp 251-285, IRL Press, Oxford, New York, Tokyo.

[51] Wong, K. P. and Tanford, C. (1973) Denaturation of bovine carbonic anhydrase B by guanidine hydrochloride. A process involving separable sequential conformational transitions. *J Biol Chem 248*, 8518-23.
[52] Robson, B. and Pain, R. H. (1976) The mechanism of folding of globular proteins. Equilibria and kinetics of conformational transitions of penicillinase from Staphylococcus aureus involving a state of intermediate conformation. *Biochem J 155*, 331-44.
[53] McCoy, L. F., Jr., Rowe, E. S. and Wong, K. P. (1980) Multiparameter kinetic study on the unfolding and refolding of bovine carbonic anhydrase B. *Biochemistry 19*, 4738-43.
[54] Goto, Y., Calciano, L. J. and Fink, A. L. (1990) Acid-induced folding of proteins. *Proc Natl Acad Sci U S A 87*, 573-7.
[55] Goto, Y. and Fink, A. L. (1990) Phase diagram for acidic conformational states of apomyoglobin. *J Mol Biol 214*, 803-5.
[56] Goto, Y., Takahashi, N. and Fink, A. L. (1990) Mechanism of acid-induced folding of proteins. *Biochemistry 29*, 3480-8.
[57] Goto, Y. and Fink, A. L. (1989) Conformational states of beta-lactamase: molten-globule states at acidic and alkaline pH with high salt. *Biochemistry 28*, 945-52.
[58] Stryer, L. (1965) The interaction of a naphthalene dye with apomyoglobin and apohemoglobin. A fluorescent probe of non-polar binding sites. *J Mol Biol 13*, 482-95.
[59] Zhou, H. M. and Tsou, C. L. (1986) Comparison of activity and conformation changes during refolding of urea-denatured creatine kinase. *Biochim Biophys Acta 869*, 69-74.
[60] Horton, H. R. and Koshland, D. E. (1967) *Methods Enzymol 11*, 856-870.
[61] Schelman, J. A. and Schelman, C. (1962) in *The Proteins* (Neurath, H., Ed.), Academic Press, New York.
[62] Miyazawa, T. (1960) Perturbation treatment of the characteristic vibrations of polypeptide chains in various configurations. *J Chem Phys 32*, 1647-1656.
[63] Chirgadze, Y. N., Shestopalov, B. V. and Venyaminov, S. Y. (1973) Intensities and other spectral parameters of infrared amide bands of polypeptides in the beta- and random forms. *Biopolymers 12*, 1337-51.
[64] Chirgadze, Y. N., Fedorov, O. V. and Trushina, N. P. (1975) Estimation of amino acid residue side-chain absorption in the infrared spectra of protein solutions in heavy water. *Biopolymers 14*, 679-94.
[65] Dolgikh, D. A., Abaturov, L. V., Brazhnikov, Lebedev Iu, O. and Chirgadze Iu, N. (1983) Acid form of carbonic anhydrase: "molten globule" with a secondary structure. *Dokl Akad Nauk SSSR 272*, 1481-4.
[66] Lockhart, D. J. and Kim, P. S. (1992) Internal stark effect measurement of the electric field at the amino terminus of an alpha helix. *Science 257*, 947-51.
[67] Stoutland, P. O., Dyer, R. B. and Woodruff, W. H. (1992) Ultrafast infrared spectroscopy. *Science 257*, 1913-7.
[68] Lockhart, D. J. and Kim, P. S. (1993) Electrostatic screening of charge and dipole interactions with the helix backbone. *Science 260*, 198-202.
[69] Eftink, M. R. (1995) Use of multiple spectroscopic methods to monitor equilibrium unfolding of proteins. *Methods Enzymol 259*, 487-512.
[70] Middaugh, C. R., Mach, H., Ryan, J. A., Sanyal, G. and Volkin, D. B. (1995) Infrared spectroscopy. *Methods Mol Biol 40*, 137-56.

[71] Phillips, C. M., Mizutani, Y. and Hochstrasser, R. M. (1995) Ultrafast thermally induced unfolding of RNase A. *Proc Natl Acad Sci U S A 92*, 7292-6.
[72] Gilmanshin, R., Williams, S., Callender, R. H., Woodruff, W. H. and Dyer, R. B. (1997) Fast events in protein folding: relaxation dynamics and structure of the I form of apomyoglobin. *Biochemistry 36*, 15006-12.
[73] Gilmanshin, R., Dyer, R. B. and Callender, R. H. (1997) Structural heterogeneity of the various forms of apomyoglobin: implications for protein folding. *Protein Sci 6*, 2134-42.
[74] Gilmanshin, R., Williams, S., Callender, R. H., Woodruff, W. H. and Dyer, R. B. (1997) Fast events in protein folding: relaxation dynamics of secondary and tertiary structure in native apomyoglobin. *Proc Natl Acad Sci U S A 94*, 3709-13.
[75] Cabiaux, V., Oberg, K. A., Pancoska, P., Walz, T., Agre, P. and Engel, A. (1997) Secondary structures comparison of aquaporin-1 and bacteriorhodopsin: a Fourier transform infrared spectroscopy study of two-dimensional membrane crystals. *Biophys J 73*, 406-17.
[76] Oberg, K. A. and Fink, A. L. (1998) A new attenuated total reflectance Fourier transform infrared spectroscopy method for the study of proteins in solution. *Anal Biochem 256*, 92-106.
[77] Chittur, K. K., Fink, D. J., Leininger, R. I. and Hutson, T. B. (1986) FTIR/ATR studies of protein adsorption in flowing systems. Approaches for bulk correction and compositional analysis in mixtures. *J Colloid Interface Sci 111*, 419-433.
[78] Wasacz, F. M., Olinger, J. M. and Jakobsen, R. J. (1987) Fourier transform infrared studies of proteins using nonaqueous solvents. Effects of methanol and ethylene glycol on albumin and immunoglobulin G. *Biochemistry 26*, 1464-70.
[79] Goormaghtigh, E., Cabiaux, V. and Ruysschaert, J. M. (1990) Secondary structure and dosage of soluble and membrane proteins by attenuated total reflection Fourier-transform infrared spectroscopy on hydrated films. *Eur J Biochem 193*, 409-20.
[80] Swedberg, S. A., Pesek, J. J. and Fink, A. L. (1990) Attenuated total reflectance Fourier transform infrared analysis of an acyl-enzyme intermediate of alpha-chymotrypsin. *Anal Biochem 186*, 153-8.
[81] Buchet, R., Varga, S., Seidler, N. W., Molnar, E. and Martonosi, A. (1991) Polarized infrared attenuated total reflectance spectroscopy of the Ca(2+)-ATPase of sarcoplasmic reticulum. *Biochim Biophys Acta 1068*, 201-16.
[82] Uchida, K., Harada, I., Nakauchi, Y. and Maruyama, K. (1991) Structural properties of connectin studied by ultraviolet resonance Raman spectroscopy and infrared dichroism. *FEBS Lett 295*, 35-8.
[83] Singh, B. R., Fuller, M. P. and DasGupta, B. R. (1991) Botulinum neurotoxin type A: structure and interaction with the micellar concentration of SDS determined by FT-IR spectroscopy. *J Protein Chem 10*, 637-49.
[84] Baenziger, J. E., Miller, K. W. and Rothschild, K. J. (1993) Fourier transform infrared difference spectroscopy of the nicotinic acetylcholine receptor: evidence for specific protein structural changes upon desensitization. *Biochemistry 32*, 5448-54.
[85] Oberg, K., Chrunyk, B. A., Wetzel, R. and Fink, A. L. (1994) Nativelike secondary structure in interleukin-1 beta inclusion bodies by attenuated total reflectance FTIR. *Biochemistry 33*, 2628-34.

[86] Safar, J., Roller, P. P., Ruben, G. C., Gajdusek, D. C. and Gibbs, C. J., Jr. (1993) Secondary structure of proteins associated in thin films. *Biopolymers 33*, 1461-76.
[87] de Jongh, H. H., Goormaghtigh, E. and Ruysschaert, J. M. (1996) The different molar absorptivities of the secondary structure types in the amide I region: an attenuated total reflection infrared study on globular proteins. *Anal Biochem 242*, 95-103.
[88] Singh, B. R., Fuller, M. P. and Schiavo, G. (1990) Molecular structure of tetanus neurotoxin as revealed by Fourier transform infrared and circular dichroic spectroscopy. *Biophys Chem 36*, 155-66.
[89] Bauer, H. H., Muller, M., Goette, J., Merkle, H. P. and Fringeli, U. P. (1994) Interfacial adsorption and aggregation associated changes in secondary structure of human calcitonin monitored by ATR-FTIR spectroscopy. *Biochemistry 33*, 12276-82.
[90] Kauppinen, J. K., Moffat, D. J., Mantsch, H. H. and Cameron, D. G. (1981) Fourier self-deconvolution: A method for resolving intrinsically overlapped bands. *Appl Spectrosc 35*, 271-276.
[91] Susi, H. and Byler, D. M. (1983) Protein structure by Fourier transform infrared spectroscopy: second derivative spectra. *Biochem Biophys Res Commun 115*, 391-7.
[92] Surewicz, W. K. and Mantsch, H. H. (1988) New insight into protein secondary structure from resolution-enhanced infrared spectra. *Biochim Biophys Acta 952*, 115-30.
[93] Cooper, E. A. and Knutson, K. (1995) Fourier transform infrared spectroscopy investigations of protein structure. *Pharm Biotechnol 7*, 101-43.
[94] Cooper, E. A. and Knutson, K. (1995) Fourier transform infrared spectroscopy investigations of protein structure. in *Physical methods to characterize pharmaceutical proteins* (Herron, J. N., Jiskoot, W. and Crommelin, D. J. A., Eds.) pp 101-143, Plenum Press, New York.
[95] Fink, A. L. (1998) Protein aggregation: folding aggregates, inclusion bodies and amyloid. *Fold Des 3*, R9-23.
[96] Khurana, R., Gillespie, J. R., Talapatra, A., Minert, L. J., Ionescu-Zanetti, C., Millett, I. and Fink, A. L. (2001) Partially folded intermediates as critical precursors of light chain amyloid fibrils and amorphous aggregates. *Biochemistry 40*, 3525-35.
[97] Seshadri, S., Khurana, R. and Fink, A. L. (1999) Fourier transform infrared spectroscopy in analysis of protein deposits. *Methods Enzymol 309*, 559-76.
[98] Seshadri, S., Oberg, K. A. and Fink, A. L. (1994) Thermally denatured ribonuclease A retains secondary structure as shown by FTIR. *Biochemistry 33*, 1351-5.
[99] Goers, J., Permyakov, S. E., Permyakov, E. A., Uversky, V. N. and Fink, A. L. (2002) Conformational prerequisites for alpha-lactalbumin fibrillation. *Biochemistry 41*, 12546-51.
[100] Souillac, P. O., Uversky, V. N., Millett, I. S., Khurana, R., Doniach, S. and Fink, A. L. (2002) Elucidation of the molecular mechanism during the early events in immunoglobulin light chain amyloid fibrillation. Evidence for an off-pathway oligomer at acidic pH. *J Biol Chem 277*, 12666-79.
[101] Souillac, P. O., Uversky, V. N., Millett, I. S., Khurana, R., Doniach, S. and Fink, A. L. (2002) Effect of association state and conformational stability on the kinetics of immunoglobulin light chain amyloid fibril formation at physiological pH. *J Biol Chem 277*, 12657-65.
[102] Uversky, V. N., Li, J., Souillac, P., Millett, I. S., Doniach, S., Jakes, R., Goedert, M. and Fink, A. L. (2002) Biophysical properties of the synucleins and their propensities to

fibrillate: inhibition of alpha-synuclein assembly by beta- and gamma-synucleins. *J Biol Chem 277*, 11970-8.

[103] Nielsen, L., Khurana, R., Coats, A., Frokjaer, S., Brange, J., Vyas, S., Uversky, V. N. and Fink, A. L. (2001) Effect of environmental factors on the kinetics of insulin fibril formation: elucidation of the molecular mechanism. *Biochemistry 40*, 6036-46.

[104] Li, J., Uversky, V. N. and Fink, A. L. (2001) Effect of familial Parkinson's disease point mutations A30P and A53T on the structural properties, aggregation, and fibrillation of human alpha-synuclein. *Biochemistry 40*, 11604-13.

[105] Ahmad, A., Uversky, V. N., Hong, D. and Fink, A. L. (2005) Early events in the fibrillation of monomeric insulin. *J Biol Chem 280*, 42669-75.

[106] Uversky, V. N., Yamin, G., Munishkina, L. A., Karymov, M. A., Millett, I. S., Doniach, S., Lyubchenko, Y. L. and Fink, A. L. (2005) Effects of nitration on the structure and aggregation of alpha-synuclein. *Brain Res Mol Brain Res 134*, 84-102.

[107] Dong, A., Prestrelski, S. J., Allison, S. D. and Carpenter, J. F. (1995) Infrared spectroscopic studies of lyophilization- and temperature-induced protein aggregation. *J Pharm Sci 84*, 415-24.

[108] Prestrelski, S. J., Tedeschi, N., Arakawa, T. and Carpenter, J. F. (1993) Dehydration-induced conformational transitions in proteins and their inhibition by stabilizers. *Biophys J 65*, 661-71.

[109] Kendrick, B. S., Cleland, J. L., Lam, X., Nguyen, T., Randolph, T. W., Manning, M. C. and Carpenter, J. F. (1998) Aggregation of recombinant human interferon gamma: kinetics and structural transitions. *J Pharm Sci 87*, 1069-76.

[110] Surewicz, W. K., Leddy, J. J. and Mantsch, H. H. (1990) Structure, stability, and receptor interaction of cholera toxin as studied by Fourier-transform infrared spectroscopy. *Biochemistry 29*, 8106-11.

[111] Byler, D. M. and Susi, H. (1986) Examination of the secondary structure of proteins by deconvolved FTIR spectra. *Biopolymers 25*, 469-87.

[112] Susi, H. and Byler, D. M. (1986) Resolution-enhanced Fourier transform infrared spectroscopy of enzymes. *Methods Enzymol 130*, 290-311.

[113] Englander, S. W. and Mauel, C. (1972) Hydrogen exchnge studies of respiratory proteins. II. Detection of discrete, ligand-induced changes in hemoglobin. *J Biol Chem 247*, 2387-94.

[114] Molday, R. S., Englander, S. W. and Kallen, R. G. (1972) Primary structure effects on peptide group hydrogen exchange. *Biochemistry 11*, 150-8.

[115] Englander, S. W. and Englander, J. J. (1972) Hydrogen-tritium exchange. *Methods Enzymol 26 PtC*, 406-13.

[116] Englander, S. W., Downer, N. W. and Teitelbaum, H. (1972) Hydrogen exchange. *Annu Rev Biochem 41*, 903-24.

[117] Dolgikh, D. A., Abaturov, L. V., Bolotina, I. A., Brazhnikov, E. V., Bychkova, V. E., Gilmanshin, R. I., Lebedev Yu, O., Semisotnov, G. V., Tiktopulo, E. I., Ptitsyn, O. B. and et al. (1985) Compact state of a protein molecule with pronounced small-scale mobility: bovine alpha-lactalbumin. *Eur Biophys J 13*, 109-21.

[118] Brazhnikov, E. V., Chirgadze Yu, N., Dolgikh, D. A. and Ptitsyn, O. B. (1985) Noncooperative temperature melting of a globular protein without specific tertiary structure: acid form of bovine carbonic anhydrase B. *Biopolymers 24*, 1899-907.

[119] Souillac, P. O., Uversky, V. N. and Fink, A. L. (2003) Structural transformations of oligomeric intermediates in the fibrillation of the immunoglobulin light chain LEN. *Biochemistry 42*, 8094-104.
[120] Raman, C. V. and Krishnan, K. S. (1928) A new class of spectra due to secondary radiation. Part I. *Indian J Phys 2*, 399-419.
[121] Thomas, G. J., Jr. (1999) Raman spectroscopy of protein and nucleic acid assemblies. *Annu Rev Biophys Biomol Struct 28*, 1-27.
[122] Xu, M., Ermolenkov, V. V., He, W., Uversky, V. N., Fredriksen, L. and Lednev, I. K. (2005) Lysozyme fibrillation: deep UV Raman spectroscopic characterization of protein structural transformation. *Biopolymers 79*, 58-61.
[123] Lednev, I. K., Ermolenkov, V. V., He, W. and Xu, M. (2005) Deep-UV Raman spectrometer tunable between 193 and 205 nm for structural characterization of proteins. *Anal Bioanal Chem 381*, 431-7.
[124] Asher, S. A. (2001) UV Resonance Raman Spectroscopy. in *Handbook of Vibrational Spectroscopy* pp 557-571, John Wiley and Sons, Ltd, Chichester, UK.
[125] Adler, A. J., Greenfield, N. J. and Fasman, G. D. (1973) Circular dichroism and optical rotatory dispersion of proteins and polypeptides. *Methods Enzymol 27*, 675-735.
[126] Provencher, S. W. and Glockner, J. (1981) Estimation of globular protein secondary structure from circular dichroism. *Biochemistry 20*, 33-7.
[127] Yang, J. T., Wu, C. S. and Martinez, H. M. (1986) Calculation of protein conformation from circular dichroism. *Methods Enzymol 130*, 208-69.
[128] Chou, P. Y. and Fasman, G. D. (1974) Conformational parameters for amino acids in helical, beta-sheet, and random coil regions calculated from proteins. *Biochemistry 13*, 211-22.
[129] Chou, P. Y. and Fasman, G. D. (1974) Prediction of protein conformation. *Biochemistry 13*, 222-45.
[130] Schulz, G. E., Barry, C. D., Friedman, J., Chou, P. Y., Fasman, G. D., Finkelstein, A. V., Lim, V. I., Pititsyn, O. B., Kabat, E. A., Wu, T. T., Levitt, M., Robson, B. and Nagano, K. (1974) Comparison of predicted and experimentally determined secondary structure of adenyl kinase. *Nature 250*, 140-2.
[131] Chou, P. Y., Adler, A. J. and Fasman, G. D. (1975) Conformational prediction and circular dichroism studies on the lac repressor. *J Mol Biol 96*, 29-45.
[132] Chou, P. Y. and Fasman, G. D. (1978) Empirical predictions of protein conformation. *Annu Rev Biochem 47*, 251-76.
[133] Chou, P. Y. and Fasman, G. D. (1978) Prediction of the secondary structure of proteins from their amino acid sequence. *Adv Enzymol Relat Areas Mol Biol 47*, 45-148.
[134] Bolotina, I. A. and Lugauskas, V. (1985) [Determination of the secondary structure of proteins from circular dichroism spectra. IV. Contribution of aromatic amino acid residues into circular dichroism spectra of proteins in the peptide region]. *Mol Biol (Mosk) 19*, 1409-21.
[135] Bolotina, I. A., Chekhov, V. O., Lugauskas, V. and Ptitsyn, O. B. (1981) [Determination of protein secondary structure from circular dichroism spectra. III. Protein-derived base spectra of circular dichroism for antiparallel and parallel beta-structures]. *Mol Biol (Mosk) 15*, 167-75.
[136] Bolotina, I. A., Chekhov, V. O., Lugauskas, V., Finkel'shtein, A. V. and Ptitsyn, O. B. (1980) [Determination of the secondary structure of proteins from their circular

dichroism spectra. I. Protein reference spectra for alpha-, beta- and irregular structures]. *Mol Biol (Mosk) 14*, 891-902.

[137] Bolotina, I. A., Chekhov, V. O., Lugauskas, V. and Ptitsyn, O. B. (1980) [Determination of the secondary structure of proteins from their circular dichroism spectra. II. Estimation of the contribution of beta-pleated sheets]. *Mol Biol (Mosk) 14*, 902-9.

[138] Vassilenko, K. S. and Uversky, V. N. (2002) Native-like secondary structure of molten globules. *Biochim Biophys Acta 1594*, 168-77.

[139] Brahms, S. and Brahms, J. (1980) Determination of protein secondary structure in solution by vacuum ultraviolet circular dichroism. *J Mol Biol 138*, 149-78.

[140] Bolotina, I. A. (1987) [Secondary structure of proteins from circular dichroism spectra. V. Secondary structure of proteins in a "molten globule" state]. *Mol Biol (Mosk) 21*, 1625-35.

[141] Manning, M. C. and Woody, R. W. (1989) Theoretical study of the contribution of aromatic side chains to the circular dichroism of basic bovine pancreatic trypsin inhibitor. *Biochemistry 28*, 8609-13.

[142] Jagannadham, M. V. and Balasubramanian, D. (1985) The molten globular intermediate form in the folding pathway of human carbonic anhydrase B. *FEBS Lett 188*, 326-30.

[143] Rodionova, N. A., Semisotnov, G. V., Kutyshenko, V. P., Uverskii, V. N. and Bolotina, I. A. (1989) [Staged equilibrium of carbonic anhydrase unfolding in strong denaturants]. *Mol Biol (Mosk) 23*, 683-92.

[144] Uverskii, V. N., Semisotnov, G. V. and Ptitsyn, O. B. (1993) [Molten globule unfolding by strong denaturing agents occurs by the "all or nothing" principle]. *Biofizika 38*, 37-46.

[145] Uversky, V. N., Semisotnov, G. V., Pain, R. H. and Ptitsyn, O. B. (1992) 'All-or-none' mechanism of the molten globule unfolding. *FEBS Lett 314*, 89-92.

[146] Bychkova, V. E., Berni, R., Rossi, G. L., Kutyshenko, V. P. and Ptitsyn, O. B. (1992) Retinol-binding protein is in the molten globule state at low pH. *Biochemistry 31*, 7566-71.

[147] Abramov, V. M., Vasiliev, A. M., Khlebnikov, V. S., Vasilenko, R. N., Kulikova, N. L., Kosarev, I. V., Ishchenko, A. T., Gillespie, J. R., Millett, I. S., Fink, A. L. and Uversky, V. N. (2002) Structural and functional properties of Yersinia pestis Caf1 capsular antigen and their possible role in fulminant development of primary pneumonic plague. *J Proteome Res 1*, 307-15.

[148] Abramov, V. M., Vasiliev, A. M., Vasilenko, R. N., Kulikova, N. L., Kosarev, I. V., Khlebnikov, V. S., Ishchenko, A. T., MacIntyre, S., Gillespie, J. R., Khurana, R., Korpela, T., Fink, A. L. and Uversky, V. N. (2001) Structural and functional similarity between Yersinia pestis capsular protein Caf1 and human interleukin-1 beta. *Biochemistry 40*, 6076-84.

[149] Roberts, G. C. and Jardetzky, O. (1970) Nuclear magnetic resonance spectroscopy of amino acids, peptides, and proteins. *Adv Protein Chem 24*, 447-545.

[150] Wuthrich, K. (1981) Nuclear magnetic resonance studies of internal mobility in globular proteins. *Biochem Soc Symp*, 17-37.

[151] Wuthrich, K. (1989) Determination of three-dimensional protein structures in solution by nuclear magnetic resonance: an overview. *Methods Enzymol 177*, 125-31.

[152] Wuthrich, K. (1989) Protein structure determination in solution by nuclear magnetic resonance spectroscopy. *Science 243*, 45-50.

[153] Wuthrich, K. (1990) Protein structure determination in solution by NMR spectroscopy. *J Biol Chem 265*, 22059-62.

[154] LeMaster, D. M., Kay, L. E., Brunger, A. T. and Prestegard, J. H. (1988) Protein dynamics and distance determination by NOE measurements. *FEBS Lett 236*, 71-6.

[155] LeMaster, D. M. (1988) Protein NMR resonance assignment by isotropic mixing experiments on random fractionally deuterated samples. *FEBS Lett 233*, 326-30.

[156] McDonald, C. C. and Phillips, W. D. (1969) Proton magnetic resonance spectra of proteins in random-coil configurations. *J Am Chem Soc 91*, 1513-21.

[157] Semisotnov, G. V., Kutyshenko, V. P. and Ptitsyn, O. B. (1989) [Intramolecular mobility of a protein in a "molten globule" state. A study of carbonic anhydrase by 1H-NMR]. *Mol Biol (Mosk) 23*, 808-15.

[158] Kuwajima, K., Harushima, Y. and Sugai, S. (1986) Influence of Ca2+ binding on the structure and stability of bovine alpha-lactalbumin studied by circular dichroism and nuclear magnetic resonance spectra. *Int J Pept Protein Res 27*, 18-27.

[159] Baum, J., Dobson, C. M., Evans, P. A. and Hanley, C. (1989) Characterization of a partly folded protein by NMR methods: studies on the molten globule state of guinea pig alpha-lactalbumin. *Biochemistry 28*, 7-13.

[160] Dobson, C. M. (1991) NMR spectroscopy and protein folding: studies of lysozyme and alpha-lactalbumin. *Ciba Found Symp 161*, 167-81; discussion 181-9.

[161] Dyson, H. J. and Wright, P. E. (2004) Unfolded proteins and protein folding studied by NMR. *Chem Rev 104*, 3607-22.

[162] Dyson, H. J. and Wright, P. E. (2002) Insights into the structure and dynamics of unfolded proteins from nuclear magnetic resonance. *Adv Protein Chem 62*, 311-40.

[163] Shortle, D. R. (1996) Structural analysis of non-native states of proteins by NMR methods. *Curr Opin Struct Biol 6*, 24-30.

[164] Dill, K. A. and Shortle, D. (1991) Denatured states of proteins. *Annu Rev Biochem 60*, 795-825.

[165] Rico, M., Santoro, J., Gonzalez, C., Bruix, M., Neira, J. L., Nieto, J. L. and Herranz, J. (1991) 3D structure of bovine pancreatic ribonuclease A in aqueous solution: an approach to tertiary structure determination from a small basis of 1H NMR NOE correlations. *J Biomol NMR 1*, 283-98.

[166] Wand, A. J., Di Stefano, D. L., Feng, Y. Q., Roder, H. and Englander, S. W. (1989) Proton resonance assignments of horse ferrocytochrome c. *Biochemistry 28*, 186-94.

[167] Redfield, C. and Dobson, C. M. (1988) Sequential 1H NMR assignments and secondary structure of hen egg white lysozyme in solution. *Biochemistry 27*, 122-36.

[168] Redfield, C. and Dobson, C. M. (1990) 1H NMR studies of human lysozyme: spectral assignment and comparison with hen lysozyme. *Biochemistry 29*, 7201-14.

[169] LeMaster, D. M. and Richards, F. M. (1988) NMR sequential assignment of Escherichia coli thioredoxin utilizing random fractional deuteriation. *Biochemistry 27*, 142-50.

[170] De Wolf, E., Gill, R., Geddes, S., Pitts, J., Wollmer, A. and Grotzinger, J. (1996) Solution structure of a mini IGF-1. *Protein Sci 5*, 2193-202.

[171] Muhlhahn, P., Czisch, M., Morenweiser, R., Habermann, B., Engh, R. A., Sommerhoff, C. P., Auerswald, E. A. and Holak, T. A. (1994) Structure of leech derived tryptase inhibitor (LDTI-C) in solution. *FEBS Lett 355*, 290-6.
[172] Bartik, K., Dobson, C. M. and Redfield, C. (1993) 1H-NMR analysis of turkey egg-white lysozyme and comparison with hen egg-white lysozyme. *Eur J Biochem 215*, 255-66.
[173] Penkett, C. J., Redfield, C., Dodd, I., Hubbard, J., McBay, D. L., Mossakowska, D. E., Smith, R. A., Dobson, C. M. and Smith, L. J. (1997) NMR analysis of main-chain conformational preferences in an unfolded fibronectin-binding protein. *J Mol Biol 274*, 152-9.
[174] Zuiderweg, E. R. and Fesik, S. W. (1989) Heteronuclear three-dimensional NMR spectroscopy of the inflammatory protein C5a. *Biochemistry 28*, 2387-91.
[175] Fesik, S. W., Gampe, R. T., Jr., Zuiderweg, E. R., Kohlbrenner, W. E. and Weigl, D. (1989) Heteronuclear three-dimensional NMR spectroscopy applied to CMP-KDO synthetase (27.5 kD). *Biochem Biophys Res Commun 159*, 842-7.
[176] Ohkubo, T., Taniyama, Y. and Kikuchi, M. (1991) 1H and 15N NMR study of human lysozyme. *J Biochem (Tokyo) 110*, 1022-9.
[177] Redfield, C., Smith, L. J., Boyd, J., Lawrence, G. M., Edwards, R. G., Smith, R. A. and Dobson, C. M. (1991) Secondary structure and topology of human interleukin 4 in solution. *Biochemistry 30*, 11029-35.
[178] Venters, R. A., Farmer, B. T., 2nd, Fierke, C. A. and Spicer, L. D. (1996) Characterizing the use of perdeuteration in NMR studies of large proteins: 13C, 15N and 1H assignments of human carbonic anhydrase II. *J Mol Biol 264*, 1101-16.
[179] Udgaonkar, J. B. and Baldwin, R. L. (1988) NMR evidence for an early framework intermediate on the folding pathway of ribonuclease A. *Nature 335*, 694-9.
[180] Roder, H., Elove, G. A. and Englander, S. W. (1988) Structural characterization of folding intermediates in cytochrome c by H-exchange labelling and proton NMR. *Nature 335*, 700-4.
[181] Hamilton, C. L. and Mcdonnell, H. M. (1968) in *Structural Chemistry and Molecular Biology* (Rich, A. and Davidson, N., Eds.) pp 115-127, WH Freeman and Co, San Francisco.
[182] Kaivarainen, A. I. and Nezlin, R. S. (1976) Spin-label approach to conformational properties of immunoglobulins. *Immunochemistry 13*, 1001-8.
[183] Hustedt, E. J. and Beth, A. H. (1999) Nitroxide spin-spin interactions: applications to protein structure and dynamics. *Annu Rev Biophys Biomol Struct 28*, 129-53.
[184] Berliner, L. H. (1976) *Spin Labeling Theory and Application*, Academic Press, New York.
[185] Dodd, G. H., Barratt, M. D. and Rayner, L. (1970) Spin probes for binding site polarity. *FEBS Lett 8*, 286-288.
[186] Roberts, G. C., Hannah, J. and Jardetzky, O. (1969) Noncovalent binding of a spin-labeled inhibitor to ribonuclease. *Science 165*, 504-6.
[187] Schmidt, P. G. and Kuntz, I. D. (1984) Distance measurements in spin-labeled lysozyme. *Biochemistry 23*, 4261-6.
[188] Kuntz, I. D. and Crippen, G. M. (1979) Protein densities. *Int J Pept Protein Res 13*, 223-8.

[189] Megli, F. M., van Loon, D., Barbuti, A. A., Quagliariello, E. and Wirtz, K. W. (1985) Chemical modification of methionine residues of the phosphatidylcholine transfer protein from bovine liver. A spin-label study. *Eur J Biochem 149*, 585-90.

[190] Carlsson, U., Aasa, R., Henderson, L. E., Jonsson, B. H. and Lindskog, S. (1975) Paramagnetic and fluorescent probes attached to "buried" sulfhydryl groups in human carbonic anhydrases. Application to inhibitor binding, denaturation and refolding. *Eur J Biochem 52*, 25-36.

[191] Semisotnov, G. V., Rodionova, N. A., Kutyshenko, V. P., Ebert, B., Blanck, J. and Ptitsyn, O. B. (1987) Sequential mechanism of refolding of carbonic anhydrase B. *FEBS Lett 224*, 9-13.

[192] Lackowicz, J. (1983) *Principles of Fluorescence Spectroscopy*, Plenum Press, New York.

[193] Eftink, M. R. and Ghiron, C. A. (1976) Fluorescence quenching of indole and model micelle systems. *J. Phys. Chem. 80*, 486-493.

[194] Eftink, M. R. and Ghiron, C. A. (1981) Fluorescence quenching studies with proteins. *Anal Biochem 114*, 199-227.

[195] Lakowicz, J. R. and Weber, G. (1973) Quenching of fluorescence by oxygen. A probe for structural fluctuations in macromolecules. *Biochemistry 12*, 4161-70.

[196] Lakowicz, J. R. and Weber, G. (1973) Quenching of protein fluorescence by oxygen. Detection of structural fluctuations in proteins on the nanosecond time scale. *Biochemistry 12*, 4171-9.

[197] Eftink, M. R. and Ghiron, C. A. (1975) Dynamics of a protein matrix revealed by fluorescence quenching. *Proc Natl Acad Sci U S A 72*, 3290-4.

[198] Eftink, M. R. and Ghiron, C. A. (1976) Exposure of tryptophanyl residues in proteins. Quantitative determination by fluorescence quenching studies. *Biochemistry 15*, 672-80.

[199] Eftink, M. R. and Ghiron, C. A. (1977) Exposure of tryptophanyl residues and protein dynamics. *Biochemistry 16*, 5546-51.

[200] Calhoun, D. B., Vanderkooi, J. M., Holtom, G. R. and Englander, S. W. (1986) Protein fluorescence quenching by small molecules: protein penetration versus solvent exposure. *Proteins 1*, 109-15.

[201] Varley, P. G., Dryden, D. T. and Pain, R. H. (1991) Resolution of the fluorescence of the buried tryptophan in yeast 3-phosphoglycerate kinase using succinimide. *Biochim Biophys Acta 1077*, 19-24.

[202] Chaffotte, A. F. and Goldberg, M. E. (1984) Fluorescence-quenching studies on a conformational transition within a domain of the beta 2 subunit of Escherichia coli tryptophan synthase. *Eur J Biochem 139*, 47-50.

[203] Weber, G. (1960) Fluorescence-polarization spectrum and electronic-energy transfer in proteins. *Biochem J 75*, 345-52.

[204] Weber, G. (1960) Fluorescence-polarization spectrum and electronic-energy transfer in tyrosine, tryptophan and related compounds. *Biochem J 75*, 335-45.

[205] Semisotnov, G. V., Zikherman, K. K., Kasatkin, S. B., Ptitsyn, O. B. and Anufrieva, E. I. (1981) *Biopolymers 20*, 2287-2309.

[206] Ingham, K. C., Tylenda, C. and Edelhoch, H. (1976) Structural studies of human chorionic gonadotropin and its subunits using tyrosine fluorescence. *Arch Biochem Biophys 173*, 680-90.

[207] Förster, T. (1951) *Fluoreszenz Organischer Verbindungen*, Vandenhoeck and Ruprecht, Gottingen, Federal Republic of Germany.

[208] Epe, B., Steinhauser, K. G. and Woolley, P. (1983) Theory of measurement of Forster-type energy transfer in macromolecules. *Proc Natl Acad Sci U S A 80*, 2579-2583.

[209] Förster, T. (1948) Intermolecular energy migration and fluorescence. *Ann Physik 2*, 55-75.

[210] Chen, R. F. (1968) Dansyl labeled proteins: determination of extinction coefficienc and number of bound residues with radioactive dansyl chloride. *Anal Biochem 25*, 412-6.

[211] Blumen, A. (1980) On the anisotropic energy-transfer to random acceptors. *J Chem Phys 74*, 6926-6933.

[212] Shore, V. G. and Pardee, A. B. (1956) Energy transfer in conjugated proteins and nucleic acids. *Arch Biochem Biophys 62*, 355-68.

[213] Weber, G. and Young, L. B. (1964) Fragmentation of Bovine Serum Albumin by Pepsin. I. The Origin of the Acid Expansion of the Albumin Molecule. *J Biol Chem 239*, 1415-23.

[214] Weber, G. and Young, L. B. (1964) Fragmentation of Bovine Serum Albumin by Pepsin. Ii. Isolation, Amino Acid Composition, and Physical Properties of the Fragments. *J Biol Chem 239*, 1424-31.

[215] Turner, D. C. and Brand, L. (1968) Quantitative estimation of protein binding site polarity. Fluorescence of N-arylaminonaphthalenesulfonates. *Biochemistry 7*, 3381-90.

[216] Semisotnov, G. V., Rodionova, N. A., Razgulyaev, O. I., Uversky, V. N., Gripas, A. F. and Gilmanshin, R. I. (1991) Study of the "molten globule" intermediate state in protein folding by a hydrophobic fluorescent probe. *Biopolymers 31*, 119-28.

[217] Ptitsyn, O. B., Pain, R. H., Semisotnov, G. V., Zerovnik, E. and Razgulyaev, O. I. (1990) Evidence for a molten globule state as a general intermediate in protein folding. *FEBS Lett 262*, 20-4.

[218] Uversky, V. N., Winter, S. and Lober, G. (1996) Use of fluorescence decay times of 8-ANS-protein complexes to study the conformational transitions in proteins which unfold through the molten globule state. *Biophys Chem 60*, 79-88.

[219] Mulqueen, P. M. and Kronman, M. J. (1982) Binding of naphthalene dyes to the N and A conformers of bovine alpha-lactalbumin. *Arch Biochem Biophys 215*, 28-39.

[220] Witz, G. and Van Duuren, B. L. (1973) Hydrophobic fluorescence probe studies with poly-L-lysine. *J Phys Chem 77*, 648-51.

[221] Lapanje, S. and Tanford, C. (1967) Proteins as random coils. IV. Osmotic pressures, second virial coefficients, and unperturbed dimensions in 6 M guanidine hydrochloride. *J Am Chem Soc 89*, 5030-3.

[222] Bucher, D., Richards, E. G. and Brown, W. D. (1970) Modifications of the Rayleigh interferometer in the ultracentrifuge for use with heme and other light-absorbing proteins. *Anal Biochem 36*, 368-80.

[223] Gast, K., Zirwer, D., Welfle, H., Bychkova, V. E. and Ptitsyn, O. B. (1986) *Int J Biol Macromol 8*, 231-236.

[224] Glatter, O. and Kratky, O. (1982) *Small Angle X-Ray Scattering*, Academic Press, New York.

[225] Feigin, L. A. and Svergu, D. I. (1987) *Structural Analysis by Small-Angle X-Ray and Neutron Scattering*, Plenum Press, New York.

[226] Eliezer, D., Chiba, K., Tsuruta, H., Doniach, S., Hodgson, K. O. and Kihara, H. (1993) Evidence of an associative intermediate on the myoglobin refolding pathway. *Biophys J 65*, 912-7.

[227] Kataoka, M., Hagihara, Y., Mihara, K. and Goto, Y. (1993) Molten globule of cytochrome c studied by small angle X-ray scattering. *J Mol Biol 229*, 591-6.

[228] Kataoka, M., Kuwajima, K., Tokunaga, F. and Goto, Y. (1997) Structural characterization of the molten globule of alpha-lactalbumin by solution X-ray scattering. *Protein Sci 6*, 422-30.

[229] Semisotnov, G. V., Kihara, H., Kotova, N. V., Kimura, K., Amemiya, Y., Wakabayashi, K., Serdyuk, I. N., Timchenko, A. A., Chiba, K., Nikaido, K., Ikura, T. and Kuwajima, K. (1996) Protein globularization during folding. A study by synchrotron small-angle X-ray scattering. *J Mol Biol 262*, 559-74.

[230] Uversky, V. N. (1993) Use of fast protein size-exclusion liquid chromatography to study the unfolding of proteins which denature through the molten globule. *Biochemistry 32*, 13288-98.

[231] Endo, S., Saito, Y. and Wada, A. (1983) Denaturant-gradient chromatography for the study of protein denaturation: principle and procedure. *Anal Biochem 131*, 108-20.

[232] Ackers, G. K. (1967) Molecular sieve studies of interacting protein systems. I. Equations for transport of associating systems. *J Biol Chem 242*, 3026-34.

[233] Ackers, G. K. (1970) Analytical gel chromatography of proteins. *Adv Protein Chem 24*, 343-446.

[234] Fish, W. W., Reynolds, J. A. and Tanford, C. (1970) Gel chromatography of proteins in denaturing solvents. Comparison between sodium dodecyl sulfate and guanidine hydrochloride as denaturants. *J Biol Chem 245*, 5166-8.

[235] Ansari, A. A. and Mage, R. G. (1976) An evaluation of effectiveness of bio-glas in molecular sieving of polypeptides in guanidine hydrochloride. *Anal Biochem 74*, 118-25.

[236] Hashimoto, T., Sasaki, H., Aiura, M. and Kato, Y. (1978) High-speed aqueous gel-permeation chromatography of proteins. *J Chromatogr 160*, 301-5.

[237] Imamura, T., Konishi, K. and Yokoyama, M. (1979) High-speed gel filtration of polypeptides in sodium dodecyl sulfate. *J Biochem (Tokyo) 86*, 639-42.

[238] Corbett, R. J. and Roche, R. S. (1984) Use of high-speed size-exclusion chromatography for the study of protein folding and stability. *Biochemistry 23*, 1888-94.

[239] Uversky, V. N. (1994) Gel-permeation chromatography as a unique instrument for quantitative and qualitative analysis of protein denaturation and unfolding. *Int. J. Bio-Chromatography 1*, 103-114.

[240] Withka, J., Moncuse, P., Baziotis, A. and Maskiewicz, R. (1987) Use of high-performance size-exclusion, ion-exchange, and hydrophobic interaction chromatography for the measurement of protein conformational change and stability. *J Chromatogr 398*, 175-202.

[241] Creighton, T. E. (1986) Detection of folding intermediates using urea-gradient electrophoresis. *Methods Enzymol 131*, 156-72.

[242] Creighton, T. E. and Pain, R. H. (1980) Unfolding and refolding of Staphylococcus aureus penicillinase by urea-gradient electrophoresis. *J Mol Biol 137*, 431-6.

[243] Creighton, T. E. (1980) Kinetic study of protein unfolding and refolding using urea gradient electrophoresis. *J Mol Biol 137*, 61-80.

[244] Chemeris, V. V., Dolgikh, D. A., Fedorov, A. N., Finkelstein, A. V., Kirpichnikov, M. P., Uversky, V. N. and Ptitsyn, O. B. (1994) A new approach to artificial and modified proteins: theory-based design, synthesis in a cell-free system and fast testing of structural properties by radiolabels. *Protein Eng 7*, 1041-52.

[245] Fedorov, A. N., Dolgikh, D. A., Chemeris, V. V., Chernov, B. K., Finkelstein, A. V., Schulga, A. A., Alakhov Yu, B., Kirpichnikov, M. P. and Ptitsyn, O. B. (1992) De novo design, synthesis and study of albebetin, a polypeptide with a predetermined three-dimensional structure. Probing the structure at the nanogram level. *J Mol Biol 225*, 927-31.

[246] Thatcher, D. R. and Hodson, B. (1981) Denaturation of proteins and nucleic acids by thermal-gradient electrophoresis. *Biochem J 197*, 105-9.

[247] Gianazza, E., Miller, I., Eberini, I. and Castiglioni, S. (1999) Low-tech electrophoresis, small but beautiful, and effective: electrophoretic titration curves of proteins. *Electrophoresis 20*, 1325-38.

[248] Kolin, A. (1958) Rapid electrophoresis in density gradients combined with pH and/or conductivity gradients. *Methods Biochem Anal 6*, 259-88.

[249] Kolin, A. (1970) pH gradient electrophoresis. *Methods Med Res 12*, 326-58.

[250] Righetti, P. G. (1988) Isoelectric focusing as the crow flies. *J Biochem Biophys Methods 16*, 99-108.

[251] Gianazza, E. (1995) Isoelectric focusing as a tool for the investigation of post-translational processing and chemical modifications of proteins. *J Chromatogr A 705*, 67-87.

[252] O'Farrell, P. H. (1975) High resolution two-dimensional electrophoresis of proteins. *J Biol Chem 250*, 4007-21.

[253] Gorg, A., Weiss, W. and Dunn, M. J. (2004) Current two-dimensional electrophoresis technology for proteomics. *Proteomics 4*, 3665-85.

[254] Hubbard, S. J., Eisenmenger, F. and Thornton, J. M. (1994) Modeling studies of the change in conformation required for cleavage of limited proteolytic sites. *Protein Sci 3*, 757-68.

[255] Hubbard, S. J., Beynon, R. J. and Thornton, J. M. (1998) Assessment of conformational parameters as predictors of limited proteolytic sites in native protein structures. *Protein Eng 11*, 349-59.

[256] Laskowski, M., Jr. and Kato, I. (1980) Protein inhibitors of proteinases. *Annu Rev Biochem 49*, 593-626.

[257] Bode, W. and Huber, R. (1992) Natural protein proteinase inhibitors and their interaction with proteinases. *Eur J Biochem 204*, 433-51.

[258] Hubbard, S. J., Campbell, S. F. and Thornton, J. M. (1991) Molecular recognition. Conformational analysis of limited proteolytic sites and serine proteinase protein inhibitors. *J Mol Biol 220*, 507-30.

[259] Fontana, A., Fassina, G., Vita, C., Dalzoppo, D., Zamai, M. and Zambonin, M. (1986) Correlation between sites of limited proteolysis and segmental mobility in thermolysin. *Biochemistry 25*, 1847-51.

[260] Novotny, J. and Bruccoleri, R. E. (1987) Correlation among sites of limited proteolysis, enzyme accessibility and segmental mobility. *FEBS Lett 211*, 185-9.

[261] Fontana, A., Zambonin, M., Polverino de Laureto, P., De Filippis, V., Clementi, A. and Scaramella, E. (1997) Probing the conformational state of apomyoglobin by limited proteolysis. *J Mol Biol 266*, 223-30.
[262] Fontana, A., Polverino de Laureto, P., De Filippis, V., Scaramella, E. and Zambonin, M. (1997) Probing the partly folded states of proteins by limited proteolysis. *Fold Des 2*, R17-26.
[263] Spolaore, B., Bermejo, R., Zambonin, M. and Fontana, A. (2001) Protein interactions leading to conformational changes monitored by limited proteolysis: apo form and fragments of horse cytochrome c. *Biochemistry 40*, 9460-8.
[264] Mihalyi, E. (1972) *Application of Proteolytic Enzymes to Protein Structure Study*, Chem. Rubber Co., Cleveland.
[265] Leontiev, V. V., Uversky, V. N. and Gudkov, A. T. (1993) Comparative stability of dihydrofolate reductase mutants in vitro and in vivo. *Protein Eng 6*, 81-4.
[266] Drabikowski, W., Kuznicki, J. and Grabarek, Z. (1977) Similarity in Ca2+-induced changes between troponic-C and protein activator of 3':5'-cyclic nucleotide phosphodiesterase and their tryptic fragments. *Biochim Biophys Acta 485*, 124-33.
[267] Drabikowski, W., Grabarek, Z. and Barylko, B. (1977) Degradation of TN-C component of troponin by trypsin. *Biochim Biophys Acta 490*, 216-24.
[268] Grabarek, Z., Drabikowski, W., Leavis, P. C., Rosenfeld, S. S. and Gergely, J. (1981) Proteolytic fragments of troponin C. Interactions with the other troponin subunits and biological activity. *J Biol Chem 256*, 13121-7.
[269] Grabarek, Z., Drabikowski, W., Vinokurov, L. and Lu, R. C. (1981) Digestion of troponin C with trypsin in the presence and absence of Ca2+. Identification of cleavage points. *Biochim Biophys Acta 671*, 227-33.
[270] Cox, J. A., Winge, D. R. and Stein, E. A. (1979) Calcium, magnesium and the conformation of parvalbumin during muscular activity. *Biochimie 61*, 601-5.
[271] Fullmer, C. S., Wasserman, R. H., Hamilton, J. W., Huang, W. Y. and Cohn, D. V. (1975) The effect of calcium on the tryptic digestion of bovine intestinal calcium-binding protein. *Biochim Biophys Acta 412*, 256-61.
[272] Pieterson, W. A., Volwerk, J. J. and de Haas, G. H. (1974) Interaction of phospholipase A2 and its zymogen with divalent metal ions. *Biochemistry 13*, 1439-45.
[273] Perrier, V., Surewicz, W. K., Glaser, P., Martineau, L., Craescu, C. T., Fabian, H., Mantsch, H. H., Barzu, O. and Gilles, A. M. (1994) Zinc chelation and structural stability of adenylate kinase from Bacillus subtilis. *Biochemistry 33*, 9960-7.
[274] Zettlmeissl, G., Rudolph, R. and Jaenicke, R. (1983) Limited proteolysis as a tool to study the kinetics of protein folding: conformational rearrangements in acid-dissociated lactic dehydrogenase as determined by pepsin digestion. *Arch Biochem Biophys 224*, 161-8.
[275] Amit, A. G., Mariuzza, R. A., Phillips, S. E. and Poljak, R. J. (1985) Three-dimensional structure of an antigen-antibody complex at 6 A resolution. *Nature 313*, 156-8.
[276] Wilson, I. A., Haft, D. H., Getzoff, E. D., Tainer, J. A., Lerner, R. A. and Brenner, S. (1985) Identical short peptide sequences in unrelated proteins can have different conformations: a testing ground for theories of immune recognition. *Proc Natl Acad Sci U S A 82*, 5255-9.

[277] Fujio, H., Takagaki, Y., Ha, Y. M., Doi, E. M., Soebandrio, A. and Sakato, N. (1985) Native and non-native conformation-specific antibodies directed to the loop region of hen egg-white lysozyme. *J Biochem (Tokyo) 98*, 949-62.

[278] Furie, B. and Furie, B. C. (1979) Conformation-specific antibodies as probes of the gamma-carboxyglutamic acid-rich region of bovine prothrombin. Studies of metal-induced structural changes. *J Biol Chem 254*, 9766-71.

[279] Delmas, P. D., Stenner, D. D., Romberg, R. W., Riggs, B. L. and Mann, K. G. (1984) Immunochemical studies of conformational alterations in bone gamma-carboxyglutamic acid containing protein. *Biochemistry 23*, 4720-5.

[280] Reichlin, M. (1974) Quantitative immunological studies on single amino acid substitution in human hemoglobin: demonstration of specific antibodies to multiple sites. *Immunochemistry 11*, 21-7.

[281] Richards, F. F., Konigsberg, W. H., Rosenstein, R. W. and Varga, J. M. (1975) On the specificity of antibodies. *Science 187*, 130-7.

[282] Blond, S. and Goldberg, M. (1987) Partly native epitopes are already present on early intermediates in the folding of tryptophan synthase. *Proc Natl Acad Sci U S A 84*, 1147-51.

[283] Goldberg, M. E., Semisotnov, G. V., Friguet, B., Kuwajima, K., Ptitsyn, O. B. and Sugai, S. (1990) An early immunoreactive folding intermediate of the tryptophan synthease beta 2 subunit is a 'molten globule'. *FEBS Lett 263*, 51-6.

[284] Pelham, H. R. (1986) Speculations on the functions of the major heat shock and glucose-regulated proteins. *Cell 46*, 959-61.

[285] Ellis, R. J. and Hemmingsen, S. M. (1989) Molecular chaperones: proteins essential for the biogenesis of some macromolecular structures. *Trends Biochem Sci 14*, 339-42.

[286] Ellis, R. J., van der Vies, S. M. and Hemmingsen, S. M. (1989) The molecular chaperone concept. *Biochem Soc Symp 55*, 145-53.

[287] Hendrix, R. W. (1979) Purification and properties of groE, a host protein involved in bacteriophage assembly. *J Mol Biol 129*, 375-92.

[288] Chen, S., Roseman, A. M., Hunter, A. S., Wood, S. P., Burston, S. G., Ranson, N. A., Clarke, A. R. and Saibil, H. R. (1994) Location of a folding protein and shape changes in GroEL-GroES complexes imaged by cryo-electron microscopy. *Nature 371*, 261-4.

[289] Martin, J. (1998) Protein folding assisted by the GroEL/GroES chaperonin system. *Biochemistry (Mosc) 63*, 374-81.

[290] Kurganov, B. I. and Topchieva, I. N. (1998) Artificial chaperone-assisted refolding of proteins. *Biochemistry (Mosc) 63*, 413-9.

[291] Martin, J., Langer, T., Boteva, R., Schramel, A., Horwich, A. L. and Hartl, F. U. (1991) Chaperonin-mediated protein folding at the surface of groEL through a 'molten globule'-like intermediate. *Nature 352*, 36-42.

[292] Uversky, V. N., Kutyshenko, V. P., Protasova, N., Rogov, V. V., Vassilenko, K. S. and Gudkov, A. T. (1996) Circularly permuted dihydrofolate reductase possesses all the properties of the molten globule state, but can resume functional tertiary structure by interaction with its ligands. *Protein Sci 5*, 1844-51.

[293] Langer, T. and Neupert, W. (1991) Heat shock proteins hsp60 and hsp70: their roles in folding, assembly and membrane translocation of proteins. *Curr Top Microbiol Immunol 167*, 3-30.

[294] Hartl, F. U., Hlodan, R. and Langer, T. (1994) Molecular chaperones in protein folding: the art of avoiding sticky situations. *Trends Biochem Sci 19*, 20-5.

[295] Hayer-Hartl, M. K., Ewbank, J. J., Creighton, T. E. and Hartl, F. U. (1994) Conformational specificity of the chaperonin GroEL for the compact folding intermediates of alpha-lactalbumin. *Embo J 13*, 3192-202.

[296] Martin, J. and Hartl, F. U. (1994) Molecular chaperones in cellular protein folding. *Bioessays 16*, 689-92.

[297] Hartl, F. U. (1994) Protein folding. Secrets of a double-doughnut. *Nature 371*, 557-9.

[298] Robinson, C. V., Gross, M., Eyles, S. J., Ewbank, J. J., Mayhew, M., Hartl, F. U., Dobson, C. M. and Radford, S. E. (1994) Conformation of GroEL-bound alpha-lactalbumin probed by mass spectrometry. *Nature 372*, 646-51.

[299] Frydman, J., Nimmesgern, E., Ohtsuka, K. and Hartl, F. U. (1994) Folding of nascent polypeptide chains in a high molecular mass assembly with molecular chaperones. *Nature 370*, 111-7.

In: Protein Structure
Editors: Lauren M. Haggerty, pp. 205-236

ISBN: 978-1-61209-656-8

Chapter 10

Steady-State Quenching of Fluorescence to Study Protein Structure and Dynamics*

Béla Somogyi[1,2], Miklós Nyitrai[2] and Gábor Hild[1]

[1]Research Group for Fluorescence Spectroscopy, Office for Academy Research Groups Attached to Universities and Other Institutions University of Pécs, Pécs, Szigeti str. 12, H-7624, Hungary

[2]Department of Biophysics, Faculty of Medicine, University of Pécs, Pécs, Szigeti str. 12, H-7624, Hungary

1. Introduction

The intrinsic fluorescence of proteins and the fluorescence emitted by fluorescent probes attached to proteins or embedded in lipid layers are extremely sensitive to different environmental parameters (like temperature, viscosity, ionic strength, electric field gradient, pH). The change in the dynamic (spatio-temporal) behaviour of the neighbouring matrix around the fluorophores can affect their fluorescence characteristics. This feature makes the fluorescence techniques to be potentially powerful tools to obtain information about the structural and dynamic parameters of macromolecules and/or supramolecular systems [1].

The aim of this chapter is to present steady-state quenching techniques applicable to obtain different kinds of information about the structure and dynamics of proteins or other macromolecular systems (for related reviews see e.g. [2-5]). Among these methods we deal here mainly with some aspects of fluorescence quenching techniques without reviewing the vast amount of experimental data accumulated in previous years. We will restrict our

* This chapter is dedicated to the memory of Professor Bela Somogyi. He had a crystal clear logic and a unique gift for seeing the essential correlation behind phenomena and making it comprehensive. His outstanding professional knowledge was accompanied by an outstanding personality. He managed to prove that being a scientist is a form of life that a real scientist can give up under no circumstances.

discussion assuming that the processes we deal with are taking place in liquid phase and the fluorescence decay is monoexponential.

Fluorescence quenching can appear by several different ways depending on the specific interaction between the excited state fluorophore and the quencher. Here we deal only with two major types of it: the collisional quenching and the fluorescence resonance energy transfer (FRET). We further limit the discussion to steady-state cases where the samples experience continuous 'not too high intensity' illumination. Regarding some basic features of the fluorescence itself and time dependent FRET methods the reader is referred to earlier chapters of this book.

2. Collisional Quenching

Collisional quenching, by definition, occurs when a quencher takes away the excitation energy of the fluorophore upon collision preventing this way the photon emission, i.e., it assumes a collision between the two reactants (fluorophore and quencher). The simple scheme of such an event in gas phase is:

$$F^* + Q \xleftrightarrow{k_+[Q]/k_-} F^* : Q \xrightarrow{k_q} F + Q \quad (1)$$

where F and F^* is the fluorophore in its ground and excited states respectively, Q is the quencher molecule, $[Q]$ is the molar concentration of the quencher, k_+ is the relative transport rate constants, k_- is the dissociation rate constant of the complex $F^* : Q$ and k_q is the rate constant characteristic for the quenching process. Dealing with a liquid phase, however, one has to consider the so-called 'cage effect' [6]. This means that those particles of the liquid forming encounter complex reside next to each other performing rotational and vibrational motion and having 10^3 to 10^4 collisions before they leave each other. Accordingly, the above scheme has to be modified in the case of liquid phase as:

$$\begin{array}{ccccc} F^* + Q & \xleftrightarrow{k_+[Q]/k_-} & F^* : Q & \xrightarrow{k_q} & F + Q \\ \updownarrow & & \updownarrow & & \\ F + Q & \xleftrightarrow{k_+[Q]/k_-} & F : Q & & \end{array} \quad (2)$$

Let τ_0 be the lifetime of the excited state F^*:

$$\tau_0 = \frac{1}{k_f + k_0} \quad (3)$$

where k_f and k_0 are the rate constants characteristic for the fluorescence transition and the sum of all the other rate constants (except the quenching) leading to the excited to ground state transitions. By the inspection of Scheme (2) it is clear that $F^* : Q$ and $F : Q$ are the encounter complexes formed between the fluorophore and quencher molecules. Depending on the relationships between the rate constants above we can observe two possibilities. The quenchers for which

$$k_q >> \tau_0^{-1} + k_- \tag{4}$$

are defined as *strong quenchers*. In the case of strong quenchers the probability that the encounter complex with the excited state fluorophore will decay by quenching is close to 1. Otherwise, when Eq. (4) does not apply for a quencher the probability that the complex decays back to the ground state by some other ways is relatively large, and these quenchers are defined as *weak quenchers*[1].

By the use of strong quencher one more phenomenon can be involved in the quenching mechanisms, which is the presence of the *dark complex* and the corresponding *static* quenching component. As scheme 2. shows there are encounter complexes formed between the ground state fluorophore and a quencher molecule. These complexes are called *dark* because due to the nature of *strong* quenchers the excitation of such a fluorophore results in an immediate quenching. Therefore, by detecting the decay of the excited state population these complexes escape notice, i.e., in the fluorescence lifetime measurements the quenching component from dark complexes is not observed. The static component than is related to the encounter complex with ground state fluorophore while the dynamic component (what we can selectively detect in time-resolved measurements) is related to the quenching sequence started with excited fluorophore in no complex with quencher at the moment of excitation. It is clear from the above that the physics behind the static and dynamic quenching is the same. The existence of static quenching in the case of liquid or quasi-liquid phase reactions is the consequence of assuming the quencher to be ideally *strong*. Since in the case of *weak* quenchers there is no such a clear distinction but, instead, there is a more complex description, the distinction between dynamic and static quenching is to some extent the consequence of the experimental limitations. The increase of the sensitivity of our instrumentation will result in the appearance of new short fluorescence lifetime component(s) related to the effect of the so-called static quenching.

3. Data Processing and Limitations

3.1. The Stern-Volmer Equation

Let us start to describe the simplest approach, which can be applied for the data processing of the quenching of steady-state fluorescence emission. This is when we assume to use an ideally strong quencher and neglect the presence of the static quenching component assuming no dark complex[2]. In such a case the fluorophore can relax back from its excited state to its ground state according to one of the three schemes below:

---

[1] For such cases the description and interpretation of the quenching becomes more difficult and therefore the application of weak quenchers is rather rare in practice.

[2] In most of the cases the static component is about 10 % or smaller of that of the dynamic one.

$$F^* \xrightarrow{k_f} F + h\upsilon \tag{5}$$

$$F^* \xrightarrow{k_0} F + heat \tag{6}$$

$$F^* + Q \xrightarrow{k_+[Q]} F^* : Q \xrightarrow{k_q} F + Q \tag{7}$$

with $k_q \approx \infty$. When continuous illumination is applied steady fluorescence intensity can be obtained. If the illumination light intensity is low enough, only a small fraction of the fluorophore population will be in excited state. Therefore, the fluorescence intensity will be proportional to the total fluorophore concentration with the proportionality factor of the rate of photon emission (scheme (5)) over the sum of all rates resulting in deexcitation. Accordingly, the fluorescence intensity with zero quencher concentration (I_0) is:

$$I_0 = \frac{k_f F^*}{(k_f + k_0)F^*} = \frac{k_f}{k_f + k_0} = k_f \tau_0 \tag{8}$$

Similarly, in the case of the presence of quencher in concentration of $[Q]$,

$$I = \frac{k_f}{k_f + k_0 + k_+[Q]} = k_f \tau \tag{9}$$

where τ and τ_0 are the fluorescence lifetime of the fluorophore in the presence and the absence of the quencher, respectively. By the use of Eqs. (8) and (9) one can arrive at the Stern-Volmer equation:

$$\frac{I_0}{I} = \frac{\tau_0}{\tau} = 1 + k_+ \tau_0 [Q] = 1 + K_{SV}[Q] \tag{10}$$

where K_{SV} is the Stern-Volmer constant obtained as the slope of the linear plot of $\frac{I_0}{I}$ *vs.* $[Q]$ (see Eq. (10)). This is the classical Stern-Volmer plot. Such a plot, however, often exhibits either upward or downward curvature in applications.

The most likely reason of the upward curvature is thought to be the presence of the static quenching due to the formation of dark complexes. Since those fluorophores forming dark complex do not contribute to the fluorescence, Eq. (10) takes the form of

$$\frac{I_0}{I} P(nd) = 1 + K_{SV}[Q] \tag{11}$$

where $P(nd)$ is the probability that a fluorophore is not in an encounter complex with a quencher molecule. The form of $P(nd)$ will depend on the way the complex between the fluorophore and quencher is formed. Assuming no chemical interaction between these reactants this probability can be described by using the Poisson distribution. Accordingly, the $P(nd)$ probability is given as

$$P(nd) = \exp(-V[Q]) \tag{12}$$

where V is the static quenching constant and this time it corresponds to a volume element around the fluorophore within which the (*strong*) quencher molecule forms an encounter complex (*dark* complex) with the fluorophore.

A different case may occur if the quencher molecule is capable of forming a special, relatively tight complex with the fluorophore or with the matrix to which the fluorophore is attached. The speciality of this complex requires the quencher to stay close enough to the fluorophore to produce immediate quenching at any time. Otherwise a complex with a relatively rare encounter between the reactants would produce dynamic quenching. In such a case the value of $P(nd)$ or the fraction of free (not complexed) fluorophores can be given as:

$$P(nd) = \frac{K_D}{K_D + [Q]} \tag{13}$$

where K_D is the equilibrium dissociation constant of the complex. In cases of weak complexes Eqs. (12) and (13) give approximately the same value with $V = K_D^{-1}$. Accordingly, in most of the cases when the static quenching is around 10 % or smaller the above two cases cannot really be distinguished and the Stern-Volmer equation takes the form of:

$$\frac{I_0}{I} = (1 + K_{SV}[Q])\exp(V[Q]) \tag{14}$$

By assuming that V $[Q]$ <<1 and therefore $\exp(V[Q]) \approx 1 + V[Q]$ and $K_{SV}V[Q]^2$ can be neglected, Eq. (14) can be approximated as:

$$\frac{I_0}{I} = (1 + K_{SV}[Q])(1 + V[Q]) \cong 1 + (K_{SV} + V)[Q] \tag{15}$$

Eq. (15) tells us that performing steady-state quenching experiment, i.e., measuring the decrease of the fluorescence intensity as a function of the actual quencher concentration, the slope of the Stern-Volmer plot, even if the plot is linear, approximately measures the sum of the dynamic, K_{SV}, and the static, V, quenching constants.

In many cases, however, the Stern-Volmer plot shows upward curvature. In such cases it is possible to use a simple fitting procedure to search for the parameters of Eq. (14). Another possibility to find an explanation for the upward curvature is to perform steady-state and time resolved measurements and compare the results. The latter provides us with the K_{SV} making it simpler to find V from the steady-state data. In a limited range of the different fluorescence parameters there is a possibility to selectively obtain the value of K_{SV} and V from steady-state anisotropy measurements [8].

3.2. The modified Stern-Volmer or Lehrer equation

In the particular case when an external quencher can extinguish only an α ($0 \leq \alpha \leq 1$) fraction of the total fluorescence, F_0, the Stern-Volmer equation can be applied for the quenchable fraction of the fluorescence intensity with the assumption that no static quenching occurs [9]. The fluorescence intensity F in the presence of quencher can then be given as:

$$F = (1-\alpha)F_0 + \frac{\alpha F_0}{1+K_{SV}[Q]} \quad (16)$$

where $[Q]$ is the quencher concentration. By rearrangement of Eq. (16) we obtain the modified Stern-Volmer (or Lehrer) equation:

$$\frac{F_0}{F_0 - F} = \frac{F_0}{\Delta F} = \frac{1}{\alpha} + \frac{1}{\alpha K_{SV}} \frac{1}{[Q]} \quad (17)$$

The linear character of Eq. (17) makes it possible to calculate the values of α and K_{SV} by the use of the slope and intercept of the straight line of $F_0 / \Delta F$ *vs.* 1/[Q]. The presence of static quenching or heterogeneity of emitters may distort Eq. (17) [10], introducing difficulties into the analysis and interpretation. By considering this possibility, Eq. (17) is applicable for the study of protein fluorescence, when part of the emitted light originates from fluorophores (such as tryptophan or tyrosine) buried inside a macromolecule.

4. Special Cases

In many practical cases the application of fluorescence quenching methods require their adaptation to the investigated system. Such adapted and specialised methods were developed in previous years to characterise protein conformation and dynamics in situations where the applied fluorophores were shielded from the solvent by parts of a protein (see e.g. [11-13]). The adaptation of the quenching method could also capitalise on special geometric arrangements in some proteins, such as in actin filaments. A special type of quenching, fluorescence resonance energy transfer (FRET) was applied in many cases as a molecular ruler, and was shown to be able to describe the dynamics of the investigated protein matrix or supramolecular systems after appropriate mathematical treatment of the data. In the following paragraphs we describe some of these methods.

4.1. Quenching the Emission of Fluorophores Buried in Proteins

In proteins the fluorophores are often located in a special environment. Thus, the interpretation of K_{SV} in Eq. (10) or (17) characteristic of the quenching of model compounds such as tryptophan in a simple solvent becomes more complicated. In the case of fluorophores buried inside the protein matrix or other macromolecule there are two different possible mechanisms to account for the collisional quenching. In both cases the quenching is produced by quencher molecules initially located outside the protein matrix. In the first mechanism the excited fluorophore and the quencher come into contact by a transient exposure of the fluorophore to the solvent [14]. Their contact then results in the quenching of the fluorescence emission. The transient exposure of the fluorophore may occur by partial unfolding of the protein or by a relatively large-scale breathing motion, which allows the quencher to reach the fluorophore. Another way to understand the collisional quenching of buried fluorophores is to consider a mechanism in which the quencher penetrates the protein body and establish the interaction with the fluorophore after the penetration. Such a mechanism was shown to be operating for oxygen [15-18] and also for acrylamide [19]. Obviously, the information content of K_{SV} will be different depending on the particular mechanism involved in the quenching process. The following two sections describe these two mechanisms and the interpretation of the quenching data in these cases.

4.1.1. The Gating Mechanism

The gated mechanism used to be discussed in terms of a reaction scheme borrowed from the literature of hydrogen exchange [15, 20]:

$$A_{cl}^{*} \xleftrightarrow{k_{cl}/k_{op}} A_{op}^{*} + Q \xrightarrow{k_D} A_{op}:Q \tag{18}$$

The subscripts *cl* and *op* represent the inaccessible (closed) and accessible (open) states, while k_D is the bimolecular quenching constant characteristic of the rate of encounter complex formation between the exposed fluorophore and the quencher molecule. Owing to the finite lifetime of the excited state, however, scheme 18. gives only an approximate description of gated quenching. For example, for the quenching constant K_{SV} for the case of scheme 18. results in:

$$K_{SV}[Q] = \frac{k_{op}k_D\tau[Q]}{k_{cl} + k_D[Q]} \tag{19}$$

with τ being the lifetime of the excited state in the absence of the quencher, i.e. it predicts the quenching reaction to be first order only in a limited range of the quencher concentration where $k_D\,[Q] << k_{cl}$. A more precise description involves the extended scheme of

$$\begin{array}{ccccc} A_{cl} & \longleftrightarrow & A_{op} & & \\ \updownarrow & & \updownarrow & & \\ A_{cl}^{*} & \longleftrightarrow & A_{op}^{*} & \xrightarrow{k_D[Q]} & A_{op}:Q \end{array} \quad (20)$$

In the general case, scheme (20) represents a given fraction, α, of the protein fluorescence. If fraction 1 - α cannot be quenched by the quencher, one can apply statistic considerations to calculate the probability W that the fraction α of the excited state fluorophores will not be quenched, i.e. the probability that the excited state assigned to α will decay in any other way but by quenching [21]:

$$W = \frac{1 + P_{cl} k_D [Q] \tau / (1 + \tau / \nu_{cl} + \tau / \nu_{op})}{1 + (\tau + \nu_{op}) / \nu_{cl} + k_D [Q] \tau / (1 + \tau / \nu_{cl} + \tau / \nu_{op})} \quad (21)$$

Here ν_{op} and ν_{cl} are the average time of the fluorophore in the open and the closed states, τ is the lifetime of the excited state in the absence of the quencher and

$$P_{cl} = (1 - P_{op}) = \frac{\nu_{cl}}{\nu_{cl} + \nu_{op}} \quad (22)$$

where P_{cl} and P_{op} are the probabilities that the fluorophore is in a closed and an open state, respectively. By the use of Eq. (21) one can obtain the modified Stern-Volmer equation:

$$\frac{F_0}{F_0 - F} = \frac{F_0}{\Delta F} = \frac{1}{\alpha_{obs}} + \frac{1}{\alpha_{obs} K_{SV}} \frac{1}{[Q]} \quad (23)$$

with the parameters

$$\alpha_{obs} = \alpha \alpha_g \quad (24)$$

and

$$K_{SV} = \frac{P_{op} k_D \tau}{\alpha_g} \quad (25)$$

where

$$\alpha_g = P_{op} \left(1 + \frac{P_{cl}}{P_{op}} \frac{\tau}{\tau + \nu_{cl}} \right) \quad (26)$$

As it is seen by inspection of Eqs. (23-26) the model describes the quenching to be first order in the reciprocal value of the quencher concentration. Further, the gating affects the apparent accessibility α_{obs} and the Stern-Volmer constant K_{SV} depending on the ratio ν_{cl} / τ (Table 1).

Table 1. The quenching parameters of gated quenching in the limiting cases when the gating is very fast ($\nu_{cl} << \tau$) or very slow ($\nu_{cl} >> \tau$) on the time scale of τ

	α_{obs}	K_{SV}
$\nu_{cl}<<\tau$	α	$P_{op}k_D\tau$
$\nu_{cl}>>\tau$	$P_{op}\alpha$	$k_D\tau$

The product $\alpha_{obs} K_{SV}$, however, remains equal to $\alpha P_{op} k_D \tau$ independently of the ratio ν_{cl} / τ. This feature of the model has a diagnostic power. For example, by changing the viscosity, which certainly alters the ratio ν_{cl} / τ but leaves the value of P_{op} unchanged (altering both ν_{cl} and ν_{op} in the same way), the above product ($\alpha_{obs} K_{SV}$) decreases linearly with increasing viscosity, while the change in K_{SV} itself shows a different profile depending on the change in ν_{cl} / τ (see Eqs. (25) and (26)). This model was applied for the RNase T_1-acrylamide system. The acrylamide quenching of the fluorescence emitted by the single tryptophan residue of RNase T_1 was studied in the presence of an increasing amount of glycerol as the viscosity-elevating agent [21]. The model and the experimental data set proved to be incompatible, suggesting that the main quenching mechanism in this particular case was not a gating type. In our model we disregarded the presence of static quenching. This shortcoming of the model became important since further refinement of experimental data [22] showed a non-negligible contribution of static quenching in this experimental system. Accordingly, we further developed our model including the static quenching process as well [23]. This extension of the model resulted in the following forms for the two parameters of the modified Stern-Volmer equation:

$$\alpha_{obs} = 1 + \frac{VP_{op}(\nu_{op}\nu_{cl} + \nu_{op}\tau + 2\nu_{cl}\tau) - K_{SV}P_{op}\nu_{cl}^2}{K_{SV}(\nu_{op}\nu_{cl} + \nu_{op}\tau) + V\tau^2} \tag{27}$$

and

$$K_{SV}^{obs} = \frac{K_{SV}(\nu_{op}\nu_{cl} + \nu\tau) + V\tau^2}{\nu_{op}\nu_{cl} + \nu_{op}\tau + \nu_{cl}\tau} \tag{28}$$

where V is the static quenching constant [23]. For an experimental test of the model a third (although not independent) characteristic parameter, namely the inverse slope σ of the modified Stern-Volmer equation, has a more attractive form:

$$\sigma = P_{op}\tau\left[k_D + V\left(\frac{1}{\tau} + \frac{1}{\nu_{op}}\right)\right] = P_{op}\left[K_{SV} + V\tau\left(\frac{1}{\tau} + \frac{1}{\nu_{op}}\right)\right] \tag{29}$$

Besides its relatively simple form (compared with the expressions for α_{obs} and $K_{SV}{}^{obs}$), σ has the additional advantage that its value can be determined with higher precision than those of the other two (because of the higher error in determining the intercept). The most attractive feature of σ is that all parameters but k_D and ν_{op} in Eq. (29) are independent of viscosity η, while in the gating model k_D is inversely and ν_{op} is linearly proportional to η. Multiplying both sides by η we see that $d(\sigma\eta)/d\eta = V P_{op}$, i.e. by plotting $\sigma\eta$ *vs.* η we should, if Eq. (29) is applicable, see a straight line with a slope of $V P_{op}$. To prove the linear relationship we have to produce a substantial variation in viscosity, which can usually be done (at constant temperature) by adding cosolvents, such as glycerol, in higher concentrations. Such an addition of a cosolvent, however, makes the system non-ideal and introduces several side effects. These effects characterizing the actual system have to be discussed separately. The study of the viscosity dependence of σ in the case of the RNase T_1-acrylamide system showed that even the extended version of the gated quenching model is incompatible with the experimental data [23].

4.1.2. The Penetration Model

In many cases, the gating model apparently fails to describe the fluorescence quenching processes for the buried tryptophan residues of proteins when neutral external quenchers are applied. The alternative model that can possibly account for the quenching data assumes that the neutral quencher molecules (such as molecular oxygen and acrylamide) can enter (penetrate) the protein matrix. The detailed description of such a model, in the general case, arrives at a rather difficult form even if the protein interior is regarded as a homogeneous and isotropic phase. Such an attempt was made to describe the quenching by molecular oxygen of the porphyrin emission of the iron-free derivatives of myoglobin and haemoglobin [16, 18]. The model starts with the detailed description of the time-dependent behaviour of a simple quenching system [16]:

$$Q + F^* \xleftrightarrow{k^+/k^-} Q \cdot F^* \tag{30}$$

If F^* and $Q\,.F^*$ are denoted by x and y respectively, the model gives $x(t)$ and $y(t)$, the time-dependent functions of x and y, as:

$$x(t) = \alpha_{0x}\exp(-m_0 t) + \alpha_{1x}\exp(-m_1 t) \tag{31}$$

$$y(t) = \alpha_{0y}\exp(-m_0 t) + \alpha_{1y}\exp(-m_1 t) \tag{32}$$

with

$$m_{0,1} = \frac{(\Gamma_0 + \Gamma_1) \pm \{(\Gamma_0 - \Gamma_1)^2 + 4k^- k^+ [Q]\}^{1/2}}{2} \quad (33)$$

and

$$\alpha_{0x} = \frac{x_0(\Gamma_0 - m_1) - k^- y_0}{m_0 - m_1} \quad (34)$$

$$\alpha_{1x} = \frac{k^- y_0 - x_0(\Gamma_0 - m_0)}{m_0 - m_1} \quad (35$$

$$\alpha_{0y} = \frac{y_0(\Gamma_1 - m_1) - k^+ [Q] x_0}{m_0 - m_1} \quad (36)$$

$$\alpha_{1y} = \frac{k^+ [Q] x_0 - y_0(\Gamma_1 - m_0)}{m_0 - m_1} \quad (37)$$

where x_0 and y_0, are the concentrations of the excited species x and y respectively right after excitation and:

$$\Gamma_0 = \Gamma + k^+ [Q] \quad (38a)$$

$$\Gamma_1 = \Gamma + \chi + k^- \quad (38b)$$

where Γ is the single-exponential radiative decay rate of the excited state and χ is the non-radiative decay rate due to quenching. Accordingly, the time-dependent fluorescence emission $F(t)$ is described by two exponential components characterized by rates m_0 and m_1:

$$F(t) = F_0(t) + F_1(t) \quad (39)$$

where

$$F_0(t) = (\alpha_{0x} + \alpha_{0y}) \exp(-m_0 t) \quad (40)$$

$$F_1(t) = (\alpha_{1x} + \alpha_{1y}) \exp(-m_1 t) \quad (41)$$

The steady state fluorescence emission is then given by:

$$\langle F \rangle = \frac{\alpha_{0x} + \alpha_{0y}}{m_0} + \frac{\alpha_{1x} + \alpha_{1y}}{m_1} \tag{42}$$

In the case of strong quenching, i.e. when:

$$\chi >> k^{-} \tag{43}$$

Eq. (42) becomes:

$$\langle F \rangle = \frac{x_0}{\Gamma_0} \tag{44}$$

Gratton and colleagues apply the above formalism for the quenching of fluorescence emitted by the buried fluorophore of a protein [16]. In the above treatment they assign the species y to the protein molecule containing quencher molecule(s) in the protein interior. Accordingly, x is the rate constant of the reaction bringing such a protein molecule to a state where the tryptophan residue has a quencher molecule within contact distance. Without giving any further details here we refer to the experimental paper of Jameson and colleagues [18] where they arrive at the conclusion that, in the case of molecular oxygen as the quencher, 'both the rate of penetration of oxygen in the protein interior and its further migration to the fluorophore' need to be considered depending on the fluorescence lifetime. For compounds having short lifetimes, such as tryptophan, the latter process dominates the quenching. Owing to the experimental difficulties caused by the low solubility and small size of oxygen, considerable attention has been paid to the quenching by acrylamide, an uncharged solute showing a high quenching efficiency. The description of the mechanism by which this molecule quenches the fluorescence emitted by buried tryptophan residues of proteins often seems ambiguous. Although many experimental data have been collected, the mechanism of action for this quencher has not been elucidated. A gating-type mechanism cannot dominate the quenching process in the case of RNase T_1 [23]. The Förster-type energy transfer mediated by dipole-dipole interaction between the tryptophan residue and acrylamide can also be excluded as a main mechanism owing to the presence of static quenching component observed for the RNase T_1-acrylamide system [22, 24]. The remaining possibility can be described by a penetration model assuming that acrylamide can enter (and make a "random walk" inside) the protein matrix. Such a simple model was reported by Eftink and Hagaman [22]. They assumed that the acrylamide molecules residing outside the protein body at the moment of excitation are mainly responsible for the quenching. Accordingly, the overall quenching process is regarded as a two-step sequential reaction. First, the quencher molecules diffuse in the solvent to the protein (characterized by k_1). After reaching the protein surface, they enter the body and reach the tryptophan residue via a second diffusion process (within the protein matrix) characterized by a different rate constant k_2. Therefore the overall rate constant k^1 for the quenching process can be described as

$$k^1 = \frac{k_1 k_2}{k_1 + k_2} \tag{45}$$

This model explains the experimental data obtained for the viscosity-dependent variation in the acrylamide quenching of the fluorescence emitted by the single tryptophan residue of RNase T_1. At low viscosities, the rate-limiting step is the internal diffusion of acrylamide (k_2) and there is only a little viscosity dependence of k^1. At higher viscosities the external viscosity makes k_1 comparable with k_2 (k_1 is inversely proportional to the viscosity) and therefore a slight but definite viscosity dependence will occur. Above this range (at the high viscosity limit), k_1 becomes rate limiting and the system exhibits a definite viscosity dependence (k^1 varies according to the change in k_1). The above model is in conflict with the presence of a definite static quenching component observed for the RNase T_1-acrylamide system [23]. Therefore, the assumption that the quenching is caused by quencher molecules outside the protein matrix at the moment of excitation is not acceptable. This conclusion was also supported by Jameson and colleagues [18] stating that even in the case of oxygen quenching fluorescence emitted by buried tryptophan residues the oxygen molecules already residing inside the protein at the excitation play a dominant role. Acrylamide molecules, because of their size, are expected to travel a shorter distance in the liquid phase during the lifetime of the excited tryptophan residues. Therefore it is likely that those acrylamide molecules situated outside the protein matrix at the moment of excitation of a buried tryptophan residue play little direct role in the quenching process. This assumption makes it possible to use a 'two-phase' model [19], which is just the opposite of that described earlier [22] in stating that quenching is due to the acrylamide molecules present inside the protein matrix at the moment of excitation. The acrylamide molecules can penetrate the structural matrix of the protein via transient holes formed by the fluctuations in volume, free energy and bond configuration of the protein. The acrylamide molecule can then move around within the protein phase (including the hydration shell) by a motion characterized by the inhomogeneous and anisotropic local rearrangements of the protein matrix. Of course, the acrylamide molecules will eventually leave the protein phase and re-enter the aqueous phase. During their migration inside the protein phase they have the chance of contacting or colliding with the excited fluorophore. In this way acrylamide exerts its quenching effect in a similar manner to oxygen and the process can be described accordingly by a general penetration model [16, 18]. The use of acrylamide as a quencher, however, allows us to introduce a greatly simplified penetration model, which can be directly tested over a wide range of quencher concentrations and viscosity values. This model has the advantage that it also deals with certain aspects of the non-ideality of the system, i.e. it considers activity changes in glycerol-water mixtures [19].

The quenching mechanism to be considered assumes that protein segments do not unfold transiently into the solution but that the native structure of the protein represents a separate phase and that the quencher and the solvent also have finite solubilities in that phase. The premise above allows the assumption that the acrylamide molecules outside the protein phase at $t = 0$ (when the excitation occurs) cannot reach the tryptophan residue during the interval of τ and thus do not contribute to the observed quenching. Accordingly, the model deals only with the acrylamide molecules with the concentration $[Q_i]$ situated in the protein phase. In fact the quenching process is now formally comparable with acrylamide quenching of free indole. The classical derivation of the Stern-Volmer equation is applicable. The apparent dynamic and static quenching constants, however, both contain the distribution constant describing the acrylamide distribution between the two phases (the protein phase and the solvent). The presence of a cosolvent such as glycerol introduces further factors specific for the effects of the cosolvent on the activity of the quencher and the protein.

To obtain the true quenching constant K_{SV} and V, both steady state and time-resolved quenching experiments should be performed (usually the steady state measurements are sensitive to both constants while the time-resolved measurements are sensitive only to the dynamic quenching constant).

4.2. The Method of Double Quenching

The fluorescence behaviour of tryptophan residues in a native protein is not uniform. Usually each residue is characterized by unique microenvironmental properties. Depending on their location in the protein matrix they may be fully or partially exposed to (or buried from) the solvent [25, 26]. Ionic quenchers, because of their charged and strongly hydrated character, can extinguish only fluorescence emitted by fluorophores located at, or near to, the surface of the protein [2, 9]. Oxygen (a non-ionic quencher), in contrast, can penetrate the protein matrix and quench all the emission of the tryptophan residues, buried or exposed [15-18, 20]. In the case of another non-ionic quencher, acrylamide, there was for long time no clear evidence as to whether it penetrates the protein matrix (thereby giving information on the internal fluctuation of the macromolecule) or whether it quenches the fluorescence of the buried fluorophores by some other mechanism(s) [15-18, 20, 21, 23, 27]. Similarly, the oxygen quenching data of proteins presenting both solvent-exposed and buried fluorophores contain mixed information. Such considerations indicate that the separation of quenching parameters associated with buried and exposed fluorophores is of special importance for the further understanding of protein fluctuations studied by fluorescence or phosphorescence quenching. To determine the acrylamide quenching constant for the buried tryptophan (Trp-314) of horse liver dehydrogenase, Eftink and Selvidge [28] used iodide in a relatively high concentration (0.45 M) to quench about 95 % of the fluorescence of the exposed tryptophan. Subsequently, acrylamide was added to the mixture; and the decrease in fluorescence then characterized the acrylamide quenching of the buried tryptophan. A similar approach was used to study the fluorescence quenching of the same tryptophan residue by oxygen [17] and to resolve the fluorescence emitted by the buried tryptophan in yeast 3-phosphoglycerate kinase [29]. Although this approach seems attractive and simple, there are several problems associated with it. For example, it is impossible to achieve 100 % quenching of the exposed fluorophore. The remaining fluorescence then might generate a considerable error. Further, it is difficult to generalize the method owing to the uncertainty of the quencher concentration necessary to quench 'totally' the fluorescence of the exposed fluorophores. Another difficulty might also arise from the potentially high concentration of the 'selective' quencher since it could result in gross conformational transitions, or even denaturation, of the protein [28].

To overcome this problem separate quenching parameters are required to characterize the two classes. Such a method will be described here as 'double quenching'. With one independent variable (the quencher concentration) a linear relationship, presented by Eq. (17), yields two parameters (α and K_{SV}). If we desire to resolve the mechanism in further detail by characterizing the quenching constants and accessibility for the two main classes of fluorophores (external and internal) we must increase the number of variables. One way to do this is to apply two quenchers simultaneously which show selectivity in their ability to quench [30]. One of the quenchers should be specific (such as I^- or Cs^+), being able to quench only the external fluorescence, while the other should be non-specific (such as oxygen or acrylamide). According to the above let us consider a protein having three classes of fluorophores: (a) external, (b) internal and quenchable by non-specific quencher only and (c) non-quenchable. The fluorescence intensities $F_{0e}(=(\alpha F_0)$, $F_{0i}(=\beta F_0)$ and $F_{0n}(=\gamma F_0)$ then add up to F_0 the total intensity. Therefore:

$$\alpha+\beta+\gamma=1 \tag{46}$$

By using the proportionality between the lifetime and the fluorescence intensity and the additivity of fluorescence one can arrive at:

$$\frac{F_{1,2}}{F_0}=\frac{\alpha}{1+K_{1e}[Q_1]+K_2[Q_2]}+\frac{\beta}{1+K_{1i}[Q_1]}+\gamma \tag{47}$$

where $F_{1,2}$ is the total fluorescence intensity in the presence of the two quenchers with a concentration of the non-specific quencher and a concentration of the specific quencher. K_{1e}, K_{1i} and K_2 are the appropriate Stern-Volmer constants. It is then obvious that, for the total fluorescence intensities F_1 measured in the presence of Q_1 and the absence of Q_2, one obtains (by inserting $[Q_2]=0$ into Eq. (47)):

$$\frac{F_1}{F_0}=\frac{\alpha}{1+K_{1e}[Q_1]}+\frac{\beta}{1+K_{1i}[Q_1]}+\gamma \tag{48}$$

Combining Eqs. (47) and (48) gives:

$$\frac{F_0}{\Delta F}=\frac{F_0}{F_1-F_{1,2}}=\frac{1+K_{1e}[Q_1]}{\alpha}+\frac{(1+K_{1e}[Q_1])^2}{\alpha K_2}\frac{1}{[Q_2]} \tag{49}$$

Equation (49) shows that by keeping $[Q_1]$ constant and varying $[Q_2]$, the plot of $F_0/\Delta F$ versus $1/[Q_2]$ results in a straight line with intercept and slope depending on $[Q_2]$. When $[Q_1]=0$, Eq. (49) reduces to the modified Stern-Volmer plot for quencher 2. Therefore two parameters (α and K_2) out of six can be obtained from a titration in the absence of Q_1. To proceed further a secondary plot can be used by plotting the square roots of slopes against $[Q_1]$:

$$\xi^{1/2}=\frac{1}{(\alpha K_2)^{1/2}}+\frac{K_{1e}}{(\alpha K_2)^{1/2}}[Q_1] \tag{50}$$

The value of K_{1e} can be determined either by using the slope and intercept of Eq. (50) or by using the already-known value of αK_2 and the slope.

The parameters then remaining to evaluate are β, γ and K_{1i}. For this, one can define X and calculate its value from the already-known parameters as

$$X = \frac{F_1}{F_0} - \frac{\alpha}{1 + K_{1e}[Q_1]} = \frac{\beta}{1 + K_{1i}[Q_1]} + \gamma \tag{51}$$

Denoting the value of X at $[Q_1] = 0$ by X_0, we write:

$$\frac{1}{X_0 - X} = \frac{1}{\beta} + \frac{1}{\beta K_{1i}} \frac{1}{[Q_1]} \tag{52}$$

Plotting $1 / (X_0 - X)$ against $1 / [Q_1]$ should result in a straight line, and the values of β and K_{1i} can accordingly be calculated from the slope and intercept. The value of γ then becomes $\gamma = 1 - \alpha - \beta$. The above method was applied to resolve the quenching parameters of lysozyme fluorescence by the use of iodide as the selective quencher and acrylamide as the non-selective quencher. The determination of the acrylamide quenching constant associated with the internal fluorophore Trp-108 proved to be about twice ($K_i = 3.5$ M^{-1}) that belonging to the more exposed residue Trp-62 ($K_0 = 1.6$ M^{-1}) [30]. The reliability of the method was later proved by independent measurements on single-tryptophan variants of lysozyme [31].

4.3. Methods Based on the Application of FRET

A special case of the fluorescence quenching when an excited fluorophore (fluorescent donor) can release its excess energy by transferring it to an appropriate fluorophore (fluorescent acceptor) located in the near vicinity of it. Appropriate means here that the emission spectrum of the donor fluorophore overlaps with the absorption spectra of the acceptor molecule. The phenomenon is called fluorescence resonance energy transfer and occurs without photon emission through dipole-dipole interactions. The rate constant characteristic for the energy transfer (k_t) depends on the inverse 6th power of the distance between the donor and acceptor molecules. The distance dependence of k_t made FRET a powerful tool to study intramolecular and intermolecular distances in molecular systems, i.e. it can be used as a 'spectroscopic ruler' [32]. Since energy is transferred from the excited donor to the acceptor, the lifetime (τ), quantum efficiency (Q) and fluorescence intensity (F) of the donor decrease in the presence of an acceptor (Eq. 53).

As a consequence the fluorescence intensity of the acceptor increases if the donor is present which process is the sensitisation (see Eq. 54). Accordingly, the efficiency of the energy transfer (E) can be determined in different ways depending on the optical properties of the investigated system. If one measures the fluorescence lifetime (τ), intensity (F) or quantum yield (Q) of the donor:

$$1 - E = \frac{\tau_D^A}{\tau_D} = \frac{F_D^A}{F_D} = \frac{Q_D^A}{Q_D} \tag{53}$$

Depending on the spectral properties of the donor and acceptor molecules, in special cases it is favorable to determine the transfer efficiency via the acceptor fluorescence. Due to the transfer, the fluorescence intensity of the acceptor will increase and the efficiency of the transfer could be calculated using the following formula:

$$\frac{F_A^D}{F_A} = 1 + \left(\frac{\varepsilon_D C_D}{\varepsilon_A C_A} \right) E \tag{54}$$

In equations (53) and (54) the lower indices indicate the donor (*D*) or acceptor (*A*), while the upper indices refer to the presence of the donor (*D*) or the acceptor (*A*). C_D and C_A are the molar concentrations, ε_D and ε_A are the molar absorption coefficients of the donor and the acceptor, respectively. The efficiency (*E*) of the transfer process can be expressed by using the rate constants of the contributing de-excitation processes as:

$$E = \frac{k_t}{k_t + k_f + k_0} \tag{55}$$

where k_t and k_f are the rate constants for the energy transfer and the donor emission respectively, while k_0 is the sum of the rate constants of all other non-radiative de-excitation processes of the excited donor molecules. The transfer rate constant can be written as:

$$k_t = d\, J\, n^{-4}\, k_f R^{-6}\, \kappa^2 \tag{56}$$

where *d* is a constant depending on the choice of units and *J* is the spectral overlap integral. The overlap integral is the overlap between the normalized emission spectrum of the donor and the absorption spectrum of the acceptor. In Eq. (56) *n* is the refractive index of the medium between the fluorophores participating in the energy transfer. *R* is the distance between the two chromophores, while κ^2 is a measure of the relative orientation of the electronic transition moments of the emitting state of the donor and the absorbing state of the acceptor. The Förster critical distance (R_0) is defined as the donor-acceptor distance at which the transfer efficiency is 0.5. R_0 can be written as:

$$R_0 = c \left(J n^{-4} \frac{k_f}{k_f + k_0} \kappa^2 \right)^{1/6} \tag{57}$$

The value of R_0 is characteristic for a donor-acceptor pair and its value can be determined separately in an independent experiment provided that κ^2 is unchanged. The combination of the above equations leads to the expression:

$$E= \frac{R^{-6}}{R^{-6}+R_0^{-6}} \tag{58}$$

This equation shows that by knowing the value of R_0 the experiment allowing the calculation of E makes it possible to evaluate the actual distance R between the donor- and acceptor. The next sections describe special FRET based applications, which apart from distance determination can provide additional information regarding the studied macromolecular system.

4.3.1. Protein Flexibility as Resolved by FRET

By using special experimental strategy and adequate mathematical formalism the method of FRET can be applied to characterise the dynamic properties of proteins. The application is based on the non-linear distance dependence of the k_t. Although there were many studies dealing with the determination of intramolecular and intermolecular distances by the use of FRET [33-43], the built-in distortion of the method was often disregarded. Both theoretical and experimental data indicated that the relative motion of the donor-acceptor pair on the time scale of excited-state lifetime could alter the rate constant of the energy transfer. The faster the relative motion is the greater k_t and thus the shorter donor-acceptor distance determined by the method. This distortion of the method is disadvantageous for donor-acceptor distance measurements. At the same time it can be used to follow the change in the internal dynamics of macromolecules [44]. Let us consider two fluorescent labels associated with a protein molecule and forming a donor-acceptor pair. Labelling of proteins with extrinsic fluorophores cannot usually be carried out in a perfectly uniform fashion. Even for proteins with intrinsic fluorophores the homogeneity of the sample is usually not perfect, for example due to the presence of conformational sub-states of proteins. The transfer rates (k_t) of the molecular population can be denoted by a distribution of k_{ti} and in any experimental study the determination of the transfer rate constant should always involve averaging:

$$\langle k_t \rangle = ck_f \langle R^{-6}\,\kappa^2 \rangle \tag{59}$$

Regarding to the averaging involved in this equation three different cases has to be considered [44]: (a) "static" case, when the fluctuation is much slower than the transfer rate; (b) the "dynamic" case, when the rate of fluctuation is much faster than the rate of energy transfer; (c) the "intermediate" case, when the two processes have comparable rates. For each of these cases there are different mathematical methods involved in the averaging of k_t values. Any real (*e.g.* macromolecular) system will involve all these modalities and thus any experiment performed with proteins entails all kinds of fluctuations under the usual analytical conditions. Accordingly, the experimentally obtained value of k_t will necessarily include all the modes of averaging superposed. Thus the mean value $\langle k_t \rangle$ seems an appropriate parameter for monitoring local fluctuations (of different relaxation times) of a macromolecule under changing environmental conditions. For a homogeneous energy transfer system in an ideally "frozen" state the quantum yield Φ_D of the donor in the presence of the acceptor can be expressed as:

$$\Phi_D = \frac{k_f}{k_t + k_f + k_0} \quad (60)$$

we introduce the factor f as:

$$f = \frac{E}{\Phi_D} = \frac{k_t/(k_f + k_t + k_0)}{k_f/(k_f + k_t + k_0)} = \frac{k_t}{k_f} \quad (61)$$

In a real experiment, however, fluctuations result in variability for the values of k_t and k_0. Therefore one can measure only the average values of E and Φ_D. In these cases,

$$f = \frac{\langle E \rangle}{\langle \Phi_D \rangle} = \frac{\langle k_t/(k_f + k_{ti} + k_{0i}) \rangle}{\langle k_f/(k_f + k_{ti} + k_{0i}) \rangle} \quad (62)$$

On the assumption of no change in k_f during the fluctuation this equation can be rearranged as:

$$f = \left\langle \frac{k_{ti}}{k_f} \frac{1/(k_f + k_{ti} + k_{0i})}{1//(k_f + k_{ti} + k_{0i})} \right\rangle = \left\langle \frac{k_{ti}}{k_f} \frac{\tau_{Di}}{\langle \tau_D \rangle} \right\rangle \quad (63)$$

where τ_{Di} is the lifetime of the ith donor population taking a momentary picture of the system, and $\langle \tau_D \rangle$ is the average lifetime. Applying Eq. (59) we have

$$f = c\, \langle \alpha_i R_i^{-6} \kappa_i^2 \rangle \quad (64)$$

where

$$\alpha_i = \frac{\tau_{Di}}{\langle \tau_D \rangle} \quad (65)$$

Considering the characteristic features of α_i it can be shown that

$$\langle \alpha_i \rangle = 1 \quad (66)$$

We must emphasize that without further assumptions and qualifications the parameter f cannot be used to yield information about the frequency and amplitude distribution of the fluctuations. To put Eq. (64) into a more useful form, let us assume that

$$k_{ti} << k_f + k_{0i} \tag{67}$$

which means that the rate constant for energy transfer is much smaller than the composite for all other deactivation processes. One can approximate this relationship from the experimentally determined value of the transfer efficiency. In this case, the value of k_{ti} in the expression for α_i can be neglected (i.e. k_{ti} and α_i can be treated as independent variables). Thus Eq. (63) simplifies, by virtue of Eq. (66), as follows:

$$f = \left\langle \frac{k_{ti}}{k_f} \right\rangle \left\langle \frac{\tau_{Di}}{\tau} \right\rangle = \left\langle \frac{k_{ti}}{k_f} \right\rangle \tag{68}$$

and Eq. (64) becomes:

$$f = c \left\langle R_i^{-6} \kappa_i^2 \right\rangle \tag{69}$$

From Eq. (69) the change in the parameter f can be related directly to a change in fluctuations or to a conformational change, which alters the static 'disorder' (or fluctuation spectrum) of the donor-acceptor system. In experimental applications the temperature profile of this parameter provides information about the average flexibility of the protein matrix located between the labels.

Viewing the change in f, it may be more convenient to use another parameter, f ', whose definition can be given similarly to that of f, except with the use of the fluorescence intensity instead of the quantum yield of the donor (therefore the parameters f and f ' are proportional to each other):

$$f' = \frac{\langle E \rangle}{\langle F_D \rangle} \tag{70}$$

where F_D is the fluorescence intensity of the donor molecule in the presence of the acceptor.

This temperature dependent FRET method was originally tested by using the single tryptophan residue in RNase T_1 as donor and pyridoxamine phosphate as acceptor attached to the lysine side group of the protein [44]. A 40 % increase in f ' was observed while the temperature increased from 293 to 314 K. The conclusion was that the temperature induced increase in f ' was due to the increase in the amplitudes of the fluctuation inside the protein matrix. O'Hara and coworkers [45] have also used the method to study thermally induced fluctuations in calmodulin (energy transfer from a tyrosine donor to a Tb(III) acceptor and transferrin (energy transfer from a Tb(III) donor to an Fe(III) acceptor). They observed similar percentage increases in f ' for both systems as in the case of the RNase T_1. This allowed them to conclude that the change in f ' is rather insensitive to the nature of the protein environment. Since all their systems are small proteins, the fluctuation patterns have similar responses to the increase in temperature. This reasoning seems to be reinforced by the results of Jona and coworkers [46] who investigated the thermally induced structural fluctuations in Ca^{2+}-adenosine-5'-triphosphatase (ATPase) of sarcoplasmic reticulum using f ' method. Their data indicate the existence of a relatively flexible structure in the region of the ATPase molecule linking the cysteins 670 and 674 and the lysine 515 (shown by a relatively high increase in f' over the temperature range 10-37 °C). In contrast, the region between the cystein cluster (cystein 670 and 674) and the Ca^{2+} site proved to be relatively rigid as revealed by the normalized energy transfer from IAEDANS to Pr^{3+}. Subsequently the temperature dependence of the FRET efficiency was applied in numerous studies to macromolecules and their complexes [47-52] and these applications resolved characteristic changes in protein dynamics in many cases. These observations seem to prove that the magnitude of thermally induced change in f' is not unified for all the proteins. It is capable of sensing even directional inhomogeneity of internal fluctuations in the protein matrix.

Although this FRET method is unique in monitoring the change in relative fluctuations of two or more specifically defined points on a macromolecule, it has to be emphasized that gross conformational or structural changes (e.g. denaturation) might also affect the value of f (or f'). Regarding the effect of conformational changes on f', this might be identified separately in many cases. Therefore the method should be characterized as capable of monitoring both gross conformational transitions and changes in fluctuations inside the macromolecular matrix as well. It was also shown that in fluorophore systems where more than one donor can transfer energy to more than one acceptor the normalised FRET efficiency defined above is the linear combination of the normalised transfer efficiencies attributed to individual donor-acceptor pairs of the system [53]. Therefore, the temperature dependence of the normalised FRET efficiency is informative regarding the changes of the intramolecular flexibility and/or conformation of proteins even in these multiple fluorophore systems.

4.3.2. The Application of FRET for Protein Systems with Special Symmetry

In special cases FRET methods can be applied to provide information regarding distance and thus conformational distribution within proteins and supramolecular protein complexes. One of these specialized methods was developed based on the work of Taylor and his colleagues [54]. Although previously many studies applied FRET to characterize different conformational states of actin filaments and their interactions with partner proteins [55-69], the FRET approach described here can provide more detailed information by capitalizing on the helical symmetry of actin filaments and thus of the complex of actin filaments and actin-binding proteins bound to it. Although the method was applied originally for the actin filament system, it is also applicable for any filamentous system with appropriate geometrical symmetry. For simplicity here the interpretation of the FRET data with actin and actin-binding proteins are described.

The application of FRET can provide three dimensional coordinates of a labeled point on an actin-binding protein in a coordinate system confined to the actin filament axis [70]. By assuming a distribution of the donor-acceptor distances and thus the coordinates, the three dimensional distributions characteristic for the location of the probes can be described with this approach. The method therefore allows the detailed description of the conformational heterogeneity of a labeled segment of an actin-binding protein.

For a single donor-single acceptor system the transfer efficiency can be calculated from Eq. (58), or in a modified form from the following equation:

$$E = R_0^6 / \left(R_0^6 + R^6\right) \tag{71}$$

When the donor is located in an actin-binding protein, and the actin protomers are labelled with acceptors, the energy transfer within this fluorophore system can occur from a single donor to more then one acceptors. Due to the complexity of the FRET processes the calculation becomes more complicated and Eq. (71) transforms to [10, 71]:

$$E = \sum (R_0 / R_i)^6 / [1 + \sum (R_0 / R_i)^6] \tag{72}$$

where R_i denotes the individual donor-acceptor distances. For the analysis of experimental data one should assume that the donor molecule in the actin-binding protein can transfer energy to acceptors located on the five nearest actin monomers. A three-dimensional Descartean coordinate system can be fixed to the actin filament, while the analyses should also consider the geometry of the actin filament. The geometry of a filament is described by the radial coordinate of the labeled residue (r), and also by the pitch of the actin monomer in the actin filament, and the relative rotation of the neighboring monomer along the genetic helix, which can be taken to be 55 Å and 166 Å, respectively. The coordinates (x, y and z) of the labeled actin residues on the five actin monomers considered can be then calculated by using the following equations:

$$x_i = r\cos((3-i)166°) \tag{73a}$$

$$y_i = r\sin((3-i)166°) \tag{73b}$$

$$z_i = (3-i)27.5\,Å \tag{73c}$$

where i takes the integer value from 1 to 5 and refers to individual actin monomers. If the acceptor molar ratio (g) is known, the probability (p_k) of the individual arrangements could be obtained from the binomial distribution as follows:

$$p_k = g^k (1-g)^{5-k} \tag{74}$$

Knowing the positions of the potential acceptor labeling sites on the five actin monomers, the cumulative transfer efficiency (E_g) at a given acceptor molar ratio (g) can be calculated using Eqs. (72) and (73) as follows:

$$E_g = \sum_k p_k E_k \tag{75}$$

where E_k is the FRET efficiency between the donor and the acceptors in the kth arrangement of actin monomers (Eq. 72), whereas p_k is the probability of the kth arrangement (Eq. 74). By applying these equations, one can compare simulated curves with the experimental data and determine the physically veritable position of the donor in the actin binding protein.

With a more complex analysis the method can also give information regarding the conformational heterogeneity of the protein matrix. To accommodate conformational distribution the donor-acceptor distance distributions were taken into account. It is assumed in the analysis that the transitions between the conformational states are slow on a nanosecond time scale. The distance distribution characteristic for a donor-acceptor system ($\omega(R)$) can be approximated with a Gaussian function, with a mean value of R_c and the full width at half-maximum of σ:

$$\omega(R) = \frac{1}{\sigma\sqrt{2\pi}} \exp\left[-\frac{(R - R_c)^2}{2\sigma^2}\right] \tag{76}$$

In this case the summation of the individual transfer efficiencies is carried out as follows.

$$E_g = \sum_{jk} p_k \omega(R_j) E_{jk} \tag{77}$$

where $\omega(R_j)$ is the probability that the donor-acceptor distance is R_j (at the jth position of the distance distribution).

The acceptor molar ratio dependence of the FRET efficiency was measured in the case of the actin-myosin protein complex system [70]. The comparison of the experimental FRET data with the expectations from the theory assuming different values for the three coordinates of the donor resolved the three dimensional positional distribution of the donor attached to the myosin. The method therefore appeared to be suitable to obtain information regarding the positional distribution of a labeled residue in a protein bound to a filamentous macromolecule.

4.3.3. FRET Measurements in Flow Cytometry

Investigation of protein-protein associations is essential in understanding structure and function relationships in living cells. Fluorescence techniques are widely used to quantify molecular parameters of biochemical and biological processes because of their inherent sensitivity, specificity and temporal resolution. Combination of fluorescence resonance energy transfer with flow cytometry - flow cytometric energy transfer (FCET) - provided a solid basis for improvements of these technologies [37, 72]. Applying donor or acceptor labelled antibodies FRET technique can be used to determine inter- and intramolecular distances of cell surface proteins. With the help of FRET, molecular dimensions can be measured in live cells providing information inaccessible with other approaches e.g. electron microscopy [73, 74].

FRET efficiency can be determined by measuring the fluorescence characteristics of the donor or the acceptor. Quantum yield, fluorescence lifetime or intensity of the donor change upon FRET. Quantum yield has not been measured in flow cytometry so far. Measuring fluorescence lifetime in a flow cytometer requires sophisticated and expensive instrumentation, therefore it is not widely applied. The simplest way to measure energy transfer is to determine the decrease in the fluorescence intensity of the donor in the presence of the acceptor. The fractional decrease in the donor fluorescence with the acceptor present is equal to the efficiency of FRET. In flow cytometry, however, donor quenching cannot be used to determine transfer efficiency on a cell-by-cell basis, because the expression levels of various proteins have broad distributions, and we cannot measure the donor intensity in the absence and presence of acceptor on the same cell. For mean values donor quenching can also be used providing a single mean FRET efficiency for the whole cell population. In addition appropriate control samples should be prepared and measured in order to tell apart whether the decrease in the donor fluorescence intensity was caused by competition or FRET between the donor and acceptor labelled molecules.

Sensitised emission of the acceptor, however, can be applied to calculate the transfer efficiency on a cell-by-cell basis since with multiple excitations the direct and sensitised emissions of the acceptor can be determined on the same cell. In flow cytometric measurements there are three unknown parameters: efficiency of FRET, the unquenched donor intensity and the non-enhanced acceptor intensity. In order to determine these parameters, three independent fluorescence intensities for the same cell must be measured. The three independent fluorescence signals differ from each other in the wavelength of the excitation or the spectral range of the detection. Because of the different wavelength dependence in the absorption and emission spectra of the donor and acceptor, the corrected transfer signal can be evaluated from the three independently measured parameters using correction factors determined on singly labelled cells. In this case biological variance in the level of expression of the investigated protein has no effect on the accuracy. The generated distribution histogram of FRET efficiency will provide information about the heterogeneity of the cell population with high statistical accuracy. Technical details of flow cytometric energy transfer measurements applying fluorescein and rhodamine as donor-acceptor pair can be found in several reviews [75-78]. Briefly, in such a system the 488 nm and the 514 nm lines of an argon ion laser are used for excitation. Spectral ranges of fluorescence are detected around the emission maximum of fluorescein (535 nm) and usually above 590 nm (emission of rhodamine). The calculation of energy transfer efficiency is performed by a computer using the equations below. In these equations the measured intensities are I_1, I_2 and I_3 and the excitation and emission wavelengths are given in parentheses. I_D and I_A stand for the theoretical (unquenched and nonenhanced) intensities of the donor (excited at 488 nm, emission detected at 535 nm) and of the acceptor (excited at 514 nm, detected at >590 nm), respectively. Since the emission spectra of the fluorescein and rhodamine overlap, and both molecules can be excited by both laser beams, correction factors have to be introduced. These factors are S_1, S_2 and S_3. The definition of these parameters is:

$S_1 = I_2 / I_1$ determined using only donor labeled cells
$S_2 = I_2 / I_3$ determined using only acceptor labeled cells
$S_3 = I_3 / I_1$ determined using only donor labeled cells.

The three detected intensities can be expressed as:

$$I_1(488 \rightarrow 535) = I_D(1-E) \qquad (78)$$

$$I_2(488 \rightarrow > \ 590) = I_D(1-E)S_1 + I_A S_2 + I_D E\alpha \qquad (79)$$

$$I_3(514 \rightarrow > 590) = I_D(1-E)S_3 + I_A + \frac{S_3}{S_1} I_D E\alpha \qquad (80)$$

I_1 is smaller than I_D because the energy transfer causes donor quenching. I_2 consists of three additive terms: 1. the overlapping fraction of the quenched fluorescein intensity, 2. the direct contribution of rhodamine and 3. Sensitized emission due to energy transfer. The proportionality factor α is the ratio of I_2 for a given number of rhodamine molecules and I_1 for the same number of fluorescein molecules. α is constant for each experimental setup, and has to be determined for every defined case. I_3 is a sum of 1. a fraction of the quenched fluorescein intensity, 2. the rhodamine intensity and 3. the sensitized emission of rhodamine due to energy transfer corrected for the lower molar extinction coefficient of fluorescein at 514 nm than at 488 nm.

From the Eqs. (78-80) the following equation can be derived:

$$\frac{E}{1-E} = \frac{1}{\alpha}\left[\frac{(I_2 - S_2 I_3)}{\left(1 - \frac{S_3 S_2}{S_1}\right) I_1} - S_1\right] \qquad (81)$$

All the parameters in Eq. (81) can be determined experimentally. If we substitute B in the right side of the Eq. (81), E can be expressed as follows:

$$E = \frac{B}{1+B} \qquad (82)$$

It is assumed that cellular autofluorescence is negligible compared to the specific fluorescence. If autofluorescence is substantial, corrections should be done. In this case, however, the correction for autofluorescence can be done using the average autofluorescence intensities of the cell population. Since in most cell types there is a good correlation between the autofluorescence detected at different spectral regions, a more elaborate correction method is also possible. Here another independent parameter should be detected, a fourth fluorescence intensity, and the autofluorescence can be calculated on a cell-by-cell basis. Naturally high autofluorescence may decrease the precision of the measurements [79]. A versatile computer program for evaluating flow cytometric energy transfer data has been elaborated recently [80].

Under special circumstances a simplified technique – using single laser excitation – can be used [81]. Using this approach, flow cytometry can provide quantitative measurements on of individual cells, allowing convenient determination of the distribution of energy transfer efficiency values in a population. In most flow cytometric energy transfer experiments fluorescein is used as donor and rhodamine as acceptor, the above considerations apply. However the equations are valid for other donor acceptor pairs, just the different spectral characteristics have to be considered. Applying Cy3 and Cy5 donor acceptor pair has several advantages over the classical fluorescein – rhodamine pair: *i.* the Cy3 and Cy5 have higher absorption coefficients, better photostability, *ii.* both the donor and the acceptor emissions are shifted towards the red region, causing autofluorescence much less of a problem, *iii.* even in commercial flow cytometers, the acceptor (Cy5) can be excited independently from Cy3, so that S_3 value would be zero and the equations become less complicated. In the future we expect to see more and more results obtained with Cy3 and Cy5 dye pair or Alexa dyes.

Many biologically important cell surface protein associations have been clarified in the last decade using flow cytometric energy transfer measurements. Among others the cell surface organization of the Major Histocompatibility Complex Class I [7, 82-87], that of the interleukin 2 receptor [88, 89] or the ErbB receptors [90, 91]. Several reviews are available about FRET measurements in biological membranes [75-77]. These reviews deal with associations of various membrane proteins, structures of receptors, conformational changes in transmembrane proteins evoked by ligands and membrane potential changes.

Concluding Remarks

The Stern-Volmer equation provides a straightforward approach to describe the quenching processes of simple model systems. The study of more complex systems, such as the intrinsic fluorescence of proteins or externally labeled macromolecules or supramolecular systems, results in a different kind of complexity, which can only be partially overcome by the application of different models. For such systems the interpretation of the data is model dependent. This feature is generally characteristic of data sets from complex systems and requires independent methods to verify the applied model. Apart from measuring molecular distances in isolated proteins the methods based on the application of a special quenching process, FRET, can also be informative regarding the dynamics of proteins and their complexes, and with the current techniques can be applied in living cells. All these considerations indicate that fluorescence quenching is a useful and widely applicable group of methods to obtain adequate information about the structural and dynamic details of biological systems.

References

[1] Lakowicz, J. R. (2004) *Principles of Fluorescence Spectroscopy*, Kluwer Academic / Plenum Publishers.

[2] Bushueva, T. L., Busel, E. P., and Burstein, E. A. (1980) Some regularities of dynamic accessibility of buried fluorescent residues to external quenchers in proteins. *Arch Biochem Biophys 204*, 161-6.

[3] Eftink, M. R., and Ghiron, C. A. (1981) Fluorescence quenching studies with proteins. *Anal Biochem 114*, 199-227.

[4] Somogyi, B., Lakos, Z., Damjanovich, S., and Rosenberg, A. (1989) Steady-State Fluorescence Quenching as a Tool to Study Protein Dynamics. *Journal of Molecular Liquids 42*, 45-58.

[5] Somogyi, B., and Lakos, Z. (1993) Protein dynamics and fluorescence quenching. *J Photochem Photobiol B 18*, 3-16.

[6] North, A. M. (1964) The Collision Theory of Chemical Reactions in Liquids, Wiley, London.

[7] Damjanovich, S., Vereb, G., Schaper, A., Jenei, A., Matko, J., Starink, J. P., Fox, G. Q., Arndt-Jovin, D. J., and Jovin, T. M. (1995) Structural hierarchy in the clustering of HLA class I molecules in the plasma membrane of human lymphoblastoid cells. *Proc Natl Acad Sci U S A 92*, 1122-6.

[8] Lakos, Z., Szarka, A., Koszorus, L., and Somogyi, B. (1995) Quenching-resolved emission anisotropy: a steady-state fluorescence method to evaluate dynamic and static quenching components. *J. Photochem. Photobiol. B:Biol. 27*, 55-60.

[9] Lehrer, S. S. (1971) Solute perturbation of protein fluorescence. The quenching of the tryptophyl fluorescence of model compounds and of lysozyme by iodide ion. *Biochemistry 10*, 3254-63.

[10] Bay, Z., and Pearlstein, R. M. (1963) A Theory of Energy Transfer in the Photosynthetic Unit. *Proc Natl Acad Sci U S A 50*, 1071-8.

[11] Somogyi, B., Matko, J., and Rosenberg, A. (1988) Evaluation of ligand induced relative change in the bimolecular quenching constant of protein fluorescence: applications to lysozyme and adenosine deaminase. *Acta Biochim Biophys Hung 23*, 135-48.

[12] Somogyi, B., Matko, J., and Rosenberg, A. (1988) Evaluation of ligand induced relative change in the bimolecular quenching constant of protein fluorescence: a steady-state model. *Acta Biochim Biophys Hung 23*, 125-33.

[13] Lakos, Z., Szarka, A., and Somogyi, B. (1995) Fluorescence quenching in membrane phase. *Biochem Biophys Res Commun 208*, 111-7.

[14] Englander, S. W., and Kallenbach, N. R. (1983) Hydrogen exchange and structural dynamics of proteins and nucleic acids. *Q Rev Biophys 16*, 521-655.

[15] Calhoun, D. B., Vanderkooi, J. M., Woodrow, G. V., 3rd, and Englander, S. W. (1983) Penetration of dioxygen into proteins studied by quenching of phosphorescence and fluorescence. *Biochemistry 22*, 1526-32.

[16] Gratton, E., Jameson, D. M., Weber, G., and Alpert, B. (1984) A model of dynamic quenching of fluorescence in globular proteins. *Biophys J 45*, 789-94.

[17] Hagaman, K. A., and Eftink, M. R. (1984) Fluorescence quenching of Trp-314 of liver alcohol dehydrogenase by oxygen. *Biophys Chem 20*, 201-7.

[18] Jameson, D. M., Gratton, E., Weber, G., and Alpert, B. (1984) Oxygen distribution and migration within Mbdes Fe and Hbdes Fe. Multifrequency phase and modulation fluorometry study. *Biophys J 45*, 795-803.

[19] Somogyi, B., Punyiczki, M., Hedstrom, J., Norman, J. A., Prendergast, F. G., and Rosenberg, A. (1994) Coupling between external viscosity and the intramolecular dynamics of ribonuclease T1: a two-phase model for the quenching of protein fluorescence. *Biochim Biophys Acta 1209*, 61-8.

[20] Calhoun, D. B., Vanderkooi, J. M., and Englander, S. W. (1983) Penetration of small molecules into proteins studied by quenching of phosphorescence and fluorescence. *Biochemistry 22*, 1533-9.

[21] Somogyi, B., Norman, J. A., and Rosenberg, A. (1986) Gated quenching of intrinsic fluorescence and phosphorescence of globular proteins. An extended model. *Biophys J 50*, 55-61.

[22] Eftink, M. R., and Hagaman, K. A. (1986) Viscosity dependence of the solute quenching of the tryptophanyl fluorescence of proteins. *Biophys Chem 25*, 277-82.

[23] Somogyi, B., Norman, J. A., Punyiczki, M., and Rosenberg, A. (1992) Viscosity dependence of acrylamide quenching of ribonuclease T1 fluorescence. The gating mechanism. *Biochim Biophys Acta 1119*, 81-9.

[24] Keizer, J. (1983) Non-Linear Fluorescence Quenching and the Origin of Positive Curvature in Stern-Volmer Plots. *Journal of the American Chemical Society 105*, 1494-1498.

[25] Lee, B., and Richards, F. M. (1971) The interpretation of protein structures: estimation of static accessibility. *J Mol Biol 55*, 379-400.

[26] Richards, F. M. (1977) Areas, volumes, packing and protein structure. *Annu Rev Biophys Bioeng 6*, 151-76.

[27] Englander, J. J., Kallenbach, N. R., and Englander, S. W. (1972) Hydrogen exchange study of some polynucleotides and transfer RNA. *J Mol Biol 63*, 153-69.

[28] Eftink, M. R., and Selvidge, L. A. (1982) Fluorescence quenching of liver alcohol dehydrogenase by acrylamide. *Biochemistry 21*, 117-25.

[29] Varley, P. G., Dryden, D. T., and Pain, R. H. (1991) Resolution of the fluorescence of the buried tryptophan in yeast 3-phosphoglycerate kinase using succinimide. *Biochim Biophys Acta 1077*, 19-24.

[30] Somogyi, B., Papp, S., Rosenberg, A., Seres, I., Matko, J., Welch, G. R., and Nagy, P. (1985) A double-quenching method for studying protein dynamics: separation of the fluorescence quenching parameters characteristic of solvent-exposed and solvent-masked fluorophors. *Biochemistry 24*, 6674-9.

[31] Harris, D. L., and Hudson, B. S. (1990) Photophysics of tryptophan in bacteriophage T4 lysozymes. *Biochemistry 29*, 5276-85.

[32] Stryer, L. (1978) Fluorescence energy transfer as a spectroscopic ruler. *Annu Rev Biochem 47*, 819-46.

[33] Stryer, L., Thomas, D. D., and Carlsen, W. F. (1982) Fluorescence energy transfer measurements of distances in rhodopsin and the purple membrane protein. *Methods Enzymol 81*, 668-78.

[34] Tron, L., Szollosi, J., Damjanovich, S., Helliwell, S. H., Arndt-Jovin, D. J., and Jovin, T. M. (1984) Flow cytometric measurement of fluorescence resonance energy transfer on cell surfaces. Quantitative evaluation of the transfer efficiency on a cell-by-cell basis. *Biophys J 45*, 939-46.

[35] Tron, L., Szollosi, J., and Damjanovich, S. (1987) Proximity measurements of cell surface proteins by fluorescence energy transfer. *Immunol Lett 16*, 1-9.

[36] Szollosi, J., Damjanovich, S., Goldman, C. K., Fulwyler, M. J., Aszalos, A. A., Goldstein, G., Rao, P., Talle, M. A., and Waldmann, T. A. (1987) Flow cytometric resonance energy transfer measurements support the association of a 95-kDa peptide termed T27 with the 55-kDa Tac peptide. *Proc Natl Acad Sci U S A 84*, 7246-50.

[37] Szollosi, J., Tron, L., Damjanovich, S., Helliwell, S. H., Arndt-Jovin, D., and Jovin, T. M. (1984) Fluorescence energy transfer measurements on cell surfaces: a critical comparison of steady-state fluorimetric and flow cytometric methods. *Cytometry 5*, 210-6.

[38] Szollosi, J., Damjanovich, S., Mulhern, S. A., and Tron, L. (1987) Fluorescence energy transfer and membrane potential measurements monitor dynamic properties of cell membranes: a critical review. *Prog Biophys Mol Biol 49*, 65-87.

[39] Matko, J., Szollosi, J., Tron, L., and Damjanovich, S. (1988) Luminescence spectroscopic approaches in studying cell surface dynamics. *Q Rev Biophys 21*, 479-544.

[40] Beechem, J. M., and Haas, E. (1989) Simultaneous determination of intramolecular distance distributions and conformational dynamics by global analysis of energy transfer measurements. *Biophys J 55*, 1225-36.

[41] Jovin, T. M., and Arndt-Jovin, D. J. (1989) Luminescence digital imaging microscopy. *Annu Rev Biophys Biophys Chem 18*, 271-308.

[42] Kawata, Y., and Hamaguchi, K. (1991) Use of fluorescence energy transfer to characterize the compactness of the constant fragment of an immunoglobulin light chain in the early stage of folding. *Biochemistry 30*, 4367-73.

[43] Szaba, G., Jr., Pine, P. S., Weaver, J. L., Kasari, M., and Aszalos, A. (1992) Epitope mapping by photobleaching fluorescence resonance energy transfer measurements using a laser scanning microscope system. *Biophys J 61*, 661-70.

[44] Somogyi, B., Matko, J., Papp, S., Hevessy, J., Welch, G. R., and Damjanovich, S. (1984) Forster-type energy transfer as a probe for changes in local fluctuations of the protein matrix. *Biochemistry 23*, 3403-11.

[45] O'Hara, P. B., Gorski, K. M., and Rosen, M. A. (1988) Energy transfer as a probe of protein dynamics in the proteins transferrin and calmodulin. *Biophys J 53*, 1007-13.

[46] Jona, I., Matko, J., and Martonosi, A. (1990) Structural dynamics of the Ca2(+)-ATPase of sarcoplasmic reticulum. Temperature profiles of fluorescence polarization and intramolecular energy transfer. *Biochim Biophys Acta 1028*, 183-99.

[47] Nyitrai, M., Hild, G., Lakos, Z., and Somogyi, B. (1998) Effect of Ca2+-Mg2+ exchange on the flexibility and/or conformation of the small domain in monomeric actin. *Biophys J 74*, 2474-81.

[48] Nyitrai, M., Hild, G., Belagyi, J., and Somogyi, B. (1999) The flexibility of actin filaments as revealed by fluorescence resonance energy transfer. The influence of divalent cations. *J Biol Chem 274*, 12996-3001.

[49] Nyitrai, M., Hild, G., Hartvig, N., Belagyi, J., and Somogyi, B. (2000) Conformational and dynamic differences between actin filaments polymerized from ATP- or ADP-actin monomers. *J Biol Chem 275*, 41143-9.

[50] Nyitrai, M., Hild, G., Bodis, E., Lukacs, A., and Somogyi, B. (2000) Flexibility of myosin-subfragment-1 in its complex with actin as revealed by fluorescence resonance energy transfer. *Eur J Biochem 267*, 4334-8.

[51] Hild, G., Nyitrai, M., and Somogyi, B. (2002) Intermonomer flexibility of Ca- and Mg-actin filaments at different pH values. *Eur J Biochem 269*, 842-9.

[52] Visegrady, B., Lorinczy, D., Hild, G., Somogyi, B., and Nyitrai, M. (2004) The effect of phalloidin and jasplakinolide on the flexibility and thermal stability of actin filaments. *FEBS Lett 565*, 163-6.

[53] Somogyi, B., Lakos, Z., Szarka, A., and Nyitrai, M. (2000) Protein flexibility as revealed by fluorescence resonance energy transfer: an extension of the method for systems with multiple labels. *J Photochem Photobiol B 59*, 26-32.

[54] Taylor, D. L., Reidler, J., Spudich, J. A., and Stryer, L. (1981) Detection of actin assembly by fluorescence energy transfer. *J Cell Biol 89*, 362-7.

[55] Dos Remedios, C. G., and Cooke, R. (1984) Fluorescence energy transfer between probes on actin and probes on myosin. *Biochim Biophys Acta 788*, 193-205.

[56] Miki, M., and Wahl, P. (1985) Fluorescence energy transfer between points in G-actin: the nucleotide-binding site, the metal-binding site and Cys-373 residue. *Biochim Biophys Acta 828*, 188-95.

[57] Miki, M., Barden, J. A., Hambly, B. D., and dos Remedios, C. G. (1986) Fluorescence energy transfer between Cys-10 residues in F-actin filaments. *Biochem Int 12*, 725-31.

[58] Miki, M., Hambly, B. D., and dos Remedios, C. G. (1986) Fluorescence energy transfer between nucleotide binding sites in an F-actin filament. *Biochim Biophys Acta 871*, 137-41.

[59] Barden, J. A., and dos Remedios, C. G. (1987) Fluorescence resonance energy transfer between sites in G-actin. The spatial relationship between Cys-10, Tyr-69, Cys-374, the high-affinity metal and the nucleotide. *Eur J Biochem 168*, 103-9.

[60] Miki, M. (1990) Resonance energy transfer between points in a reconstituted skeletal muscle thin filament. A conformational change of the thin filament in response to a change in Ca2+ concentration. *Eur J Biochem 187*, 155-62.

[61] Miki, M. (1991) Detection of conformational changes in actin by fluorescence resonance energy transfer between tyrosine-69 and cysteine-374. *Biochemistry 30*, 10878-84.

[62] Miki, M., O'Donoghue, S. I., and Dos Remedios, C. G. (1992) Structure of actin observed by fluorescence resonance energy transfer spectroscopy. *J Muscle Res Cell Motil 13*, 132-45.

[63] O'Donoghue, S. I., Hambly, B. D., and dos Remedios, C. G. (1992) Models of the actin monomer and filament from fluorescence resonance-energy transfer. *Eur J Biochem 205*, 591-601.

[64] Miki, M., and Kouyama, T. (1994) Domain motion in actin observed by fluorescence resonance energy transfer. *Biochemistry 33*, 10171-7.

[65] Moens, P. D., Yee, D. J., and dos Remedios, C. G. (1994) Determination of the radial coordinate of Cys-374 in F-actin using fluorescence resonance energy transfer spectroscopy: effect of phalloidin on polymer assembly. *Biochemistry 33*, 13102-8.

[66] dos Remedios, C. G., and Moens, P. D. (1995) Fluorescence resonance energy transfer spectroscopy is a reliable "ruler" for measuring structural changes in proteins. Dispelling the problem of the unknown orientation factor. *J Struct Biol 115*, 175-85.

[67] Moens, P. D., and dos Remedios, C. G. (2001) Analysis of models of F-actin using fluorescence resonance energy transfer spectroscopy. *Results Probl Cell Differ 32*, 59-77.

[68] Prochniewicz, E., Walseth, T. F., and Thomas, D. D. (2004) Structural dynamics of actin during active interaction with myosin: different effects of weakly and strongly bound myosin heads. *Biochemistry 43*, 10642-52.

[69] Fajer, P. (2004) Conformational switching in muscle. *Adv Exp Med Biol 547*, 61-80.

[70] Nyitrai, M., Hild, G., Lukacs, A., Bodis, E., and Somogyi, B. (2000) Conformational distributions and proximity relationships in the rigor complex of actin and myosin subfragment-1. *J Biol Chem 275*, 2404-9.

[71] Gennis, R. B., and Cantor, C. R. (1972) Use of nonspecific dye labeling for singlet energy-transfer measurements in complex systems. A simple model. *Biochemistry 11*, 2509-17.

[72] Tron, L., Szollosi, J., and Damjanovich, S. (1984) New trends in studying slow diffusion processes of the cytoplasmic membrane components. *Acta Biochim Biophys Acad Sci Hung 19*, 247-57.

[73] Clegg, R. M. (1995) Fluorescence resonance energy transfer. *Curr Opin Biotechnol 6*, 103-10.

[74] Clegg, R. M. (2002) FRET tells us about proximities, distances, orientations and dynamic properties. *J Biotechnol 82*, 177-9.

[75] Matyus, L. (1992) Fluorescence resonance energy transfer measurements on cell surfaces. A spectroscopic tool for determining protein interactions. *J Photochem Photobiol B 12*, 323-37.

[76] Szollosi, J., Damjanovich, S., and Matyus, L. (1998) Application of fluorescence resonance energy transfer in the clinical laboratory: routine and research. *Cytometry 34*, 159-79.

[77] Szollosi, J., Nagy, P., Sebestyen, Z., Damjanovicha, S., Park, J. W., and Matyus, L. (2002) Applications of fluorescence resonance energy transfer for mapping biological membranes. *J Biotechnol 82*, 251-66.

[78] Vereb, G., Matko, J., and Szollosi, J. (2004) Cytometry of fluorescence resonance energy transfer. *Methods Cell Biol 75*, 105-52.

[79] Sebestyen, Z., Nagy, P., Horvath, G., Vamosi, G., Debets, R., Gratama, J. W., Alexander, D. R., and Szollosi, J. (2002) Long wavelength fluorophores and cell-by-cell correction for autofluorescence significantly improves the accuracy of flow cytometric energy transfer measurements on a dual-laser benchtop flow cytometer. *Cytometry 48*, 124-35.

[80] Szentesi, G., Horvath, G., Bori, I., Vamosi, G., Szollosi, J., Gaspar, R., Damjanovich, S., Jenei, A., and Matyus, L. (2004) Computer program for determining fluorescence resonance energy transfer efficiency from flow cytometric data on a cell-by-cell basis. *Comput Methods Programs Biomed 75*, 201-11.

[81] Szollosi, J., Matyus, L., Tron, L., Balazs, M., Ember, I., Fulwyler, M. J., and Damjanovich, S. (1987) Flow cytometric measurements of fluorescence energy transfer using single laser excitation. *Cytometry 8*, 120-8.

[82] Matyus, L., Bene, L., Heiligen, H., Rausch, J., and Damjanovich, S. (1995) Distinct association of transferrin receptor with HLA class I molecules on HUT-102B and JY cells. *Immunol Lett 44*, 203-8.

[83] Bene, L., Szollosi, J., Balazs, M., Matyus, L., Gaspar, R., Ameloot, M., Dale, R. E., and Damjanovich, S. (1997) Major histocompatibility complex class I protein conformation altered by transmembrane potential changes. *Cytometry 27*, 353-7.

[84] Jenei, A., Varga, S., Bene, L., Matyus, L., Bodnar, A., Bacso, Z., Pieri, C., Gaspar, R., Jr., Farkas, T., and Damjanovich, S. (1997) HLA class I and II antigens are partially co-clustered in the plasma membrane of human lymphoblastoid cells. *Proc Natl Acad Sci U S A 94*, 7269-74.

[85] Nagy, P., Matyus, L., Jenei, A., Panyi, G., Varga, S., Matko, J., Szollosi, J., Gaspar, R., Jovin, T. M., and Damjanovich, S. (2001) Cell fusion experiments reveal distinctly different association characteristics of cell-surface receptors. *J Cell Sci 114*, 4063-71.

[86] Bodnar, A., Bacso, Z., Jenei, A., Jovin, T. M., Edidin, M., Damjanovich, S., and Matko, J. (2003) Class I HLA oligomerization at the surface of B cells is controlled by

exogenous beta(2)-microglobulin: implications in activation of cytotoxic T lymphocytes. *Int Immunol 15*, 331-9.

[87] Panyi, G., Bagdany, M., Bodnar, A., Vamosi, G., Szentesi, G., Jenei, A., Matyus, L., Varga, S., Waldmann, T. A., Gaspar, R., and Damjanovich, S. (2003) Colocalization and nonrandom distribution of Kv1.3 potassium channels and CD3 molecules in the plasma membrane of human T lymphocytes. *Proc Natl Acad Sci U S A 100*, 2592-7.

[88] Vereb, G., Matko, J., Vamosi, G., Ibrahim, S. M., Magyar, E., Varga, S., Szollosi, J., Jenei, A., Gaspar, R., Jr., Waldmann, T. A., and Damjanovich, S. (2000) Cholesterol-dependent clustering of IL-2Ralpha and its colocalization with HLA and CD48 on T lymphoma cells suggest their functional association with lipid rafts. *Proc Natl Acad Sci U S A 97*, 6013-8.

[89] Vamosi, G., Bodnar, A., Vereb, G., Jenei, A., Goldman, C. K., Langowski, J., Toth, K., Matyus, L., Szollosi, J., Waldmann, T. A., and Damjanovich, S. (2004) IL-2 and IL-15 receptor alpha-subunits are coexpressed in a supramolecular receptor cluster in lipid rafts of T cells. *Proc Natl Acad Sci U S A 101*, 11082-7.

[90] Nagy, P., Bene, L., Balazs, M., Hyun, W. C., Lockett, S. J., Chiang, N. Y., Waldman, F., Feuerstein, B. G., Damjanovich, S., and Szollosi, J. (1998) EGF-induced redistribution of erbB2 on breast tumor cells: flow and image cytometric energy transfer measurements. *Cytometry 32*, 120-31.

[91] Nagy, P., Vereb, G., Sebestyen, Z., Horvath, G., Lockett, S. J., Damjanovich, S., Park, J. W., Jovin, T. M., and Szollosi, J. (2002) Lipid rafts and the local density of ErbB proteins influence the biological role of homo- and heteroassociations of ErbB2. *J Cell Sci 115*, 4251-62.

Index

C

D

E

F

G

O

P

R

S

T

U

V

W

Y

Z